Combining and Reporting Analytical Results

Edited by

A. Fajgelj
International Atomic Energy Agency, Agency, Vienna, Austria

M. Belli
Italian Environmental Protection Agency, Rome, Italy

U. Sansone
International Atomic Energy Agency, Agency, Vienna, Austria

RSCPublishing

The proceedings of the International Workshop Combining and Reporting Analytical Results: The Role of Traceability and Uncertainty for Comparing Analytical Results, held in Rome on 6-8 March 2006.

Special Publication No. 307

ISBN: 978-0-85404-848-9

A catalogue record for this book is available from the British Library

Published by The Royal Society of Chemistry,
Thomas Graham House, Science Park, Milton Road,
Cambridge CB4 0WF, UK

Registered Charity Number 207890

For further information see our web site at www.rsc.org

Printed by Henry Ling Ltd, Dorchester, Dorset. UK

Combining and Reporting Analytical Results

The international workshop on Combining and Reporting Analytical Results — The Role of (metrological) Traceability and (measurement) Uncertainty was organized by:

Italian Agency for Environmental Protection and Technical Services (Rome, Italy) and

Interdivisional Working Party on Harmonization of Quality Assurance of the International Union of Pure and Applied Chemistry (IUPAC)

and sponsored by:

Centro Sviluppo Materiali S.p.A., Rome, Italy

Co-Operation on International Traceability in Analytical Chemistry

CIPM Consultative Committee for Amount of Substance – Metrology in Chemistry (CCQM)

International Bureau of Weights and Measures (BIPM), Sevres, France

International Atomic Energy Agency
Vienna, Austria

Committee on Reference Materials (REMCO) of the International Organization for Standardization (ISO), Geneva, Switzerland

United Nations Industrial Development Organization (UNIDO), Vienna, Austria

PREFACE

This book contains lectures presented at the international workshop on "Combining and Reporting Analytical Results — The Role of (metrological) Traceability and (measurement) Uncertainty for Comparing Analytical Results", held from 6 to 8 March 2006 in Rome, Italy. The IUPAC Interdivisional Working Party on Harmonization of Quality Assurance and the Italian Agency for Environmental Protection and Technical Services have cooperated in organization of this event.

The idea for this workshop originated from editors' long term experience and observations at various meetings of different technical/scientific organizations and groups, where analytical (measurement) results are considered. Almost all organizations face similar problems when the analytical results are to be combined and reported, regardless of their "metrological level", being a field laboratory, producer of reference materials, proficiency testing organizer or even a national metrology institute. Combining measurement results obtained by one analyst in one laboratory employing one measurement procedure, using one measurement technique is the starting point, and the two questions: "How to report the associated measurement uncertainty and how to establish and demonstrate the metrological traceability of combined results?" are the major points of concern. The complexity of these questions expands with the increasing number of measurement procedure/techniques, number of laboratories and measurement results that need to be considered. The problems described are not at all new. In recent years, however, in following the latest developments on the international measurement scene, in fulfilling requirements for laboratory accreditation according to international standards, e.g. ISO/IEC 17025, as well as ISO Guides 34, and 43 related to formal demonstration of competence of reference materials producers and proficiency testing organizers, more attention has been paid to metrological traceability and measurement uncertainty as two important quality components of measurement results. A wide international interest in the topics covered by the workshop program is reflected in the number of cosponsoring organizations, namely: Centro Svillupo Materiali. S.p.A (CMS), the International Atomic Energy Agency (IAEA), the Consultative Committee for Amount of Substance — Metrology in Chemistry (CCQM), International Bureau of Weights and Measures (BIPM), the Co-operation on International Traceability in Analytical Chemistry (CITAC), the ISO Committee on Reference Materials (ISO REMCO) and the United Nations Industrial Development Organization (UNIDO). All these organizations were represented by speakers at the workshop and have contributed papers for this book.

The contributions in this book provide an overview of current practices used in different laboratories from different scientific fields to combine and report measurement results, at the same time describing some basic scientific considerations as well as discussions related to legislative aspects. Practical examples from, environmental monitoring laboratories, reference material producers, clinical chemistry, as well as from the top metrological level, e.g. key comparisons and pilot studies organized in support and to enable comparability of measurement and calibration results at world-wide level, are included. The contributions in this book were prepared by scientists from laboratories dealing daily with sets of measurement results, their combination and reporting. We hope that the reader will be able to extract information for her/his own specific case and be able to apply it in practice. Taking into account that there will never be only one way of combining and reporting measurement results, this book is not a classical text book and critical reading is required.

This is the sixth book in a series based on the work of the IUPAC Interdivisional Working Party on Harmonization of Quality Assurance. Also this time the IUPAC's role in achieving consensus was followed by bringing together technical experts so that they could share their experience and expertise. Co-operation with other organizations is essential for making any significant progress on an international scale. Besides the above listed co-sponsoring organizations, substantial support and contribution to the successful organization of the workshop came this time from the IUPAC Secretariat and the Analytical Chemistry Division and locally from the Italian Agency for Environmental Protection and Technical Services (APAT) and Centro Sviluppo Materiali S.p.A (CSM). We are especially grateful to Ms. S. Rosamilia, Mr. F. Falcioni, Mr. P. de Zorzi, Ms. G. Gelati, Ms. T. Guagnini and Mr. F. Babalini for the excellent conditions provided during the workshop and Mr. A. De Maio, the Director of APAT Environmental Department for his support.

Aleš Fajgelj   Maria Belli   Umberto Sansone

Contents

ENVIRONMENTAL METROLOGY IN ITALY: THE ROLE OF APAT

M. Belli

Agenzia per la Protezione dell'Ambiente e per i Servizi Tecnici (APAT), Servizio Laboratori, Misure ed Attività di Campo, Via di Castel Romano, 100, 00128 Roma, Italy

1 INTRODUCTION

The objective of environmental monitoring is to quantify the condition of ecological systems in spatial and temporal differentiation. The organized and systematic measurement of selected variables provides the establishment of baseline data and the identification of both natural and human-induced changes in the environment. Therefore, spatially and temporally measured values have to correspond to real conditions and not to different measurement methods. In addition, the environmental monitoring enables to guide the formulation and the implementation of environmental management policies designed to protect human health and well-being, which includes ecological well-being. In general, the objectives of environmental monitoring programs can be summarized as follows;

- to verify the compliance with national or international environmental quality standards;
- to provide a basis for the implementation of environmental legislation;
- to assess human population and ecosystem exposure to pollution.

To reach these objectives, both at national or international level, a combination of a high amount of environmental data are necessary, collected in different period and coming from different sources. In the case of data used to provide information on the status of natural values and threatening processes, and to determine the type and magnitude of trends over time in assessing long-term trends, quality assurance, high precision and consistency of data, are of the utmost importance. The same considerations are valid if the monitoring data are used to evaluate the pattern of environmental contamination across a country or Europe. In addition, the assessment of compliance with national or international environmental quality standards, is vitally dependent on reliable environmental data. In this frame, the analytical laboratories have a great role, as the results of analytical measurements may be the basis upon which economic, legal or environmental management decisions are made, and they are essential in international trade, environmental protection, safe transportation, law enforcement, consumer safety and the preservation of human health. It is essential that such measurements are accurate, reliable, cost effective and defensible to ensure that correct decisions are made. This requires data determined with standardized methods, measurement results traceable to national or international standards with a stated measurements uncertainty and implementation of quality assurance/quality control (QA/QC) systems.

The environmental monitoring in Italy is performed at regional level and the data are then collected and combined by the Italian National Environmental Protection Agency

(APAT), in order to assess trends and human and ecosystem exposure to pollution. Analytical data are produced by around one hundred laboratories belonging to the Italian Regional and Provincial Environmental Protection Agencies (ARPAs/APPAs) that in some cases use different approaches and different analytical methods. As focal point for the European Environmental Agency, APAT has the role to harmonize the environmental monitoring activities within the Italian territory, and to assure the comparability of the environmental data produced at regional level. To this end the following initiatives were promoted by the Environmental Metrology Service (Servizio Laboratori, Misure ed Attività di Campo) of APAT:

- the provision of matrix reference materials, similar to the test sample being measured in the regional laboratories (soil, sediments, compost, water, wastes, etc.), widely used for internal and external quality control activities. The analytes of interest are metals (for liquid and solid matrices), pesticides, organic pollutants and anions/cations (for liquid matrices);
- the establishment at national level of a permanent advisory group (GTP), according to the ISO guide 43-1[1] on proficiency testing scheme. GTP includes representatives from each regional agency. Quality managers, user of analytical data and laboratories are represented in the permanent advisory group;
- organization of proficiency tests and inter-laboratory comparison exercises;
- the establishment of a network of reference laboratories at national level (expert laboratories) for the different analytical field, in order to harmonise the different analytical methods used in environmental monitoring activities;
- the establishment of a link with the Italian National Institute of Metrology to disseminate the concepts of traceability in the Italian Regional and Provincial Environmental Protection Agencies laboratories, trough dedicated courses and equipment calibration.

2 REFERENCE MATERIAL PRODUCTION AND CHARACTERIZATION

Reference materials (RM) are one of the tools used to obtain comparability of analytical results. In recent times, there has been increasing interest worldwide in the accuracy, traceability and comparability of analytical measurements and the role that matrix reference materials play in the process. The ultimate goal of any measurement process is to ensure accuracy and to establish traceability to common universal reference points (preferably the SI) through an unbroken chain of comparisons. The use of reference materials is essential for method validations, calibration and internal/external quality control activities. The reference materials produced by APAT are closely matching in terms of matrix to the samples analyzed in the Italian environmental laboratories and are used mainly for the organization of inter-laboratory exercises. The organization of inter-laboratory exercises provides a continuous check on the comparability of environmental analytical results across Italy and identifies the determinations for which improvements are required. In addition, the reference materials produced by APAT may be used by the regional laboratories, as internal quality control materials, to check routinely if a measurement method is under statistical control.

Solid and liquid reference materials are produced following ISO Guide 35[2]. As an example for solid materials, the raw material is dried at a constant temperature of +40°C in a ventilated oven. The material is then sieved through a 2 mm mesh sieve and the resulting fraction above 2mm is discarded. The fraction below 2mm is milled into powder (<90 micrometers) and homogenized over two weeks by mixing into a cylindrical drum placed on a roll-bed. The bulk homogeneity of the sample is checked by measuring the C and N concentrations on 10 sub-samples (10-15 g each), taken directly from the cylindrical drum. If the data of C and N content does not show any measurable heterogeneity (coefficient of variation below 1% for C and N), the material is bottled. The bottling is carried out in one day. Bottling is carried out in a way to prevent the possible segregation of fine particles (20 samples, each of about 30 g, are taken from the centre of the cylindrical drum immediately after stopping the rotation and placed into 10 pre-cleaned brown glass bottles). The drum is again rotated for a further 2 minutes and again 20 samples are taken in the same way and bottled. The sampling from the cylindrical drum and the bottling of the samples continued following this procedure until the material is finished. Each batch of RM is formed by about 1000 bottles.

The main characteristics of reference materials produced by APAT are homogeneity and stability of the analytes of interest. The within- and between-bottles variability is estimated by one-way analysis of variance (ANOVA) for the analytes of interest, as required by ISO Guide 35[2]. The homogeneity of the material, or better the fitness for purpose of the material, is then confirmed using the results of the inter-laboratory exercises[3].

A short term stability study is carried out to demonstrate that the contents of the analytes of interest are not changed in the period of the inter-laboratory exercises. The stability study is carried out at different temperatures to study the effect of different temperatures on the properties of the material. Generally, the stability study is carried out at 40 and 20 °C and normally last 3 months. At the start of the stability test, 25 bottles are stored at a reference temperature (-18 °C) at which it is assumed that no instability can occur. Additional 5 bottles are stored at +20 °C and 5 bottles at +40 °C. After 1, 2 and 3 months, 5 bottles are transferred from -18 °C to +20 °C and 5 bottles from -18 °C to +40 °C. After three months 3 sub-samples collected from each bottle are measured in one run (under repeatability conditions). As homogeneity tests, stability tests are carried out on the analytes of interest. For each temperature (+20 and +40 °C), the following parameters are assessed:

- standard deviations between the bottles stored at the same temperature for the same time interval, the mean value of concentration and the coefficient of variation (CV %);
- standard deviations between the mean values of concentration of bottles stored for different time periods, the mean values and the coefficient of variation (CV %);
- ratios of the mean values of measurements on bottles stored at +20 °C and +40 °C, respectively, and the mean values of measurements on samples stored at -18 °C for the same period;
- linear regression of the above mentioned ratios, the uncertainty contribution due to the material stability;
- analysis of variance (ANOVA) to assess the influence of the storage period at +20 °C and +40 °C on the stability of the material.

The characterization of the reference materials, defined as the assignment of concentration data to the interested analytes which approaches as closely as possible the "true value", together with uncertainty limits, is another step of prime importance for the use of reference materials. The key characteristic of a reference material is that the properties of interest are measured and assigned on the basis of the accuracy. The goal is the arrival at the best possible estimate of the unknown "true value". It implies the reliable assignment of a value to a property of a material. It encompasses selection of measurands, appropriate analytical methodologies, adequately calibrated and properly used.

The assignment of property value to the reference materials produced in the APAT laboratories is carried out following different strategies: measurements in one or two expert laboratories working independently using different analytical methods or more likely, collaborative analysis in a group of expert laboratories using one definitive method. Robust mean and robust standard deviation are calculated for each set of analytical results. The assigned values are then determined as the robust average of all laboratory mean values, while the expanded uncertainty range is calculated as reported in ISO 13528[3].

3 ORGANIZATION OF INTERLABORATORY EXERCISES

As above reported, one of the main purpose of the reference materials prepared in APAT Laboratory is the organization of inter-laboratory comparisons, to check the overall quality of the analytical results of the environmental laboratory across Italy. The main aims of the inter-laboratory exercises organized by APAT are: to get information on the comparability of the environmental analytical results across the Country, to define the analytical methods that are fitness-for-purpose for environmental monitoring activities and to assist the ARPA/APPA laboratories in meeting the requirements of the monitoring activities. Literature reports many different schemes to carry out inter-laboratory comparisons and proficiency testing in chemical and biological analyses, specially for scores assignment[4]. Furthermore, environmental monitoring involves the analysis of a vast range of matrices with a wide number of analytes to be determined. In order to design inter-laboratory exercises and proficiency testing that are fit for purpose for environmental monitoring, a Technical Advisory Group (GTP) was established by APAT to ensure a smooth operation and the success of all inter-laboratory exercises. The support of an advisory group to the PT and inter laboratory comparison organisers is suggested also by the ISO Guide 43. The functions and the activities of the GTP coordinated by APAT are reported in another paper on this proceedings[5].

4 NETWORK OF REFERENCE LABORATORIES

In 2004 APAT has established a first network of reference laboratories to support APAT and ARPA/APPA system in environmental monitoring activities. The core task of this first network is the technical evaluation and validation of methods for pesticides, furan, dioxin and asbestos. Results achieved in method validation of a multiresidue analytical method for the determination of pesticide residues in water samples are reported elsewhere in this book[6].

The 10 reference laboratories for pesticide residue determinations participated in an inter-laboratory comparison organised by APAT in 2005 on the determination of pesticide residues in water. The exercise has been carried out on a new material (solid material "pill" easy soluble in water spiked with the analytes of interest) developed by the Institute Pasteur de Lille[7]. Participants were required to measure the analytes under investigation applying their routine measurement procedures and analytical technique. The analysis of the forms filled in by each laboratory showed that all laboratories applied the same measurement procedure for the determination of pesticide residues[8]. Figure 1 compare the result of chlorpyriphos content in water, obtained by the reference laboratories (a) with those obtained by all laboratories participating to the inter-laboratory exercise (b). The results show that almost all reference laboratories are in agreement with the consensus value. The consensus value was obtained applying the robust statistics on the mean values obtained by all laboratories participating to the inter-laboratory comparison. The good results achieved from the reference laboratories are attributable to the harmonization of all procedure steps between the group. This achievement show clearly the importance of a continuous activity of harmonization between laboratories across the Country.

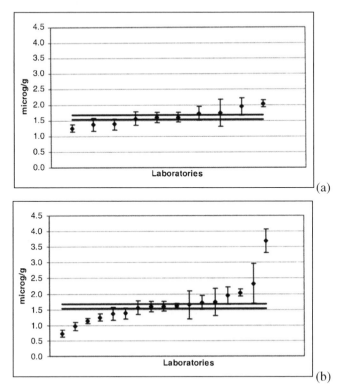

Figure 1 *Chlorpyriphos consensus value 1.6±0.1 µg/g. Reference laboratories results (a), all laboratories participating to the inter-laboratory exercise (b).*

5 COLLABORATION WITH THE NATIONAL INSTITUTE OF METROLOGY

In recent times, there has been increased demand for environmental data of demonstrated quality. In this respect, the traceability concept is now extensively discussed in the light of the specificities of chemical and biological measurements and with regards to environmental measurements. Discussions generally highlight that a direct application of theoretical metrology concepts to environmental measurements is impossible, because there are major differences between chemical/biological and physical measurement processes (e.g., chemical analysis results often strongly depend upon the nature of samples, whereas physical measurements are less or not affected). In addition a wide variety of analytical problems are encountered in relation to different parameters and matrices, and some necessary preliminary steps as sampling and sample pre-treatment, may affect the final analytical result[9]. In this respect, environmental studies and monitoring based on analytical measurement can only be valid if data are obtained under a reliable quality assurance regime (demonstration of data traceability to stated reference together with the use of validated measurement methods).

To improve the traceability of environmental analysis a strong collaboration has been established in Italy among the National Environmental Protection Agency (APAT), the Regional and Provincial Environmental Protection Agencies (ARPA/APPA) and the National Institute of Metrological Research (INRIM). In this framework, the following activities has been developed:

- development of a traceability chain for ozone measurements[10];
- theoretical and practical training courses on traceability for the following measurements:
- mass determination;
- pH;
- conductivity;
- gas measurements (air quality);
- ozone measurements.

6 FUTURE PLAN

The environmental metrology service of APAT, established to support the Italian regional environmental laboratories, to enhance the quality of their analytical measurement data through the provision of matrix reference materials, validated procedures, training in the implementation of quality control, and through the evaluation of measurement performance by organization of proficiency tests and inter-laboratory comparison exercises, is certified according to ISO 9001:2000.

In the next year the APAT laboratories for the production and characterization of reference material will be accredited ISO 17025:2005 and ISO Guide 34[11]. The network of reference laboratories will be extended to measurements on atmospheric particulate matter (PM10, PM2,5), characterization of waste and environmental ecotoxicology. The link between APAT and INRIM will continue in order to fully develop the traceability chain for ozone and extend the traceability chain to the measurements of the other parameters of the air quality.

References

1. ISO Guide 43-1, Proficiency testing by interlaboratory comparison. Part 1, International Standardization Organization, 1997
2. ISO Guide 35, Certification of reference materials – general and statstical principles, International Standardization Organization, 2006
3. ISO 13528 Statistical methods for use in proficiency testing by interlaboratory comparisons, International Standardization Organization, 2005
4. M. Thompson, S. L. R. Ellison, R. Wood, The international harmonized protocol for the proficiency testing of analytical chemistry laboratories, Pure Applied Chemistry, Vol. 78, n. 1
5. P. de Zorzi, S. Barbizzi, S. Gaudino, A. Pati, S. Rosamilia, Interlaboratory comparison: the APAT approach, this proceedings
6. M. Antoci, S. Barbizzi, B. Bencivegna, D. Centioli, S. Finocchiaro, M. Fiore, F. Fiume, V. Giudice, M. Lorenzin, M.C. Manca, M. Morelli, E. Sesia, M. Volante, Collaborative study for pesticide residues in water samples, this proceedings.
7. Final technical report of EC project n° GMA2-2001-52010, Preparation of a CRM: Pesticide in water-feasibilty study, 2004

8. S. Barbizzi, M. Belli, D. Centioli, P. de Zorzi, R. Mufato, H. Muntau, G. Sartori, G. Stocchero Rapporto conclusivo "Studio collaborativo APAT-SC001 pesticidi in acqua sotterranea", in publication as APAT report

9. Ph. Quevauviller, Traceability of environmental chemical measurements, Trend in Analytical Chemistry, Vol. 23, N.3, 2004.

10. G. Castrofino, M. Sassi, S. Curci, A. Di Leo, Catena di riferibilità per la misura della frazione molare di ozono in atmosfera, APAT CTN-ACE 2004 www.sinanet.apat.it

11. ISO Guide 34, General requirements for the competence of reference material producers, International Standardization Organization, 2003.

COMPARABILITY AND QUALITY OF EXPERIMENTAL DATA UNDER DIFFERENT QUALITY SYSTEMS

S. Caroli

Istituto Superiore di Sanità, Viale Regina Elena 299, 00161 Rome, Italy

1 INTRODUCTION

Important decisions are often taken on the basis of experimental data. Hence, it is crucial that such data be comparable, reliable and valid. No laboratory can in fact be run without a fit-for-purpose quality system in place. Quality has been defined by ISO as "*The totality of features and characteristics of a product or service that bear on its ability to satisfy stated or implied needs*". To date, quality systems are basically inspired either by the Good Laboratory Practice (GLP) principles or by the accreditation criteria.

Laboratory work may be of two different types: *i) the* outcome of the investigation are exact figures, to which precision and reproducibility are expected to be attached; *ii)* the outcome of the investigation is, in a general sense, complex information which should be credible, reliable and comparable. In the former case, what matters more are the experimental measurements. In this context quality is assessed in terms of precision and reproducibility of the numerical data obtained. The ability of the laboratory to generate such data is thus of primary importance. Quality systems based on accreditation criteria are ideal in this respect. In the latter case, the focus is on the overall study as such. Third parties should be enabled to reconstruct the whole course of the study and to check its integrity so that confidence can be gained in the way the study results have been obtained. Under such circumstances, quality systems based on the GLP principles do apply. Which approach is to be preferred depends only on the scope and goals of the activities performed in the laboratory, although it should not be overlooked that accreditation is basically voluntary, whereas the GLP system is prescribed by law for those Test Facilities (TFs) undertaking non-clinical safety studies.

There is still some confusion surrounding the terms of accreditation and certification. As this may well misleading, consensus has been reached on the following definitions: *accreditation* is a means used to identify competent testing laboratories, whereas *certification* is the official approval granted by a given authority.

2. KEY ASPECTS OF A QUALITY SYSTEM BASED ON THE ACCREDITATION CRITERIA

As set forth by the IUPAC, "*The international scientific community recognizes that a laboratory must take appropriate measures to ensure that it is capable of providing data of the required quality. Such measures include: i) internal quality control procedures; ii)*

participation in proficiency testing schemes; iii) validated methods of analysis; iv) accreditation to an international standard." (1).

Accreditation-based quality systems are governed by the international Standard ISO/IEC 17025 (2). This standard exploits the extensive experience gained in implementing the ISO/IEC Guide 25 and EN 45001 norms and replaces them both. The ISO/IEC Standard sets forth the requirements a laboratory has to meet to be recognized as competent to carry out tests and/or calibrations, including sampling. The pillars of an accreditation system are listed in Table 1.

Method validation is central to the accreditation process as reliability and comparability of data are crucial to perform experimental meaningful tests and to achieve credible results which can be profitably used by the client, *i.e.*, the end-user. It should be noted that the overall validation process covers all of the pivotal phases of an experimental measurement and not only the mere quantification step, as illustrated in Fig. 1. In turn, method validation as such should at least cover the parameters given in Table 2.

3 KEY ASPECTS OF A QUALITY SYSTEM BASED ON THE PRINCIPLES OF GOOD LABORATORY PRACTICE

In the early 1960's the US Food and Drug Administration (FDA) became aware that some studies on the safety of new chemicals performed by TFs for regulatory purposes were basically unreliable. Evidence was in fact provided of major adverse effects of such substances which had not been reported at the time when the authorization to production and commerce was granted. In the early 1970's the US Congress undertook the re-assessment of studies submitted by some TFs to Regulatory Authorities (RAs) and suspected to be fraudolent. Under such conditions thousands and thousands of safety studies on industrial chemicals, pesticides, herbicides, drugs, cosmetics, and food and feed additives were conducted for years (about 35 – 40 % of all toxicological studies authorized in the USA in that period). As an example, an article published by *The Washington* Post in 1997 is shown in Fig. 2.

Senator Edward Kennedy declared at the US Congress of January 20, 1976, that "...*unreliable, undocumented and fraudolent research is the most frightening menace to the health and safety of people. That research be wrong because of technical problems or because of the lack of competence or even due to criminal negligence is less important than the very fact that it is wrong...*"

The principles GLP were conceived to harmonize the conduct of non-clinical safety studies and to minimize the risk of fraud. Since the early years, this matter became a priority for the Organisation for Economic Co-operation and Development (OECD) which set up the GLP principles in order to promote and manage the mutual acceptance of non-clinical safety studies in the Member Countries. According to OECD, the principles of GLP are a quality system concerned with the organizational process and the conditions under which safety studies are planned, conducted, controlled, recorded, reported and archived. In practice, they form a body of reciprocally dependent documented items that make the falsification of a study more time-consuming and expensive than its actual correct performance.

The three major acts of OECD in the field of GLP are as follows: *i*) Decision of the Council concerning the Mutual Acceptance of Data (MAD) in the Assessment of Chemicals [C(81)30(Final)]; *ii*) Council Decision-Recommendation on Compliance with Principles of Good Laboratory Practice; *iii*) [C(89)87(Final)] Council Decision concerning the Adherence of Non-member Countries to the Council Acts related to the Mutual

Acceptance of Data in the Assessment of Chemicals [C(97)114(Final)]. As a part of the permanent activities of its *Environment, Health and Safety* Programme, the OECD also prepares and publishes Test Guidelines for Chemical Substances to be used when performing GLP studies and thus enhance their reliability.

The Series on the GLP principles and compliance monitoring consists at present of 14 monographs, as detailed in Table 3, whereas Table 4 details the pillars of a GLP system (3, 4). These guides form the core of the legal provisions of the European Union in the field of GLP (5, 6).

4 CONCLUSIVE REMARKS

The two quality systems have been conceived to meet quite different needs. In other words, the accreditation criteria are designed to manage activities in a laboratory where routine quantitative measurements (such as   analytical determinations) are carried out, whereas the GLP principles are intended to guarantee the integrity of data generated in non-clinical safety studies. Their respective fundamental characteristics are summarized in Table 5. From this standpoint, it is worth mentioning that, *e.g.*, the GLP system prescribes that the Director of the TF, the person responsible of the Quality Assurance Unit, the Study Director and the Archivist be all independent of each other to fully guarantee the fair conduct of the study, while in the case of the accreditation system the first two functions can coincide and the third one does not exist. On the other hand, in the accreditation system, it is imperative to have a quality manual, which in turn is not formally requested in the GLP system, although in the latter the Standard Operating Procedures play basically the same role. Moreover, a study plan, mandatory in the GLP system, is not needed in the accreditation one, not to speak of the fact that management of complaints and participation in proficiency testing is mandatory in the latter, but not necessary in the former. As regards validation of methods, the GLP system requires that validated methods are in place, but does not impose that such methods are set up according to the GLP principles, any other fit-for-purpose quality system being acceptable to this end.

All this provides clear evidence of the profound diversity in the approaches and goals of the two systems, although some common aspects are present. In this regard, in recent years, the OECD has established a dialogue group to verify where the two systems can actually interact, thus minimizing useless duplication of efforts. In conclusion, the selection of the quality system to be adopted should be carefully made on the basis of the prevailing activities carried out in the laboratory. Quality is inescapable, but it has a cost: a wrong decision can only lead to failure.

Table 1 *Key aspects of a laboratory compliant with an accreditation system.*

Service to the client	Motivation of personnel
Policy for complaints	Laboratory setting
Control of non-conformities	Validation of methods
Quality manual	Equipment
Management of records	Management reviews
Internal audits	Test and calibration items
Measurement traceability	Report of results

Table 2 *Parameters to be ascertained to validate an analytical method.*

Applicability	Limit of detection
Selectivity	Limit of quantification
Calibration and linearity	Sensitivity
Trueness	Ruggedness
Accuracy	Robustness
Precision	Fitness for purpose
Recovery	Matrix variation
Range	Measurement uncertainty

Table 3 *The OECD series on the GLP principles and compliance monitoring.*

No. 1.   OECD Principles of Good Laboratory Practice.
No. 2.   Revised Guides for Compliance Monitoring Procedures for Good Laboratory Practice (1995).
No. 3.   Revised Guidance for the Conduct of Laboratory Inspections and Study Audits (1995).
No. 4.   Quality Assurance and GLP (as revised in 1999).
No. 5.   Compliance of Laboratory Suppliers with GLP Principles (as revised in 1999).
No. 6.   The Application of the GLP Principles to Field Studies (as revised in 1999).
No. 7.   The Application of the GLP Principles to Short Term Studies (as revised in 1999).
No. 8.   The Role and Responsibilities of the Study Director in GLP Studies (as revised in 1999).
No. 9.   Guidance for the Preparation of GLP Inspection Reports (1995).
No. 10. The Application of the Principles of GLP to Computerised Systems (1995).
No. 11. The Role and Responsibilities of the Sponsor in the Application of the Principles of GLP (1999).
No. 12. Requesting and Carrying Out Inspections and Study Audits in Another Country (2000).
No. 13. The Application of the OECD Principles of GLP to the Organisation and Management of Multi-site Studies (2002).
No. 14. The Application of the OECD GLP Principles to *in vitro* Studies (2004).

Table 4 *Key aspects of a Test Facility compliant with a GLP system.*

Director of the Test Facility	Study plan
Study Director	Final report
Quality Assurance Unit	Standard Operative Procedures
Archivist	Test Site (if applicable)
Sponsor	Principal Investigator (if applicable)
Test and reference items	

Table 5 *Key elements of the accreditation and GLP system.*

Accreditation quality system	GLP quality system	Overlapping aspects
Management of complaints	Master schedule	Management
Uncertainty of measurements	Study director	Motivation
Proficiency testing	Archivist	Training
Preventive actions	Quality assurance unit	Reference materials
Service to the client	Study plan	Equipment and maintenance
Sampling	Test article	Method validation
	Reports	Chain of custody
		Quality control procedures
		Corrective action
		Audits
		Sample reception

Figure 1 *Major steps of the validation process.*

THE WASHINGTON POST   Thursday, Sept. 8, 1977   A 11

Wide Errors, Possible Fraud Found in Private Lab Testing

By Bill Richards
Washington Post Staff Writer

Federal investigators have uncovered widespread flaws and, in some cases possible fraud in private testing laoratory results that form the data base for much of the government's chemical, drug and pesticide safety standards.

Information gathered by the Food and Drug Administration on test procedures by three private laboratories has been turned over to Justice Department officials in Chicago and New Jersey, according to federal officials. The investigations began last year.

Officials involved in the investigations by both the FDA and the Environmental Protection Agency said the faulty test results could cause problems in determining whether some products already approved for the marketplace are safe.

"Based on what we've found so far, there is a serious question about the data generated by any private testing laboratory," an FDA official said yesterday. "We don't know what data is bad but at the same time we don't know what is good either."

Last month EPA announced it had found "deficiencies" in tests performed for pesticide firms by Industrial Biotest Laboratory of Northbrook, Ill. The suburban Chicago laboratory ran 3,400 pesticide tests and has done other types of testing for a wide variety of manufacturers dealing with federal agencies such as FDA and the National Cancer Institute.

Federal officials who declined to be identified, said virtually every major or "pivotal" study done by IBT has

shown serious flaws. The laboratory is one of the firms referred to the Chicago U.S. attorney's office for further investigation based on the FDA's investigation.

EPA officials said they have found at least four other testing laboratories with suspect testing results. The four firms, officials said, have conducted a total of 4,000 animal tests for pesticide firms which in turn were used by EPA to set safety standards.

"We're looking at IBT as just the tip of the iceberg," said Edwin L. Johnson, EPA's pesticides chief. "There is no indication that the rest of the industry runs its business any differently than IBT does."

The other two firms identified as having been referred to Justice officials for further action are, G. D. Searle & Co., a leading pharmaceutical manufacturer, and Biometric Testing Inc. of Princeton Cliffs, N. J. IBT and Searle have denied manipulating test data. Biometric could not be reached for comment.

The Searle firm was the subject of extensive hearings last year by the Senate Health and Administrative Practice and Procedure subcommittee. At those hearings subcommittee Chairman Sen. Edward M. Kennedy (D-Mass.) sharply questioned the entire animal research industry. The hearings touched off the FDA's investigation into testing laboratories, FDA officials said.

Federal officials estimate that there are about 700 private testing laboratories scattered around the United States. Most do only a few animal

studies on contract to commercial manufacturers. But some, such as IBT, offer a wide variety of testing services.

EPA officials said they are preparing a spot-check program of as many as 100 laboratories which do pesticide testing for commercial manufacturers. The agency has already notified 33 pesticide makers to review their own data supplied by IBT, and plans to call on as many as 100 more for similar in-house reviews.

The spot checks by the agency will be full-scale audits matching raw data compiled by testing laboratories with the final reports they submit to manufacturers, EPA officials said.

Similar audits already run on some pesticide testing labs have turned up test animals which were reported to have died one week and then been alive the next, officials said.

The officials said a large number of the pesticide products tested by labs which have turned up suspect results are used on crops grown for human consumption.

FDA officials appeared split over the seriousness of the laboratory data findings. One official who asked not to be named, said the pattern of the test data deficiencies appeared "mind boggling."

Ernest Brisson, coordinator of the FDA's toxicological laboratory inspection program, however, said the FDA's findings so far were not overly disturbing. "We expected worse," Brisson said.

In its first 40 laboratory examinations, Brisson said, FDA investigators had to notify five firms of serious deficiencies.

He acknowledged that the early inspections were superficial and were done by inexperienced inspectors. "We're learning as we go along," he said.

Figure 2 *Article published by The Washington Post in 1977 on the frauds in chemical testing.*

References

1   IUPAC Technical Report, *Harmonized Guidelines for Single-Laboratory. Validation of Methods of Analysis.*

2   ISO/IEC 17025, *General Requirements for the Competence of Testing and Calibration Laboratories.*

3   OECD Series on Principles of Good Laboratory Practice and Compliance Monitoring, Number 1, *OECD Principles on Good Laboratory Practice* (as revised in 1997), ENV/MC/CHEM (98)17.

4   S. Caroli (Ed.), *The New Principles of Good Laboratory Practice: Priorities, Problems, Perspectives*, Ann. Ist. Super. Sanità, **38** (2002), pp.110.

5   Directive 2004/9/EC of the European Parliament and Council, 11 February 2004, *Off. J. EU*, **L50** (2004), 28 -43.

6   Directive 2004/10/EC of the European Parliament and Council, 11 February 2004, *Off. J. EU*, **L50** (2004), 44 -59

THE ROLE OF TRACEABILITY IN SUSTAINABLE DEVELOPMENT: THE UNIDO APPROACH

Dr. Otto R. Loesener Diaz[1] and Dr. Lamine M. Dhaoui[2]

[1] Industrial Development Officer, Trade Capacity Building Branch, UNIDO
[2] Senior Industrial Development Officer and Deputy to the Director; Trade Capacity Building Branch, UNIDO

1   INTRODUCTION

The topic of "Traceability", as it relates to trade and sustainable development, is at the forefront of policy and technical assistance discussions. The paper provides a review of the relation between trade and sustainable development, assessing the complex link between them and traceability from the UNIDO perspective.

Metrological Traceability can be available to the extent to which, technical infrastructure that comprises several key components is existent in an economy or a region, i.e. a metrology system (M) to ensure that the results of tests and measurements are in fact traceable to the International System of Units –wherever possible- and are performed correctly, an accreditation system (A) to ensure conformity assessment systems are monitored and meet international requirements and a national documentary standards body (S) to develop or adopt requirements for products, processes and systems. This includes Proficiency Testing (PT) Schemes and Certified Reference Materials (CRMs).

Developing countries are encountering a series of trade implications for the acceptance of their products in highly regulated markets that are closely related to the Agreement on Application of Sanitary and Phyto-sanitary Measures (SPS) and on Technical Barriers to Trade (TBT). Metrological Traceability and measurement uncertainty are key for comparing analytical results, that are used by decision makers, who have the power to allow the products to enter a given market or not. Therefore, Product Traceability is increasingly becoming more important. Product Traceability needs Metrological Traceability and there is a strong link between them.

The UNIDO approach through its Trade Capacity Building Initiative and its cooperation with WTO includes:

Removal of supply-side constraints;
Establishment or Strengthening of MAS infrastructure, so as to be able to prove conformity with market requirements; and
Integration into the multilateral trading system (MTS).

Examples of trade implications for selected developing countries and current efforts of UNIDO to overcome such situations (i.e. Bolivia, Mozambique or UEMOA region) are included.

2 QUALITY INFRASTRUCTURE, TRADE AND SUSTAINABLE DEVELOPMENT

Due to increasing globalization and trade liberalization, international economic exchanges have grown exponentially over the recent years and have become a major source of economic growth and social development. Enterprises in developing countries are more and more excluded from the new production and trade patterns. This is linked to many structural impediments and supply-side obstacles the countries are encountering in achieving productive efficiency and competitiveness. Industrial enterprises are increasingly aware of the need to improve product quality and productivity, however, they typically do not have the knowledge and skills to develop and implement appropriate quality management system solutions. The lack of productive capacity, quality infrastructure, services and related skills prevents the enterprises from accessing global markets **and to** integrate into international production and supply chains.

But even where productive capacity is established, there are difficulties with accessing markets, as products have to comply with a myriad of technical standards, health and safety requirements set by the importing markets. The TBT Agreement seeks to regulate that technical regulations and standards do not create unnecessary barriers to trade, but this requires that countries be in a position to fully participate in the standard-setting processes, while having a full-fledged infrastructure and systems for metrology, accreditation, certification and testing, as well as technical support and information services for industry.[1]

Developing countries also need to assist their industries and concerned government institutions to overcome or eliminate unnecessary technical barriers to trade caused by disparities in standards, metrology and conformity assessment practices between different trading partners. This requires mutually developed and recognized systems of metrology, accreditation and standardization (MAS) which enhance market transparency for manufacturers and purchasers and perform important protective functions for consumers and employees. MAS techniques and methodologies could also be used for the establishment of an internationally recognized measuring and monitoring system to qualify and quantify mineral and agricultural exports before they are shipped to ensure revenues and preserve the natural resource base.

Figure 1 shows the complex link between the different components of the Quality Infrastructure: Metrology, Accreditation and Standardization (i.e. Standards and Conformity Assessment), the industry at the production level as precondition for sustainable development in an economy.

Many developing countries are in the process of building up their MAS / Quality Infrastructure. As a consequence, several developing countries lack full participation in technical international organizations such as

- BIPM: International Bureau of Weights and Measures,
- IAF: International Accreditation Forum,

---

[1] See paper: JCDCMAS: "Building corresponding technical infrastructure to support sustainable development and trade in developing countries and countries in transition;" Joint Committee on co-ordination of technical Assistance to Developing Countries in Metrology, Accreditation and Standardization (JCDCMAS), www.jcdcmas.org, 2005.

This Committee includes UNIDO, the International Bureau of Weights and Measures (BIPM), the International Accreditation Forum (IAF), the International Electrotechnical Commission (IEC), the International Laboratory Accreditation Cooperation (ILAC), the International Organization for Standardization (ISO), the International Telecommunication Union's International Telecommunication Standardization Sector (ITU-T) and the International Organization of Legal Metrology (OIML).

- IEC: International Electrotechnical Commission,
- ILAC: International Laboratory Accreditation Cooperation,
- OIML: International Organization of Legal Metrology,

for a variety of reasons. Main benefits of participation in these organizations is the possibility to become signatory of their MRAs as follows:

BIPM: CIPM Mutual Recognition Arrangement (Mutual recognition of national measurement standards and of calibration and measurement capability – CMC),

IAF: Multilateral Mutual Recognition Agreements (accreditation of bodies dealing with certification of Products, Quality Management Systems, Environmental Management Systems, jointly with ILAC inspection bodies),

IEC: Peer Review,

ILAC: Mutual Recognition Arrangement (laboratory accreditation, jointly with IAF inspection bodies),

OIML: Mutual Acceptance Agreement (approval of legal measuring instruments),

that are contribute to facilitate trade.

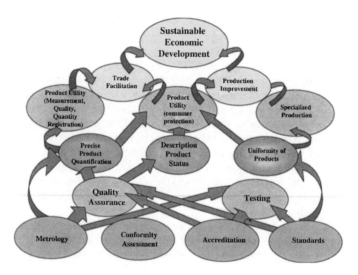

Figure 1 *MAS / Quality Infrastructure and sustainable development*

The General Council of the World Trade Organization (WTO) has been examining since 2000 the concerns raised by various developing country members regarding what they consider to be the inadequate implementation of some WTO Agreements, including the Agreement on Technical Barriers to Trade (TBT) and the Agreement on the Application of Sanitary and Phytosanitary Measures (SPS). In the context of the Triennial Reviews conducted by the TBT Committee, a survey was designed to assist developing country members of WTO to identify and prioritize their specifi needs so as to be able to better fulfill its requirements. UNIDO conducted an analysis of the responses to that questionnaire.[2]

---

[2] See UNIDO: "Relevance of UNIDO Services to the Responses to the WTO Questionnaire in Document G/TBT/W/178;" Vienna, United Nations Industrial Development Organization, Trade Capacity Building Working Paper Series, 03/05, V-05-91227, December 2005.

3 THE UNIDO APPROACH TO TRADE CAPACITY BUILDING

3.1 The UNIDO Approach

The role of UNIDO in connection with technical cooperation activities related to trade aims at building competitive industries for world markets and focuses on[3-4]:

- enabling developing countries to establish the essential quality and conformity assessment infrastructure required to increase exports
- assisting selected productive sectors with high export *potential to upgrade product* and production quality, comply with applicable standards and regulations so that they can export successfully
- assisting in cases where export products encounter technical barriers, and provide advice on technical solutions to the problem
- strengthening existing regional trade related organizations and arrangements.

The strategy adopted by UNIDO for the delivery of Trade-Related Technical Assistance and Capacity Building relies on the 3Cs approach and collaboration with UN and other Intergovernmental organizations, as well as *with technical organizations at* international and national level[5] (see Figure 2).

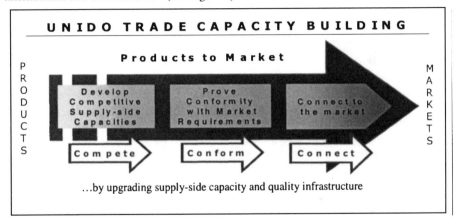

Figure 2 *3Cs approach*

Most developing countries have liberalized their markets in the hope that this would aotmatically allow them to expand their industrial base. However, the reduction and sometimes elimination of tariffs and quotas did not lead automatically to a substantial increase in their exports due to the following:

---

[3] See UN: "Review of technical cooperation in the United Nations;" Report of the Secretary-General, Fifty-eighth session (strengthening of the United Nations system), document A/58/382, New York, September 2003.

[4] See UNIDO: "Enabling Developing Countries to Participate in International Trade: Strengthening the supply capacity - A UNIDO Strategy for capacity building;" Vienna, United Nations Industrial Development Organization, March 2002.

[5] See UNIDO: "Trade Capacity Building: Supply-Side Development, Conformity and Integration in the Global Market;" UNIDO Position Paper in: "The Role of Industrial Development in the Achievement of the Millenium Development Goals;" Proceedings of the Industrial Development Forum and Associated Round Tables, Vienna, United Nations Industrial Development Organization, V.04-54862, June 2004.

- many developing countries do not have the capacity to produce goods conforming with export market requirements
- many developing countries are unable to assess and credibly demonstrate conformity of their products with technical regulations, standards and other market requirements
- many developing countries lack the capacity to connect effectively with global markets.

2.2 UNIDO Services

UNIDO's response can draw on a number of services and tools in order to service the specific requirements of the developing countries and economies in transition for developing I) competitive productive supply and export capacities and ii) MAS / Quality Infrastructure and services. In addition to integrated in-house services, such as: I) Industrial, Trade and Competitiveness Policy, ii) Investment Promotion, iii) Technology Promotion, Private Sector Development, iv) Agro-industries' technical support, v) Cleaner Production, vi) Energy and Climate Change as well as vii) Supply Chain Development, the core functions of the Industrial Competitiveness and Trade service are as follows:

3.2.1   *Strengthening the regulatory framework for conformity.*
- Assist governments with the development and effective application of a regulatory regime for metrology, accreditation, standards, testing and quality, and support for the development of an operational framework for the implementation of these regulations.

3.2.2   *Standards.*
- Establish or strengthen the capacity of existing standardization bodies by providing training, setting-up sub-sectoral technical committees, assist with the design and implementation of awareness programmes.
- Promote the adoption of standards at national and regional level, and assist participation in regional and international standards setting fora and networks.

3.2.3   *Metrology (measurement).*
- Establish or strengthen laboratory capacities for industrial and legal metrology by identifying measurement, calibration, verification and equipment requirements in accordance with the manufacturing and export needs of a country; thus, assisting in the physical set-up and start-up of laboratories, upgrading of measurement equipment according to international standards, training of technicians and assistance in networking, inter-comparison participation, mutual recognition arrangements and accreditation.
- Implement UNIDO software Measurement and Control-Chart Toolkit (MCCT) to meet the requirements related to metrological control of the ISO 9000:2000 standards.

3.2.4.   *Accreditation.*
- Establish or strengthen the capacity of accreditation bodies for the accreditation of system certifiers and laboratories.
- Assist emerging accreditation bodies to obtain international recognition from the International Accreditation Forum (IAF) or the International Laboratory Accreditation Cooperation (ILAC) by conducting pre-evaluations through the signature of Multilateral Recognition Agreements (IAF-MLA) or Arrangements (ILAC-MRA).

- Assist in networking and partnership arrangements with other national/regional institutions.

3.2.5 Testing.

- Establish or strengthen laboratory capacities for sampling, inspection, material and product testing including microbiological and chemical analysis, by specifying testing and equipment requirements. Provide technical support for the harmonization of testing procedures, training of staff through twinning arrangements, as well as assistance in networking, partnership and agreements for conformity assessment.

3.2.6 Certification.

- Assist for the development of national capability in certification, including product conformity mark schemes. Capacity building pilot projects related to product and system standards such as ISO 9000, ISO 14000 and HACCP.

3.2.7 Traceability.

- Develop national capability to comply with the EU traceability directives. Capacity building pilot projects to promote the compliance to traceability regulations.

3.2.8 Competitiveness enhancement through quality and productivity improvement.
The UNIDO Quality Approach enables the enterprises to enter into self-sustained and continuous improvements without the need for continued assistance by:

- Building institutional and human capacity at the level of Governments and Institutions for implementing quality management methodologies and systems (TQM, ISO 9000:2000, 6-Sigma, SPC, etc.) through practical demonstration in groups of pilot-enterprises for improving their quality and productivity.
- Promoting productivity by establishing Regional and National Quality and Productivity Centers. The centers act as one-stop-shop for productivity and quality improvements in manufacturing sector and associated institutions, by fostering production management upgrading at the level of enterprises, industrial sectors, supply chains, technology institutes and policy-related government bodies. Furthermore, through process and competitiveness benchmarking services the Center will be able to help identifying, adapt and promote best manufacturing practices.
- Improving capability to monitor and increase business performance through the implementation of UNIDO Business Excellence Software packages: Pharos (Business Navigator) suitable for SMEs, PRODUCE-Plus for production management and productivity measurement, BEST (Business Environment Strategic Toolkit) and FIT (Financial Improvement Toolkit) and MCCT (Measurement Control Chart Toolkit) for monitoring calibration of equipment and carry out simple Statistical Process Control.

3.2.9 Industrial Restructuring and Upgrading.

- Build capacities in public and private sector institutions to develop national industrial upgrading and restructuring policies, support mechanisms including financing schemes, restructuring and upgrading programmes. Develop national consulting capability to address the restructuring and upgrading needs.

3.2.10 Global forum functions.

- Foster linkages and cooperation among regional and international organizations such as WTO, BIPM, IAF, ILAC, ISO IUPAC, OIML and regional and international

standardization and accreditation bodies in order to facilitate networking, promotion of mutual recognition of certificates and harmonization of standards.
- Monitor global trends in standardization, conformity assessment, metrology, accreditation and testing in order to promote awareness and strengthen the trade capacity of developing countries through applied research and benchmarking studies, training and participation in international conferences.
- Carrying out and dissemination of benchmarking analyses to identify best practices of standardization bodies and laboratories.
- Develop and improve business excellence tools such as the business environment strategic toolkit (BEST), the financial improvement toolkit (FIT) and the business performance navigator (PHAROS), and printed material for upgrading of business performance such as the manual on diagnostic for restructuring and upgrading through applied research and benchmarking studies in quality management, industry modernization and productivity.

3.3 Principles

Technical cooperation services are designed and implemented to comply with UNIDO quality criteria, based on the following principles:
- *Relevance.* Services will meet mutually identified development constraints that UNIDO clients have in meeting the developmental requirements of their target groups;
- Demand- orientation. *Programmes are based on comprehensive needs assessment and the full involvement of recipients in the programming exercise.*
- *Local ownership.* Recipients of UNIDO services have clearly expressed their requirements and are ready to use UNIDO services to overcome their constraints;
- *Effectiveness / efficiency.* UNIDO and its clients have analyzed the least-cost approach to fully solve their development constraints;
- *Sustainability / impact.* The design and implementation of services ensure sustained benefits for the client and significant improvements at the level of the target group.

UNIDO provides services within the coordination mechanism of the United Nations system (the United Nations Development Assistance Framework (UNDAF), Common Country Assessment (CCA)), and complements bilateral and multilateral development cooperation programmes so as to produce synergy and enhance the impact of the entire development effort of a specific country.

3.4 Methodological Approach

The following tools and modalities are applied to deliver technical assistance in these fields:
- *Awareness campaigns.* Seminars and workshops to raise awareness on a given subject.
- *Short- and long-term missions.* International experts transfer technical and/or managerial knowledge to national experts to establish or upgrade parts of the national quality infrastructure.
- *Study tours.* Promotion of the participation of national experts in regional or international meetings, seminars, workshops, conferences, etc. as well as in training programmes in another country.
- *On-the-job training.* Practical training programmes in-situ offered by international experts in the home country to transfer technical and/or managerial knowledge to

national experts.
- *Seminars and workshops.* Training programmes addressed to different target groups to transfer specific knowledge on national quality infrastructure, national quality policy, technical regulations, standards, accreditation, certification, inspection, metrology, quality promotion and requirements of the TBT Agreement.
- *Train the Trainers.* International experts train national experts to build local capacity and human resources available in the country, thus addressing sustainability issues.
- *Twinning arrangements.* Transfer of knowledge through exchange of experts and study tours between two countries / regions.
- *Pre-Peer Evaluations.* Evaluation of accreditation bodies through experts from different countries to increase mutual confidence and enable developing countries to join international Mutual Recognition Arrangements.
- *Capacity building.* Establishment or upgrading of infrastructure in standardization bodies, metrology institutes, calibration and sector-specific testing facilities, inspection and accreditation bodies, enquiry points, as well as quality, productivity and competitiveness centers, etc.

4 THE ROLE OF TRACEABILITY

UNIDO supports technical assistance programmes at country and regional level in Standardization, Conformity Assessment, Metrology and TBT-related matters, in which traceability also plays a role.

At the country level, the main vehicle for the provision of technical assistance is Integrated Country Programmes that provide comprehensive and multi-disciplinary packages of services in response to local support requirements, and the particular structural and capacity constraints in the industry. For example, a programme may include services in the field of industrial supply-side policy, investment and technology promotion, SME development, upgrading of agro-industries, and cleaner production. Most integrated programmes include Standardization, Conformity Assessment, Metrology or TBT-related issues as a priority component requested by the recipient government. Some examples are: UEMOA (regional), Bolivia and Mozambique (national).

4.1. UEMOA / EU / UNIDO Regional Programme

The joint initiative by the European Union (EU), UNIDO and the Western African Monetary and Economic Union (UEMOA) in accreditation, standardization and quality promotion aims at facilitating trade and reducing poverty in Africa.

The "Union Economique et Monétaire Ouest Africaine" (UEMOA) counts 8 member states: Benin, Burkina Faso, Côte d'Ivoire, Guinea Bissau, Mali, Niger, Senegal and Togo. This regional organizations' objective is to establish- through accelerating the regional integration process- an economic and monetary union ensuring the creation of a regional market with the free movement of goods and people.

Despite a strong potential for economic growth, the region had not been able to take advantage of special initiatives for exports (ACP country access to EU, "Everything but Arms", AGOA...) nor rip the benefits of globalization and international trade because of several significant issues. These issues include:
- Insufficient 1) support institutions, 2) accredited laboratories and testing/analysis capacities, 3) metrology services, 4) strained staff, 5) information on standards, 6)

quality promotion activities
- High physical infrastructure costs
- Lack of necessary modern production capacities
- Low product quality that does not meet international standards for exports
- Poor capacities of institutional infrastructures.

UNIDO's Quality Programme Phase I posed itself as a viable solution to these problems and as a key element to fostering UEMOA integration into regional and international trade. The global objective of this proposal was to strengthen regional economic integration and trade in the UEMOA region through the establishment of essential quality and conformity assessment infrastructures. The programme particularly aimed at enhancing the competitiveness of high export-potential sectors in the member states by ensuring that the goods produced in the region comply with international standards for exports.

The main results of the UEMOA-UNIDO-EU are encouraging and are, namely:
- The development and upgrading of quality infrastructures in all 8 UEMOA countries
- The establishment of 4 permanent regional structures, which are the West African Accreditation System (SOAC), the Regional Standardization, Certification and Quality Promotion Secretariat (NORMCERQ), the West African Metrology Secretariat (SOAMET) and the Regional Quality Coordination Committee (CRECQ)
- The current adoption of a framework legislation, which will enhance consumers' awareness and consumers' protection
- The launching of 3 internet-accessible databases on laboratories, standardization and quality promotion
- The creation of a 'UEMOA Regional Quality Award' and of 8 'National Quality Awards'
- The lift of the ban on Benin fish products and the resumption of their exports
- The development of a West African Cotton Standard (in collaboration with the African Cotton Association, Chad and Cameroon).

Sectoral support was provided as follows:
1.   Cotton Sector:
In order to develop the cotton sector's exports and revenues in the region, UNIDO has:
- Bought HVI for Mali and Togo
- Ordered HVI for Senegal
- Restored Burkina Faso's HVI
- Promoted the UEMOA Cotton Standard
- Identified and strengthened the regional technical center for cotton
- Made available a manual on how to ensure cotton quality in the UEMOA region.
2.   Fish Sector:
In order to develop and maintain the fish sector's exports and revenues, UNIDO has:
- Introduced the HACCP system in fisheries in Togo, Benin and Cote d'Ivoire
- Trained the authorities and companies' staff of the fish sector in sanitary inspection in Benin, Togo and Cote d'Ivoire
- Helped Benin and Togo to resuming their fish exports with an estimated value of USD 200 million thanks to the UNIDO Quality Programme
3. Cashew Nut:
In order to develop the cashew nut sector's exports and revenues, UNIDO has:
- Elaborated a transformation and export strategy for the cashew nut sector

- Trained the operators of the cashew nut sector in sanitary safety in Guinea-Bissau
- Contributed in establishing partnerships between operators of the cashew nut sector in Guinea Bissau and Mozambique.

The availability of laboratory facilities close to accreditation, their participation in Proficiency Testing Schemes and the delivery of Certified Reference Materials was key to overcome the export ban imposed on some of the participating countries.

A second phase of this regional programme is currently being planned, with a view not only to consolidate the results achieved in the first phase, but also to expand the structures to the English speaking ECOWAS countries. Also as a consequence of this programme a major cotton initiative was launched recently together with WTO.

4.2. Mozambique: Enhancing The Capacities Of The Food Safety And Quality Assurance System For Trade

Like other developing countries Mozambique faces the challenge of overcoming technical barriers to trade in food products especially with the changing composition of its potential exports from the traditional primary products to high value perishable and value added food products. Global trade in high value food products has expanded enormously over the last decades facilitated in part by comparatively low and declining tariff barriers as well as pressure generated by expanding year round supplies. However, trade in these products is governed by an overly dynamic array of food safety and agricultural health standards that have been developed to address various risks such as pesticides, microbes, toxins etc. Sanitary and phytosanitary standards in the public and private sectors have proliferated in recent years and continue to evolve internationally, nationally and within individual supply chains.

The constraints in meeting TBT and SPS requirements have resulted in the closure of many food processing industries, having failed to remain competitive in terms of safety and quality in the dynamic global market. This has created serious problems in the generation of non-farm income, employment and even in the development of the support industry (packaging and labelling, food chemicals) etc. For a country with an agro-based economy, the inability to utilize its most abundant resource and take advantage of the various concessionary access to the EU and the USA markets (made available under the EU- "Everything But Arms" initiative and the US- "Africa Growth and Opportunities Act) translates into considerable loss of potential revenue from food trade and ultimately directly contributes to poverty. Realizing the importance of establishing a national food safety and quality infrastructure, Mozambique adopted a food and nutrition security strategy that amongst others seeks to increase the quality and quantity of nationally produced foods, and to increase the capacity to import and export products in line with the liberalisation of markets. In addition, the National Quality Policy (NQP) developed with assistance from UNIDO, has clear indications of objectives, priorities, implementation plan and estimated costs of the various components to improve the quality infrastructure in the country. It is a useful guide to decision makers at all levels (the government international, donor and private sector levels).

UNIDO has conducted a preliminary assessment in Mozambique with special focus on the public support institutions. The project has thus been formulated from the findings summarized in Figure 3.

As LCD, the issues applicable to Mozambique are the same as most of the UEMOA countries, i. e.:

- Insufficient 1) support institutions, 2) accredited laboratories and testing/analysis capacities, 3) metrology services, 4) strained staff, 5) information on standards, 6) quality promotion activities
- High physical infrastructure costs
- Lack of necessary modern production capacities
- Low product quality that does not meet international standards for exports
- Poor capacities of institutional infrastructures.

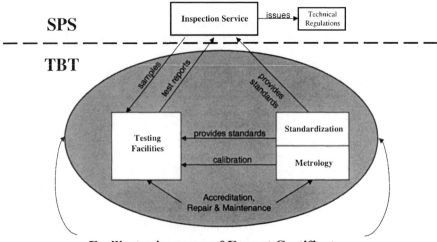

Facilitates issuance of Export Certificates

Figure 3 *Improvement of the MAS/Quality Infrastructure*

In order to increase the quantity and quality of nationally produced foods and the capacity to import/export products in line with the principle of liberalising markets, the project is focusing on:

- Harmonizing food legislation and awareness raising on food safety
- Developing a coordination mechanism between the various players in the system
- Upgrading the National Quality Infrastructure, including national metrology, calibration and food testing facilities, their accreditation, repair and maintenance system, as well as related standards and conformity assessment activities (sampling, inspection, technical regulations) and National Enquiry Points,

so as to obtain credibility while issuing certificates to export following priority products: i) cashew nuts, ii) honey, iii) fruits, iv) edible oils, without any hindrance.

4.3. Bolivia

UNIDO and WTO have joined forces to assist developing countries to overcome supply capacity constraints and the lack of Quality Infrastructure. The two organizations have signed a Memorandum of Understanding and established a concept for providing assistance to developing countries consisting of three modules:

- Develop supply-side capacity and competitiveness
- Assess and demonstrate conformity to market requirements

- Develop connectivity to the global market and integration into the multilateral trading system.

The implementation of this concept was started on a pilot basis in nine countries, one of them being Bolivia. Brazil nuts were identified as the priority product with high and strategic export potential encountering barriers to trade and trade-capacity building weaknesses see Figure 4.

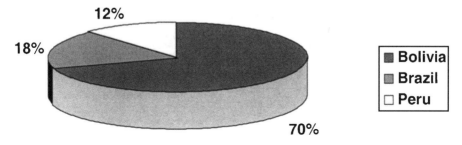

Figure 4 *Export share of Brazil nuts*

The project builds on the results of a quick assessment and proposes concrete measures to expand Bolivian exports of Bolivian nuts through an improvement of the supply-side capacity, establishment and upgrading of the conformity assessment infrastructure (test laboratories and certification bodies) as well as through enhanced awareness of the benefits of integration of the country into the multilateral trading system and implementation of WTO agreements, in particular the SPS agreement. Brazilian nuts exports from Bolivia are endangered, due to the lack of (accredited) testing facilities to check the aflatoxin level. The acceptance of Brazil nuts depends on the aflatoxin level that varies from market to market between 4 and 20 ppb (type B).

4.4 Product Vrs. Metrological Traceability

Recent regulations on Food Traceability, have put in place strict guidelines requiring that all food manufactured and sold should be safe and fully traceable "farm to fork" and back again. They affect particularly LDCs, which have to abide by these regulations in order to sell their goods. It is, therefore, crucial for them to focus on product traceability issues.

Hence, traceability regulations with regard to the product do certainly imply the need of metrological traceability / measurement uncertainty, in connection with issuance of test reports by (accredited) local laboratories. In this regard, availability of Certified Reference Materials at the laboratory facilities, the use of validated analytical methods and the participation in proficiency testing exercises are also a precondition to obtain market entry.

In the field of traceability, UNIDO has undertaken evaluation of the current situation in the three cases. Traceability chains according to the new version of the VIM are being worked out in collaboration with IUPAC, in order to ensure effective implementation of technical assistance activities, as in several cases metrological traceability in chemistry is high in demand.

QUALITY ASSURANCE OF CHEMICAL MEASUREMENTS - METROLOGICAL OR MANAGEMENT EFFORT

Mirko Prosek, Alenka Golc-Wondra, Maja Fir

National Institute of Chemistry, Hajdrihova 19, 1000 Ljubljana, Slovenia

INTRODUCTION

Results of chemical analysis are very important factor for correct decision in many fields of human activities. Every day millions of analyses are done in industry, hospitals, research institutes, control laboratories etc. Users want to be sure about the reported values and we should demonstrate quality of our work and present the confidence that can be placed on our reports. This task is not easy today and had not been easy even thirty or more years ago. There were no sophisticated instruments, computers, and metrological standards, but still analytical results were sound and decisions were correct. Where was the miracle of their work? The answer is not difficult; they were familiar with analytical management. Analytical knowledge was oriented towards understanding and elimination of sample uncertainty and not into formalistic solutions of measurement uncertainty so popular today. From the "chemical approach" described more than 30 years ago by John K. Taylor, which based on the idea that a quality assurance in an analytical laboratory is a product of two closely related tasks, **the quality control** and **the quality assessment** we now arrived to "physical approach" based on metrological parameters like **traceability** and **measuring uncertainty**.

In order to demonstrate that quality assurance in chemical analytical laboratory primarily depends on correct scientific based analytical management (assessment) and only then, in the following step on metrological parameters, we looked through classic papers from John K. Taylor (1-4) and made some new measuring experiments.

More that 30 years ago Taylor wrote down a sentence; *"The objective of quality assurance programs for analytical measurements is to reduce measurements errors to tolerable limits and to provide a means of ensuring that the measurements generated have a high probability of acceptable quality."* This statement is valid, but what has been changed. Taylor describes quality assurance in an analytical laboratory as a product of two simultaneous activities. Quality control is a mechanism established to control errors, and quality assessment verifies that the system is operating within acceptable limits. This second part is a part of the analytical management. In his article from 1981 Taylor put down, *"Quality is a subjective term. What is a high quality in one situation may be a low or unacceptable quality in another case. Clearly the tolerable limits of error must be established for each case. Along with this there must be a clear understanding of the measurement process and its capability to provide the results desired."*

With examples taken from our experiments and calculations (5,6) we want to show how important is management in analytical chemistry and to demonstrate that definitions and solutions presented 30 years ago are still present.

Results are reported with the analytical uncertainty which is a combination of sample uncertainty and measurement uncertainty. Sample uncertainty is normally much bigger then measurement uncertainty and it is in some cases uncontrollable. If we want to report the amount of analyte in a sample with a complex matrix or in an inhomogeneous sample we are faced with such a problem. Measurement uncertainty with 10% RSD is rejected, but at the same time the sample uncertainty of 50% or more is toleratted.

It is very important that we have in mind that end users of our work are concerned about results and not methods. Analytical measurements are made because it is believed that compositional information is needed for some end use in problem solving. Each analytical result has a reason to exist; we are not making and improving analytical procedure just for sport, but to solve real problems. Selection of analytical method comes from the end user and also its uncertainty should be tailored according to his expectation. Expensive, time consuming primary methods are usually bad selection in a real world; nevertheless they have very small measurement uncertainty.

The principle of measuring uncertainty is known for years, but ISO Guide, published in 1993 formally established general rules for evaluating and expressing uncertainty in measurement across a broad spectrum of measurements. In 1995 these rules were, without serious reflection applied in chemical measurements with the Eurachem guide "Quantifying Uncertainty in Analytical Measurement" published in 1995 (7) and 2nd edition of the same guide from 2000 (8). In this guideline metrological and analytical tasks are mixed and principles of physical measurements are put across chemical measurements without thinking about facts, that for most chemical measurements, significant uncertainties are not coming from the same sources as in physical measurements (9). Chemical standards differ from physical standards. Even if we can prepare sufficiently pure standards, we can not incorporate them into sample matrix correctly, to produce substances that can reliably calibrate chemical measurements. Matrix match between standard and sample is critical and difficult to achieve. In real life it is not possible to prepare representative sample. It is not possible to get "true value" but is equally not possible to get homogeneous sample.

Poor analytical results may be result of contaminated reagents, operator errors, bad instrument calibrations, poor data handling, etc. These errors can be simply eliminated with proper quality assurance but the problem of invalid sampling could not be solved at all.

In most cases sampling uncertainty is much bigger than measurement uncertainty. In **Figure 1** and equation 1 relation between relative standard deviations (RSD) of sampling s_S, and measurement s_M in HPLC and TLC are shown. In HPLC measurement uncertainty is normally less then 2% but if it is combined with 16% of sampling uncertainty (smaller amount of probes) the analytical uncertainty is 16.12%. If we take HPTLC with measurement uncertainty about 8 % and combine it with 8 % of improved sampling uncertainty (more probes) we obtain only 11.31% of analytical uncertainty. This is theoretical example, but in real life it is often possible to get better result with worse analytical techniques due to more suitable and more reliable sampling.

In order to report reliable results, we have to prepare procedures, whenever possible that measurements are conducted in such a way that each standard deviation can be evaluated separately. Unfortunately this principle is not often applicable in real life. In many cases it is possible to separate s_M from s_S but the usual price for this operation is a large increase of sampling uncertainty.

$$s_0{}^2 = s_{Sampling}{}^2 + s_{Measurement}{}^2 + s_{Other}{}^2 \qquad\qquad eq.1$$

Youden has pointed out that once the measurement uncertainty is reduced to a third or less of the sampling uncertainty, further reduction in the measurement uncertainty is of little importance. (10). Therefore, if the sampling or sample uncertainty are large and can not be reduced; a rapid approximate analytical method can be even better solution. In many cases the best solution is qualified management of more analytical procedures of low precision that permits more probes to be examined and offer the best way to reduce analytical uncertainty.

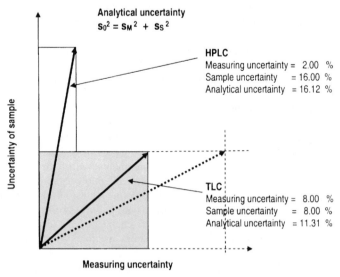

Figure 1. *Analytical uncertainties of certain samples, measured with quantitative HPLC and HPTLC.*

2   EXPERIMENTAL WORK

In order to demonstrate how a method with greater measurement uncertainty can even improve the quality of reported result by reducing the analytical uncertainty we prepared an experiment, **Figure 2**.

On a clean glass plate dimensions 40 x 40 cm, 50 ml of a solution (2 mg/mL salicylic acid in methanol) were applied with a spraying device (CAMAG) normally used for derivatization of TLC plates. Around the plate 10 cm wide glass frame was placed. Salicylic acid eventually deposit on a frame was wiped off and quantified. This value was later subtracted from theoretically applied amount in order to get correct concentration of deposited salicylic acid on the glass plate. Obtained value was used for calculation of recovery of extraction from the cottons.

After the evaporation of solvent, 64 probes were taken from the different positions of the test plate using swab method (11). Salicylic acid was quantitatively wiped out with wetted cotton (2.0 mL of solvent was applied on each swab). Cottons were inserted into a 50 ml brawn bottles and 28 ml of solvent was added in each bottle. Swabs were extracted on ultrasonic bath for 15 minutes. An aliquot of 5 ml was taken as a sample test solution.

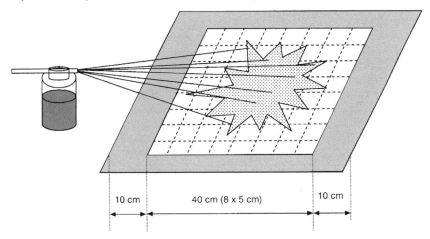

Figure 2. Preparation of test plate with spraying.

HPLC analysis

Prepared sample were analyzed with HPLC system consisting of Constametric 4100 pump (TSP), auto sampler Spectra system AS100 equipped with 10 µL fix loop and UV detector Spectramonitor 3200 (TSP). Calibration curve with 4 calibration standard in rang from 0.2 mg/L to 2 mg/L of salicylic acid were prepared. Prior the run SST (System Suitability Test) was done. Six times calibration standard 1 mg/L was injected and retention time, number of plate, peak area, and peak symmetry were calculated. After SST, randomly selected samples and three sets of calibration standards were injected in duplicate, together 158 injections in one sequence run.

Mobile phase was methanol : water 1:1, and formic acid (pH = 2.6) Flow rate was 1.5 mL/min. UV absorption was measured at 254 nm. For the whole sequence 18 hours on HPLC system were needed.

TLC analysis

Samples were analyses also with HPTLC (12), 64 samples and 4 calibration standards were applied manually with micropipettes (Camag, Mutenz CH) on five HPTLC plates (Merck Silica gel 60, Code 1.05641). Plates were developed in normal HPTLC tank, mobile phase: cyclohexane / isopropanol / chloroform / acetic acid, 60/5/5/10 and quantified with TLC scanner III (Camag). The necessary time for analysis of 5 TLC plates was 5 hours, 1 hour for application, 1 hour for development in 3 tanks and 3 hours for scanning and data processing. In **Table 1** an overview of both methods is presented.

Results

Final results expressed as mg/swab area are shown in table **2a** (HPLC) and **2b** (HPTLC). These values were further used for studying which number of swabs and swabbing positions gave the best recovery. Different number of swabs, between 1 and 12, was selected and different positions of swabbing were combined.

Table 1. Comparison of both techniques

method	HPLC	HPTLC
instrumentation	HPLC pump Auto sampler HPLC Column	Development tanks HPTLC plates Micropipettes
quantification	UV detector	TLC scanner III
Necessary time	7 minutes/injection	2 minutes/track
Number of analysis	158 injections (8/hour)	90 lanes/5 plates (18/plate)
Price (all samples)	900 Euro	300 Euro
Plate Recovery (%)	62.99 mg (87.9%)	66.21 mg (92.4%)
Analytical RSD	28.16 %	28.53 %
RSD of analytical method	0.33 %	4.66 %
Sample uncertainty	28.16 %	28.15%
Routine analysis (example)		
Sequence or **Application plan**	3 SST 4 calib. standards 2 different swabs (2x)	SST (10x scan of calib.standard no. 2 4 calib. standards 10 different swabs
N of injection/application.	11	14
Price	75 Euro	60 Euro
Bias	Small	Small
Sample Recovery (%)	88.85%	94.59%

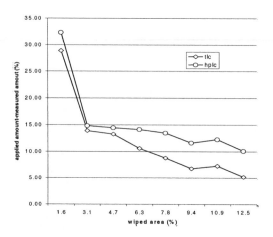

Figure 4. *Difference between true values and measured values as the function of different number of collected swabs. Swab area was 25 cm², from 1 to 8 swabs were merged in one sample. Wiped areas represented 1.6 to 12.5 % of total area. Swab positions were selected randomly. Plotted values represent the average of eight trials.*

If only one swab is taken we have 64 different results, if two swabs are combined in one sample we obtain 2016 combinations, and with three swabs there are already 41664 combinations. With more swabs, there are even more combinations. The number of combined swabs has no influence on average value and SD, as long as all the combinations are taken. This is not against the logic, there is only one true value of analyte, and there is only one distribution of analyte in the sample.

Table 2
applied amount is 71,68 mg per plate

a.) Results of TLC measurement (mg/swab)

1	2	3	4	5	6	7	8	
0.63	0.72	0.96	0.96	1.09	0.65	0.74	0.61	*1*
0.99	0.79	0.95	1.36	0.95	1.04	0.74	0.70	*2*
1.08	1.52	1.11	1.25	0.98	0.92	1.04	0.77	*3*
0.75	1.37	1.23	1.46	1.49	1.24	1.40	0.76	*4*
0.82	1.11	1.14	1.66	1.27	1.31	1.02	0.64	*5*
0.78	1.08	1.15	1.67	1.30	1.34	1.30	0.90	*6*
0.97	1.48	1.36	1.61	1.23	1.12	1.00	0.74	*7*
0.59	0.62	0.52	1.08	0.88	0.91	0.74	0.67	*8*

Measured amount 66.21 mg, Recovery = 92.4%. Average = 1.03 mg/swab, RSD = 28.53%

b.) Results of HPLC measurement (mg/swab)

1	2	3	4	5	6	7	8	
0.57	0.72	0.86	0.93	1.15	0.69	0.78	0.51	*1*
0.82	0.78	0.89	1.22	0.93	1.01	0.80	0.54	*2*
1.00	1.28	1.09	1.23	0.99	0.88	0.96	0.58	*3*
0.78	1.30	1.16	1.39	1.27	1.22	1.29	0.63	*4*
0.74	1.22	1.10	1.56	1.20	1.21	1.05	0.55	*5*
0.77	1.15	1.16	1.55	1.17	1.26	1.17	0.79	*6*
0.96	1.32	1.42	1.58	1.24	0.95	1.01	0.65	*7*
0.63	0.66	0.67	1.06	0.80	0.78	0.78	0.60	*8*

Measured amount 62.99 mg, Recovery = 87.9%. Average = 0.98 mg/swab, RSD = 28.16%

Usually it is not possible to analyze the whole sample, and it is necessary to prepare a representative sub sample, according to our skill. In **Figure 4** the relation between recoveries and swabbed area are shown. We selected average value, median, and maximum (worst case) in order to evaluate the total amount of analyte. Our samples differ from a single swab 1.6% of wiped area to combination of more swabs up to 12.5% of wiped area, **Figure 5.** Random selection of swabbing positions can be used when we have no idea about a history of a sample. If we know the origin of a sample, and want to obtain a representative (useful) result with a small sample uncertainty, we have to think over how to prepare correct procedure of sampling.

In our experiment the person who sprayed the test plate, and the person who took swabs was not the same. According to logic we can expect that the biggest concentration is in the centre of a plate and the smallest concentration is at the corners. In **Figure 6** prepared swabbing templates are shown. Results presented in **Table 3** show that templates (d) and (h) give the best recovery. The templates (g) and (c) give the smallest amounts due to swabbing at the edges, while (f) and (e) modes give the highest values due to the swabbing in the middle of the plate.

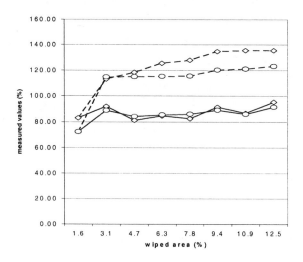

Figure 5. *Changes of median values and increasing number of swabs (wiped area), lower curves and at worst case values and number of swabs, upper curves.*

3 CONCLUSION

Our experiment shows that measurement uncertainty of in HPTLC and in HPLC method: is relatively small and does not influence on recovery. Reliable result can be obtained only with thoughtful selection of swabbing. If we want to come close to the true value we have to manage analytical procedure with well considered selection of number and positions o sampling locations. Presented results also show that median gives acceptable recovery similar to average value, while selection of maximum value (worst case principle) is no reliable and with three or more swabs gives too high values, from 110% to 140%. Thi: procedure can be useful when we need a result for safe decision, for instance in cleanin, validation.

Analysts are today occupied with traceability, uncertainty and with ideas how to find primary methods of measurements with the highest metrological quality. Physica measurements and metrology are important factor in our work, but they can not provide information about sample uncertainty. The "Error budget" procedure can not replace validation. It is possible to extract measurements uncertainty from validation results, but i is questionable if this is really necessary. When a method is used with different type o samples we have to revalidate our method. Professionally planed validation is the only wa: to get information about sample uncertainty, which is always bigger than measuremen uncertainty.

In real world it is very difficult to get two identical samples so only the analytica management can improve the confidence into the result of chemical measurements.

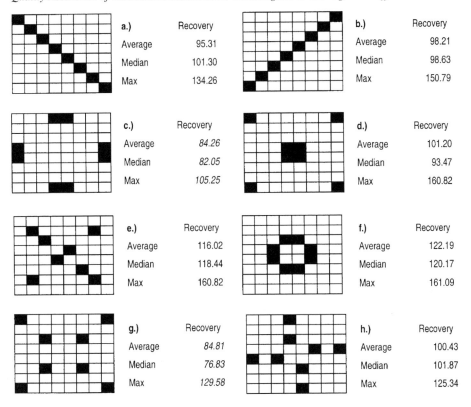

	Recovery
a.)	
Average	95.31
Median	101.30
Max	134.26

	Recovery
b.)	
Average	98.21
Median	98.63
Max	150.79

	Recovery
c.)	
Average	*84.26*
Median	*82.05*
Max	*105.25*

	Recovery
d.)	
Average	101.20
Median	93.47
Max	160.82

	Recovery
e.)	
Average	116.02
Median	118.44
Max	160.82

	Recovery
f.)	
Average	122.19
Median	120.17
Max	161.09

	Recovery
g.)	
Average	*84.81*
Median	*76.83*
Max	*129.58*

	Recovery
h.)	
Average	100.43
Median	101.87
Max	125.34

Figure 6. The difference between true value, and measured values expressed in percentage. The average value of 8 swabs was used for quantification.

References

1   Taylor JK (1981) Anal Chem 53 : 1588A-1596A
2   Taylor JK (1983) Anal.Chem 55 : 600A –605A
3   Taylor JK (1985) Handbook for SMR Users, National Bureau of Standards NBS/SP160/100
4   Taylor JK (1987) Quality Assurance of Chemical measurements, Lewis, Michigan, ISBN 0-87371-097-5
5   Prošek M, Pukl M, Miksa L, Golc-Wondra A (1993) J Planar Chromatog 6 : 62-65
6   Prošek M, Golc Wondra A, Vovk I (2001) J Planar Chromatog 14 : 62-65
7   International Organization for Standardization (1995) Guide to the Expression of Uncertainty in Measurements, ISO, Case Postal 56, CH1211 Geneva
8   EURACHEM/CITAC (2000) 2nd edn. Quantifying Uncertainty in Analytical Measurement : http://www.vtt.fi/ket/eurachem/
9   Horowitz W (1998) J AOAC 81 : 785-794
10 Youden WJ (1967) J AOAC 50 : 1007-1015
11 Cooper DW (1997) Pharm. Technol. Asia Sep/Oct : http://www.texwipe.com/
12 Agbaba D, Lazarevic R, Zivanov-Stakic D, Vladimirov S (1995) J Planar Chromatog 8 : 393-395

COMBINATION OF RESULTS FROM SEVERAL MEASUREMENTS - AN EVERLASTING PROBLEM

W. Hässelbarth and W. Bremser

Federal Institute for Materials Research and Testing (BAM), Department I "Analytical Chemistry, Reference Materials", Richard-Willstätter Str. 11, 12489 Berlin, Germany

1 INTRODUCTION

One of the most intriguing and persistent problems in the evaluation of measurement data is how to combine the results obtained in several measurements of the same quantity. Here the problem is not so much how to distill from the data a good estimate of the quantity itself – most often some kind of mean will do, and different means (unweighted, weighted etc.) will generate similar values. The big question, however, is how to estimate the uncertainty of the respective mean value. For example, when having to combine discrepant results (i.e. in case of differences beyond what would be covered by specified uncertainties), any reasonable uncertainty estimate will need to account for the bias observed, and this task requires an appropriate mathematical model as a basis for data analysis. Depending on the model and the data evaluation technique, uncertainty estimates may vary considerably, leaving the user with a difficult decision. But even when the results agree within specified uncertainties, it is often unclear whether the data are suffiently independent to justify the use of the famous factor of $1/\sqrt{N}$ in the calculation of the uncertainty of a mean value, and how to proceed if not. While the "classical" papers on this subject (treating the problem in the framework of statistical data evaluation) are by now quite old, the evaluation of measurement uncertainty according to the GUM [1] has shed some new light and triggered new investigations and publications on this problem.

This paper sets out to review the most common problems encountered when undertaking the task to combine results from different measurements of the same quantity, and the most simple approaches for solving these problems. Even with this limited scope this review is far from complete. For example, the problem how to determine a reference value and its uncertainty from the results of an interlaboratory comparison (focussed on so-called Key Comparisons of national metrology institutes) has recently attracted a lot of research and papers [2], including a contribution to this volume. The items selected for this review mirror the preferences of the authors and, of course, their limited overview.

2 GETTING STARTED: TWO MEASUREMENTS

For getting started, let us consider the most simple case of two measurement results $a \pm u(a)$ and $b \pm u(b)$ where a and b may be single values or mean values from replicate measurements, and $u(a)$, $u(b)$ are standard uncertainties obtained from an evaluation of measurement uncertainty for a and b. The task is to determine a combined result $c \pm u(c)$.

In absence any specific reasons suggesting another estimate, the default choice for c will be the common mean

$$c = \frac{a+b}{2} \tag{1}$$

Straighforward application of uncertainty propagation according to the GUM gives a combined standard uncertainty $u(c)$ as follows

$$u(c) = \frac{1}{2}\sqrt{u^2(a) + u^2(b)} \tag{2}$$

Now, what can be wrong, or require amendment, with these combination rules?

1.1 Mean Values

Obviously c should be located between a and b, i.e. be some kind of average of these data. The common mean is the democratic choice. However, there may be reasons to prefer an undemo-cratic one, giving unequal weights to a and b. For example, a may be the result of a single measurement ($n_a = 1$) while b is a mean value of 100 replicate measurements ($n_b = 100$). Then it may appear more appropriate to use a weigthed mean according to

$$c = \frac{n_a \cdot a + n_b \cdot b}{n_a + n_b} \tag{1a}$$

Also, if the uncertainties $u(a)$ and $u(b)$ are dramatically different, it may be argued that this implies different levels of confidence in the data a and b which should be accounted for by appropriate weighing. According to basic statistical considerations, this weighing should be done using the inverse variances (squared standard uncertainties) according to

$$c = \frac{v_a \cdot a + v_b \cdot b}{v_a + v_b} \quad \text{with weights} \quad v_a = \frac{1}{u^2(a)} \quad , \quad v_b = \frac{1}{u^2(b)} \tag{1b}$$

The two weighing schemes above are interrelated, because most often a mean value obtained from 100 replicate measurements will have a lower uncertainty than a value obtained from a single measurement.

As an immediate benefit, weighted means will give smaller uncertainties than the common (unweighted) mean. However, this does not necessarily imply that weighted means are closer to the truth. This will only happen if high weights (many replicates, low uncertainty) happen to correlate with low bias. Such correlation is certainly desirable but cannot be taken for granted in any case.

Weighted averaging will perform badly if the weights are inappropriate. This will happen e.g. with weighted means obtained according to equation (1b) when the uncertainty $u(a)$ was erroneously determined much lower than $u(b)$. Then a will wrongfully dominate the mean value.

In summary, nothing is wrong with the common mean as a default combination rule. In case of major differences with respect to replicate numbers or uncertainty levels, weighted means may be preferred. This will give a lower uncertainty – which may however be questionable.

1.2 Uncertainty of Mean Values

Given measurement results 100 ± 3 and 140 ± 4, the mean value $c = 120$ obtained from eq. (1) may still be a reasonable joint estimate of the measurand. But considering the difference $\Delta = b - a = 40$, the standard uncertainty $u(c) = 2.5$ obtained from eq. (2)

obviously is not. This is the *discrepancy problem*: the measurement results are not consistent within specified uncertainties. Thus at least one of the uncertainties $u(a)$ and $u(b)$ was largely underestimated. Since one or both of the data $u(a)$ and $u(b)$ are invalid, the combined uncertainty $u(c)$ is invalid, too.

Another, less commonly recognised source of trouble is the *correlation problem*: the combination rule according to eq. (2) is only valid for independent measurements. If measurements have major uncertainty sources in common, a correlation term has to be included in the combined uncertainty of a mean value. As a rule, this will give a larger uncertainty than eq. (2). More about this issue in section 1.2.2.

The uncertainty of a weighted mean requires a modification of eq. (2), and the uncertainties so obtained are often substantially smaller than for the common mean (see section 1.2.3).

Last but not least, when the task is to combine the data from two sets of replicate measurements, and uncertainty evaluation is restricted to variability, the most simple approach is to pool the data. This issue is discussed in section 1.2.4.

1.2.1 Handling of Discrepant Results.   In an ideal world the solution to the discrepancy problem would be to re-examine and amend the uncertainty budgets for the two measurement results, obtaining amended uncertainties $u_{am}(a)$ and $u_{am}(b)$ which are compatible with the observed difference $\Delta = b - a$. In practice, however, this approach is most often not applicable. The data are what they are, and the alternative is to either declare a meaningful combination impossible, or estimate a combined uncertainty from the available data, i.e. from $u(a)$, $u(b)$ and $(b - a)$. The go/no-go decision requires careful consideration – this is not the place for advice on this difficult issue. Certainly there should be limits as to the acceptable level of discrepancy. These limits should be based on considerations of the state of the art and fitness for purpose rather than on statistical considerations (e.g. significance levels).

Accommodation of the observed discrepancy by way of amended uncertainties requires some kind of model concerning issues like the level of coverage of the between-results bias by the amended uncertainty, the allocation of the between-results bias (i.e. whether the bias is attributed to both measurements or only one), and the characteristics of the error terms utilised.

Note: We assume $b > a$ throughout this section to get around the use of absolute values $|b - a|$.

   a) Minimum coverage of observed differences

Consistency of measurement results $a \pm u(a)$ and $b \pm u(b)$ may be examined by comparing the difference $\Delta = b - a$ with its expected magnitude, as given by the combined standard uncertainty of the difference.

$$u(\Delta) = \sqrt{u^2(a) + u^2(b)} \tag{3}$$

An obvious idea would be to mimick significance testing, e.g. by using $\Delta > 2u(\Delta)$ as a criterion for significant lack of consistency (at a significance level of about 95 %). However in this setting it is better to disregard significance levels and use $\Delta > u(\Delta)$ as a criterion for lack of consistency, requiring action (see [3]). More conveniently than comparing a difference Δ with its standard uncertainty $u(\Delta)$, the square $\Delta^2 = (b - a)^2$ is compared with the variance $u^2(\Delta) = u^2(a) + u^2(b)$.

If $(b - a)^2 > u^2(a) + u^2(b)$, a pragmatic approach to account for the lack of coverage by the specified uncertainties is to take the difference as a combined excess uncertainty according to $(b - a)^2 - u^2(a) - u^2(b) = u_{ex}^2(a) + u_{ex}^2(b)$. Next, amended uncertainties are defined by $u_{am}^2(a) = u^2(a) + u_{ex}^2(a)$ and $u_{am}^2(b) = u^2(b) + u_{ex}^2(b)$. Now the amended uncertainty of the mean value is calculated according to $u_{am}^2(c) = \frac{1}{4}[u_{am}^2(a) + u_{am}^2(b)] = \frac{1}{4}$

$[u^2(a) + u^2(b) + u_{ex}^2(a) + u_{ex}^2(b)] = \frac{1}{4} [u^2(a) + u^2(b) +(b - a)^2 - u^2(a) - u^2(b)] = \frac{1}{4} (b - a)^2$.
Note that the subdivision of the difference $(b - a)^2 - u^2(a) - u^2(b)$ into a sum of two distinct terms, $u_{ex}^2(a) + u_{ex}^2(b)$, is completely immaterial, but all error terms have to be uncorrelated.

In summary the result of this uncertainty evaluation is as follows:

$$u(c) = \frac{1}{2}\sqrt{u^2(a) + u^2(b)} \quad \text{for compatible results, i.e. } (b - a)^2 \leq u^2(a) + u^2(b)$$

$$(4)$$

$$u_{am}(c) = \frac{1}{2}(b - a) \qquad \text{for discrepant results, i.e. } (b - a)^2 > u^2(a) + u^2(b)$$

Following the approach by Lyons [3], the calculation above may be justified by a measurement error model as follows: Assume that both measurements are affected by an unknown bias and model this by a random shift with zero mean and standard deviation s_{shift}. Accordingly, amended uncertainties $u_{am}(a)$ and $u_{am}(b)$ are obtained as root sums of squares of $u(a)$ and s_{shift} or $u(b)$ and s_{shift}, respectively. Assuming further that the shifts for measurements a and b are uncorrelated, and equating the expected magnitude of the difference $b - a$ with the observed value, the shift standard deviation is estimated as $s_{shift}^2 = \frac{1}{2} [(b - a)^2 - u^2(a) - u^2(b)]$. This gives a final result of $u_{am}^2(c) = \frac{1}{4} (b - a)^2$.

Brute-force adaptation of one-way Analysis of Variance (ANOVA) to standard uncertainties (see section 3.1), taken to the limit of two groups of data, gives the same result as above, i.e. eq. (4).

The uncertainty estimate $u_{am}(c) = \frac{1}{2} (b - a)$ may appear inappropriate because the uncertainties $u(a)$ and $u(b)$ are not explicitly contained in the formula but only covered numerically. This is due to the fact that the amended uncertainties are tailored to effect minimum coverage of the observed difference $b - a$. As an alternative approach, following ref. [4], the amended uncertainties may be defined using the squared total difference $(b - a)^2$ instead of the excess difference $(b - a)^2 - u^2(a) - u^2(b)$.

(b) Extended coverage of observed differences
In the approach by Levenson et al. [4], the observed difference $(b - a)$ is accommodated by an additional uncertainty contribution based on a rectangular probability distribution. In the publication this is done by direct amendment of the uncertainty of the mean value according to $u_{am}^2(c) = \frac{1}{4} [u_{am}^2(a) + u_{am}^2(b)] + u_{ex}^2(c)$, where the additional contribution is taken to be the variance of a rectangular distribution with zero mean and a width of $(b - a)$. The standard deviation of this distribution is $u_{ex}(c) = (b - a) / 2\sqrt{3}$. This gives an amended uncertainty of

$$u_{am}(c) = \frac{1}{2}\sqrt{u^2(a) + u^2(b) + \frac{(b - a)^2}{3}} \qquad (5)$$

As an alternative approach, in an example given in [4] the observed difference is allocated to a residual bias in one of the measurements (b), while the other measurement (a) is taken as unbiased. Assuming this, the observed difference $(b - a)$ can be accommodated by amendment of the uncertainty of b according to $u_{am}^2(b) = u^2(b) + u_{ex}^2(b)$. In this setting, the additional contribution may be estimated by the variance of a rectangular distribution with zero mean and a half-width of $(b - a)$, i.e. $u_{ex}^2(b) = (b - a)^2 / 3$. This gives an amended variance of the mean of $u_{am}^2(c) = \frac{1}{4} [u^2(a) + u_{am}^2(b)] = \frac{1}{4} [u^2(a) + u^2(b) + u_{ex}^2(b)] = \frac{1}{4} [u^2(a) + u^2(b) +(b - a)^2 / 3]$, i.e. the same result as eq. (5).

1.2.1 Handling of Correlated Results. If measurements have major uncertainty sources in common, a correlation term has to be included in the combined uncertainty of a mean value. Otherwise the uncertainty may be substantially underestimated. For a mean of two results, this is not a major issue because the correct uncertainty is at most by a factor of $\sqrt{2}$ larger than the value calculated from eq. (2). However, for a mean of $N \geq 10$ results the correct uncertainty may be up to a factor of $\sqrt{N}$ larger than the value obtained without correlation, and this certainly is a major issue. This topic will therefore be revisited later on (section 3.6).

1.2.2   Uncertainty of Weighted Means.

For a weighted mean of two measurement results,

$$c_w = \frac{w_a \cdot a + w_b \cdot b}{w_a + w_b} \qquad \text{with weights } w_a, w_b \tag{6}$$

the combined standard uncertainty is given by

$$u(c_w) = \sqrt{\frac{w_a^2 u^2(a) + w_b^2 u^2(b)}{(w_a + w_b)^2}} \tag{7}$$

Eq. (7) also applies if the inverse variances are used as weights ($w_a = 1 / u^2(a)$, $w_b = 1 / u^2(b)$).

Note: Concerning inverse-variance weighted means, it has been proposed to attribute an uncertainty to the standard uncertainties u(a) and u(b) and include this in the calculation of the combined standard uncertainty of a weighted mean [5]. However, this is neither appropriate from a theoretical point of view: the uncertainties are what they are, and introducing a second level of uncertainty evaluation does not comply with GUM principles – nor from a practical perspective: the uncertainty of mean values is unduly inflated.

Given discrepant results $a \pm u(a)$ and $b \pm u(b)$, it may appear doubtful to use an inverse-variance weighted mean. This would only be reasonable if it could be taken for granted that smaller uncertainty correlates with smaller bias. Assuming this, we may use the weighted mean with variances amended for lack of consistency according to $u_{am}^2(a) = u^2(a) + u_{ex}^2$, $u_{am}^2(b) = u^2(b) + u_{ex}^2$ with $u_{ex}^2 = \frac{1}{2} [(b-a)^2 - u^2(a) - u^2(b)]$, see section 1.2.1. This gives

$$c_w = \frac{1}{2}\left[(a+b) - \frac{u^2(b) - u^2(a)}{b-a}\right] \tag{6a}$$

$$u(c_w) = \frac{1}{2}\sqrt{(b-a)^2 - \frac{[u^2(b) - u^2(a)]^2}{(b-a)^2}} \tag{7a}$$

Compared with the common mean $c = \frac{1}{2} (a - b)$, the weighted mean is shifted towards a if $u(b) > u(a)$ and towards b if $u(b) < u(a)$.

An even more daring approach (concerning the selection of weights) would be to use the original variances $u^2(a)$, $u^2(b)$ as weights and amend the uncertainty of the weighted mean by a numerical factor. This is the Birge approach, see section 3.3.1.

1.2.4 Pooling replicate data.

Given data $A = \{a_1, a_2, ..., a_m\}$ and $B = \{b_1, b_2, ..., b_n\}$ from two sets of replicate measurements, the most simple combination is to pool the data, take the grand mean (c_{pool}) and calculate the standard deviation of the pooled data (s_{pool}). If uncertainty evaluation is restricted to variability, a promising candidate for the standard uncertainty of the grand mean is

$$u(c_{pool}) = \frac{s_{pool}}{\sqrt{m+n}} \tag{8}$$

What can be wrong, or require amendment, with these combination rules? Firstly, it may be doubtful whether such combination makes sense at all. If the data A and B are supposed to be for the same measurand, there should be some overlap between the range of A and the range of B. Otherwise there is a bias between A and B. Similar to the case of two results $a \pm u(a)$ and $b \pm u(b)$, a decision has to be taken whether or not this bias is still acceptable and data combination, accommodating the bias, should be undertaken.

If so, taking the grand mean (c_{pool}) is an appropriate default combination. However, if there is a bias between A and B, eq. (8) will not provide a valid standard uncertainty of the grand mean.

The problems arising above – assessment of bias between A and B, and uncertainty estimation for the grand mean accounting for between-groups bias – are most conveniently addressed by one-way ANOVA. To simplify the calculation, we will assume data series of equal length ($m = n$) and comparable variability ($s_a \approx s_b$). Also we have to introduce some notation as follows:

Table 1 – *Summary statistics for data A and B*

$\overline{a}$ – mean value of data A	$\overline{b}$ – mean value of data B
s_a – standard deviation of data A	s_b – standard deviation of data B
$s(\overline{a})$ – standard deviation of the mean $\overline{a}$	$s(\overline{b})$ – standard deviation of the mean $\overline{b}$

In this scenario the variability of the pooled data may be modelled by combination of two random effects: within-group variations and between-groups variations. Accordingly the variance of the pooled data is decomposed into a sum

$$s^2_{pool} = s^2_{within} + s^2_{between} \tag{9}$$

In this expression $s^2_{within} = \frac{1}{2}[s^2_a + s^2_b]$ is the average variance of the data A and B, and $s^2_{between}$ is obtained as a difference according to $s^2_{between} = \frac{1}{2}[(\overline{b} - \overline{a})^2 - s^2(\overline{a}) - s^2(\overline{b})]$, provided that this difference is positive. Like in section 1.2.1, $(\overline{b} - \overline{a})^2 > s^2(\overline{a}) + s^2(\overline{b})$ is taken to indicate lack of consistency of the data A and B, requiring amendment by a between-groups variance $s^2_{between}$. Given a positive difference, the variance of the grand mean is obtained according to

$$u^2_{am}(c_{pool}) = \frac{s^2_{within}}{2n} + \frac{s^2_{between}}{2} \tag{10}$$

Here the different denominators of the terms in the sum are due to the fact that the between-groups effect introduces a within-group correlation which has to be accounted for in the variance of the grand mean. Evaluating eq. (10) finally gives $u^2_{am}(c_{pool}) = \frac{1}{4}(\overline{b} - \overline{a})^2$. In summary the result of this uncertainty evaluation is as follows:

$$u(c_{pool}) = \frac{s_{pool}}{\sqrt{2n}} \quad \text{for compatible results, i.e. } (\overline{b} - \overline{a})^2 \leq s^2(\overline{a}) + s^2(\overline{b}) \tag{11}$$

$$u_{am}(c_{pool}) = \frac{1}{2}(\overline{b} - \overline{a}) \text{ for discrepant results, i.e. } (\overline{b} - \overline{a})^2 > s^2(\overline{a}) + s^2(\overline{b})$$

The considerations and results presented here are largely parallel to those in section 1.2.1. The main difference is in the modelling of excess variance and the use of the average variance, $\frac{1}{2}[u^2(a) + u^2(b)]$ instead of the variance s^2_{pool} of the pooled data, which is of course not available in scenario 1.2.1. To complete the analogy, it should be noted that $s^2_{pool} \leq \frac{1}{2}[s^2_a + s^2_b]$ for compatible data A and B. This gives

$$u(c_{pool}) \leq \frac{1}{2}\sqrt{s^2(\overline{a}) + s^2(\overline{b})} \quad \text{for compatible results, i.e. } (\overline{b} - \overline{a})^2 \leq s^2(\overline{a}) + s^2(\overline{b}) \tag{12}$$

The assumptions above, data series of equal length ($m = n$) and comparable variability ($s_a \approx s_b$) are convenient but not mandatory for an ANOVA. We will come back to this issue in section 3.1.

1.3 Combination of measurements instead of measurement results

If possible, the task of combining measurements should not be deferred to final results $a \pm u(a)$ and $b \pm u(b)$ according to the demand: take the data or leave them. Instead, combination should already be addressed in the planning phase, be carried out on the complete data obtained by measurement or otherwise, and utilise method performance data from validation studies and quality control. This approach is used at the Analytical Chemistry Division of BAM, when multiple methods are employed to determine reference values, e.g. in an in-house certification of a reference material, and according to the authors´ information this is also used at the NIST. All data are given to a "data evaluation expert group" where they are analysed and processed into a mean value and an associated uncertainty – in close cooperation with the laboratories involved.

2 INTERMEZZO: UNCERTAINTY EVALUATION ACCORDING TO GUM

2.1 Procedural aspects: Uncertainty budgets versus whole-method performance characteristics

The *Guide to the Expression of Uncertainty* (GUM) almost exclusively treats a single approach for uncertainty evaluation: the "modelling approach" based on a comprehensive mathematical model of the measurement process, where every uncertainty contribution is associated with a dedicated input quantity, the uncertainty contributions are evaluated individually and combined as variances. This is therefore often (mis)conceived as being "the GUM approach" for uncertainty evaluation. Actually the GUM principles admit a variety of approaches, but this fact was buried under a plethora of papers and lectures celebrating the "modelling approach" as a new paradigm in measurement quality assurance. Only recently alternative "experimental approaches" utilising classical statistical methods have re-gained appreciation in the metrology community. These approaches are based on whole-method performance investigations, designed and conducted as to comprise the effects from as many relevant uncertainty sources as possible. The data utilised in these approaches are typically precision and bias data obtained from within-laboratory validation studies, quality control, interlaboratory method validation studies, or proficiency tests. In summary, after more than ten years of discussion about measurement uncertainty, consolidation of the "new approach" according to GUM and traditional approaches is well on the way. Given a comprehensive list of relevant effects/uncertainty sources, uncertainty evaluation may be carried out using various different approaches. They range from individual quantification and combination of input uncertainties to collective quantification, e.g. using a reproducibility standard deviation for a standard test procedure, determined according to the ISO 5725 series [6].

For ensuring that all relevant uncertainty sources are covered, error models developed in various testing fields are useful tools. A model widely used in the field of chemical analysis is the "ladder of errors" [7]: a hierarchical scheme based on a classification of measurement error according to repeatability error – run bias – laboratory bias – method bias. This scheme provides the basis for the definition and evaluation of various method performance characteristics: repeatability standard deviation, intermediate-precision standard deviation, reproducibility standard deviation and bias estimates. The main tool for the evaluation of method performance data is Analysis of Variance, but currently the tool box is re-examined, old tools are polished and new tools added to support comprehensive uncertainty evaluation, see e.g. ref [8].

2.2 Technical aspects: Implications of comprehensive uncertainty evaluation

The *Guide to the Expression of Uncertainty* (GUM) has paved the way to comprehensive measurement uncertainties including contributions beyond measurement variability. GUM principles require that

- uncertainty evaluation is comprehensive, accounting for all relevant sources of measurement error, random errors as well as residual systematic errors (left after correction);
- uncertainties arising from random and systematic effects are treated alike, i.e. are expressed and combined as variances of associated probability distributions;
- statistical evaluation of measurements (Type A) and alternative techniques, based on other data / information (Type B), are recognised and utilised as equally valid tools.

The GUM puts much emphasis on the message, that there is no principal difference between uncertainties arising from random and systematic effects, and instead puts the focus on different types of evaluation (Type A, Type B). However, when it comes to comparing and combining different measurement results, it is of paramount importance to identify which errors are shared by the measurements and which are individual. Otherwise the combined uncertainty of a sum or a difference may be substantially miscalculated.

To demonstrate this effect, consider again two measurement results $a \pm u(a)$ and $b \pm u(b)$, obtained using the same measuring instrument. Assume that the uncertainty is dominated by two sources of measurement error: random variation (between replicate measurements) and calibration error (error in calibration measurements and error in the calibration standards). Then we have

$$u^2(a) = s^2(a) + u_{cal}^2(a) \quad ; \quad u^2(b) = s^2(b) + u_{cal}^2(b) \tag{13}$$

Assume further that the repeatability standard deviations $s(a)$ and $s(b)$ are about the same, and this is also true for the calibration uncertainties $u_{cal}(a)$ and $u_{cal}(b)$. These assumptions provide for simple formulas, but they are not material for any conclusions.

Unless the measuring instrument was calibrated between measurement of a and b, the calibration error is shared by the two measurements. Therefore these measurements are not independent but correlated. As a consequence the uncertainty budget for the sum and the difference (and similar for the product or a quotient) has to include a correlation term as follows.

$$u^2(a \pm b) = u^2(a) + u^2(b) \pm 2u(a,b) \tag{14}$$

With the simplifying assumptions above, the covariance $u(a, b)$ is given by the calibration uncertainty according to $u(a, b) = u_{cal}^2$. This gives

$$u^2(a + b) = 2s^2 + 4u_{cal}^2 \quad ; \quad u^2(a - b) = 2s^2 \tag{15}$$

instead of the erroneous result $u^2(a \pm b) = 2s^2 + 2u_{cal}^2$ obtained when disregarding correlation.

These results are certainly not surprising, because it is well known that measurement bias cancels out in difference and ratio measurements. However these plain facts are only brought out in uncertainty evaluation, if the issues of shared uncertainty sources and accounting for correlation are recognised.

3 FULL SCALE: MORE THAN TWO MEASUREMENTS

With just two measurement results $a \pm u(a)$ and $b \pm u(b)$ there is only very limited scope for statistical data processing. However, if a major number N of measurement results $x_i \pm u(x_i)$ have to be combined into a mean value and an associated standard uncertainty, $\bar{x} \pm u(\bar{x})$, statistical tools for analysing and processing the data become available. For example, the variability of the results may be quantified by the standard deviation of the data x_i. A number of approaches utilising standard statistical tools are reviewed in the sections to follow. There we assume reasonably well-behaved data such that the common mean or a weighted mean can be used. Robust means, accommodating outliers or skewed data, are addressed in a later part of this chapter (section 3.4).

According to GUM principles, the expanded uncertainty U should cover a large fraction of the distribution of values that could reasonably be attributed to the measurand. As a rule, $U = k \cdot u$ is taken to effect about 95 % coverage. Given the case that none of the results $x_i \pm u(x_i)$ is obviously invalid, the interval $\bar{x} \pm U(\bar{x})$ should cover the entire range. Targetting for minimum 95 % coverage, we would put $\bar{x} = \frac{1}{2} [x_{max} + x_{min}]$, $U(\bar{x}) = \frac{1}{2} [x_{max} - x_{min}]$ and therefore estimate the standard uncertainty of the mean as

$$u(\bar{x}) = \frac{x_{max} - x_{min}}{4} \tag{16}$$

For $N = 2$ results, this uncertainty estimate is rather optimistic, compared with the results obtained in section 1.2.1. This is due to the fact that here we have targetted at 95 % coverage while previously the target was "one-standard uncertainty" coverage or better.

For large numbers N of measurement results, the uncertainty estimate above would rather be considered as unduly pessimistic, because cancellation of measurement errors is completely ignored. In other words, eq. (16) assumes total correlation among the measurements results.

An alternative approach would assume uncorrelated measurements results and introduce the famous factor $1/\sqrt{N}$ to account for error cancellation. This would give

$$u(\bar{x}) = \frac{x_{max} - x_{min}}{4\sqrt{N}} \tag{17}$$

Most often the "truth" will be somewhere in between. Assuming an average correlation coefficient of about 0.5 gives an approximate uncertainty estimate (see section 3.6) of

$$u(\bar{x}) = \frac{x_{max} - x_{min}}{4\sqrt{2}} \tag{18}$$

3.1 Approaches based on Analysis of Variance

Analysis of Variance (ANOVA) is an extremely versatile and powerful tool for data analysis and data consolidation. Among innumerous other applications, ANOVA has been, and still is used as the preferred tool for the evaluation of interlaboratory studies for value assignment of certified reference materials. Comprehensive guidance on this issue is given in ISO Guide 35 [9]. A definitive treatment of the use of ANOVA for combination of measurement results was given by Cochran in a paper entitled "The combination of estimates from different experiments" [10].

3.1.1 Classical one-way ANOVA. In the typical setting of an interlaboratory study for value assignment of certified reference materials, a number of laboratories receive samples of the candidate material and perform replicate measurements of the property value under consideration. As a rule the number of replicates is fixed. The typical result of such measurement campaign is a set of data as follows.

Table 2 *Structure of a data set for one-way ANOVA*

	M'mt 1	M'mt 2	M'mt ..	M'mt M
Lab 1	x_{11}	x_{12}	...	x_{1M}
Lab 2	x_{21}	x_{22}	...	x_{2M}
Lab ..	...	...	...	...
Lab L	x_{L1}	x_{L2}	...	x_{LM}

The standard one-way ANOVA, as summarised below, assumes equal numbers of replicates in each group (laboratory), i.e. the scheme above without empty boxes. Outlier elimination may change that. But this does not cause a major problem, because non-uniform data (unequal numbers of replicates) can also be evaluated by ANOVA, using slightly more complicated equations or (preferably) appropriate software. In addition the within-group variabilities should be comparable.

Before starting we have to specify some notation as follows.

Table 3 *Summary statistics for ANOVA data*

	Laboratory data		Summary data
M	number of replicates (same for all labs)	L	number of labs (data groups)
x_{lm}	m-th replicate of l-th group (laboratory)	$\bar{\bar{x}}$	grand mean
$\bar{x}_l$	mean of replicates of l-th lab (data group)	s_{single}	standard deviation of single values x_{lr}
s_l	standard deviation of replicates of l-th lab (data group)	s_{means}	standard deviation of lab means $\bar{x}_l$

Here the grand mean is the mean of all single values x_{lm}, but with the same number of replicates for all laboratories (data groups), the mean of lab means $\bar{x}_l$ gives the same result.

The main objectives of data analysis are (i) to decide whether the different groups of replicates are mutuall consistent and can be pooled, and (ii) to determine an appropriate estimate for the standard deviation of the grand mean, depending on the outcome of step (i).

Data analysis is based on a statistical model, where data variability is described by combination of two random effects: within-group variations and between-groups variations according to

$$x_{lm} = \mu + \beta_l + \varepsilon_{lm} \tag{19}$$

In this equation μ is the actual value of the measurand, β_l the bias of measurements made in the l-th laboratory, and ε_{lm} the random error in the m-th replicate measurement of that laboratory. The β's and ε's are supposed to be random variables with expectation zero and variance σ^2_{bias} and σ^2_{rep}, respectively. Accordingly, the variance of a single value is given by the sum of these two variances, $\text{var}(x_{lm}) = \sigma^2_{bias} + \sigma^2_{rep}$. As a consequence of the bias term β shared by measurements made in the same laboratory, replicates are correlated, with a covariance of $cov(x_{lp}, x_{lq}) = \sigma^2_{bias}$ while measurements from different laboratories are independent. Considering this, the variance of a laboratory mean is given by $var(\bar{x}_l) = \sigma^2_{bias} + \sigma^2_{rep} / M$, and the variance of the grand mean is obtained as follows:

$$var(\bar{\bar{x}}) = \frac{\sigma^2_{bias}}{L} + \frac{\sigma^2_{rep}}{L \cdot M} \tag{20}$$

The variances σ^2 above are theoretical parameters defining the quantities to be estimated by experimental variances s^2 (squared standard deviations). The variance accounting for bias between laboratories is estimated by the difference between the experimental variance s^2_{means} of the laboratory means $\bar{x}_l$ and the average repeatability variance of the laboratory means, as given by $s^2(\bar{x}_l) = s^2_l / M$.

$$s^2_{bias} = s^2_{means} - \frac{1}{L}\sum_l s^2(\bar{x}_l) \tag{21}$$

If eq. (21) returns a result of $s^2_{bias} \leq 0$, this means that the within-group variability covers the overall data variability. Hence there is no need to introduce a between-groups bias as an additional source of variability. As a consequence, the data may be pooled, s^2_{bias} may be taken as zero, and s^2_{single} may be taken as an estimate s^2_{rep} of the within-laboratory variance. The standard deviation of the grand mean is obtained from the standard deviation of the single values according to

$$s_{pool}(\bar{\bar{x}}) = \frac{s_{single}}{\sqrt{L \cdot M}} \tag{22}$$

Note: Often a significance test (F test) is used to examine whether there is a significant between-groups bias. In the setting considered here this would not be appropriate.

If eq. (21) returns a result of $s^2_{bias} > 0$, this means that the within-group variability fails to cover the overall data variability. Hence a between-groups bias is required as an additional source of variability. In this case pooling is not allowed. As a consequence the within-laboratory variance has to be estimated by the average variance of the replicate measurements made in the various laboratories. This gives $s^2(\bar{\bar{x}}) = \{s^2_{means} - [\Sigma s^2(\bar{x}_l) / M]\} / L + \{\Sigma s^2_l\} / ML = s^2_{means} / L$. Hence in the non-pooling case the standard deviation of the grand mean is obtained from the standard deviation of the laboratory means according to

$$s_{no\_pool}(\bar{\bar{x}}) = \frac{s_{means}}{\sqrt{L}} \tag{23}$$

With minor amendment, these results are also valid for unequal numbers of replicates. When pooling is allowed, the mean of single values and its standard deviation are used, otherwise the mean of laboratory means and its standard deviation. These two means take slightly different values.

Eq. (22) can only be used if the individual data x_{lm} are available. However there is a good approximation, where only the laboratory means and their standard deviations are required. If $s^2_{bias} \leq 0$, it can be shown that the average within-group variance, $\bar{s}^2_{lab} = [\Sigma s^2_l] / L$ is an upper, and most often quite close bound of s^2_{single}. This gives

$$s_{pool} = \sqrt{\frac{\bar{s}^2_{lab}}{L \cdot M}} \tag{22a}$$

as a good approximation, which also lends itself to generalisations, where experimental standard deviations are replaced by comprehensive standard uncertainties.

If the within-group variability is widely different among groups, a weighted mean should be used to account for that. Details how to perform an ANOVA on this basis are given in the paper by Cochran [10].

The application of ANOVA described above only marks the tip of an iceberg. Concerning the power of this tool for data analysis as well as for experimental design, ref. [11] provides a nice example, concerning the determination of uncertainty of analytical measurements from collaborative study data.

3.1.2 Adaptation to measurement uncertainty. Classical ANOVA deals exclusively with the variability of data obtained by multiple measurements, the main tool being decomposition of experimental variance and allocation of variance components to specified sources of variability. In this setting measurement results are typically presented as sets of replicates. In most applications these data are summarised in the usual format mean value & standard deviation, and it is these summary data that are processed in data analysis and data consolidation. That is, ANOVA can be performed without individual replicate data. Therefore, at least formally, experimental standard deviations may be substituted by GUM-type standard uncertainties. This may not be fully compatible with the statistical model used in the original ANOVA but should be good enough as a pragmatic approach. A first level of refinement is provided by Chi^2-techniques, see section 3.2. Advanced data evaluation most often makes use of Maximum Likelihood techniques, which are however beyond the scope of this paper.

For a better understanding of ANOVA in this setting, we will first focus on the examination of compatibility of the differences between the results obtained by the various laboratories, as given by the data $\bar{x}_l \pm s(\bar{x}_l)$ – laboratory mean values and associated standard deviations – and the expected values of these differences. The latter are the standard devia-tions associated with the differences. Following the approach in section 1.2.1, the magnitude of the various differences $(\bar{x}_k - \bar{x}_l)$ is compared with their standard deviation. Laboratory results $\bar{x}_k \pm s(\bar{x}_k)$ and $\bar{x}_l \pm s(\bar{x}_l)$ are deemed to be consistent if the (absolute value of) the dif-ference is below (or at most equal to) the standard deviation of the difference, and inconsistent if the difference is larger. More conveniently, this comparison is carried at the level of squares. Then consistency means $(\bar{x}_k - \bar{x}_l)^2 \leq s^2(\bar{x}_k) + s^2(\bar{x}_l)$ and $(\bar{x}_k - \bar{x}_l)^2 > s^2(\bar{x}_k) + s^2(\bar{x}_l)$ indicates inconsistency.

Most often individual statements of consistency or inconsistency will be doubtful, and a statement on average will be more appropriate. This is obtained by comparing the mean squared difference $<(\bar{x}_k - \bar{x}_l)^2>_{k \neq l}$ with the average sum of variances $<s^2(\bar{x}_k) + s^2(\bar{x}_l)>_{k \neq l}$. Evaluating the mean squared difference gives $2s^2_{means}$, where s_{means} is the standard deviation of the laboratory means $\bar{x}_l$. Evaluating the average sum of variances gives $2<s^2(\bar{x}_l)>_l$, where the bracket denotes the average variance of the laboratory means $\bar{x}_l$. Thus the outcome of an assessment of consistency on average is

$$s^2_{means} \leq \frac{1}{L} \sum_l s^2(\bar{x}_l) \qquad \text{indicates consistency of the data } \bar{x}_l \pm s(\bar{x}_l) \qquad \rceil$$

$$(24)$$

$$s^2_{means} > \frac{1}{L} \sum_l s^2(\bar{x}_l) \qquad \text{indicates inconsistency of the data } \bar{x}_l \pm s(\bar{x}_l) \qquad \rfloor$$

Given consistency, there is no need for introducing a between-results bias as an additional source of variability, and the standard deviation of the mean of laboratory means would be estimated as

$$s\left(\overline{\overline{x}}\right) = \frac{1}{L}\sqrt{\sum_l s^2\left(\overline{x}_l\right)} \tag{25}$$

Lack of consistency is accommodated by introducing an excess variance to account for between-results bias according to

$$s_{ex}^2 = s_{means}^2 - \frac{1}{L}\sum_l s^2\left(\overline{x}_l\right) \tag{26}$$

This extra variance is then added to the variance for each mean value, giving an amended variance of $s^2{}_{am}(\overline{x}_l) = s^2(\overline{x}_l) + s^2{}_{ex}$. Using this, the variance of the mean of laboratory means is obtained as $s^2{}_{am}(\overline{\overline{x}}) = [\Sigma s^2{}_{am}(\overline{x}_l)] / L^2 = \{\Sigma[s^2(\overline{x}_l) + s^2{}_{ex}]\} / L^2 = s^2{}_{means} / L$. Thus we end up with the familiar result

$$s\left(\overline{\overline{x}}\right) = \frac{s_{means}}{\sqrt{L}} \tag{27}$$

In summary, starting from the assessment of compatibility of the differences between the results obtained by the various laboratories, as given by the laboratory means and associated standard deviations, and the expected values of these differences, as given by the standard deviations associated with the differences, and introducing an average excess variance to account for lack of consistency, we have recovered the scheme used in ANOVA. The way it has been introduced above, this scheme is not restricted to data $\overline{x}_l \pm s(\overline{x}_l)$ where the uncertainties are experimental standard deviations but could also be used for data $x_l \pm u(x_l)$ with comprehensive standard uncertainties – provided that these standard uncertainties can be taken as *uncorrelated*. Otherwise covariances $u(x_k, x_l)$ have to taken into account. This will give a lower uncertainty of differences and a higher uncertainty of sums and mean values (see section 3.6).

Note: The approach above can be extended by introducing individual excess variances which are determined by way of equating squared deviations from the mean value to the amended variance of the respective deviation. The latter object is somewhat tricky [see eq. (34)], but in the end the familiar result, eq. (27), for the uncertainty of the mean value is recovered. This is also true for the Chi-squared techniques described in the next section.

According to common understanding, ANOVA assumes comparable uncertainties $u(x_l)$ in order not to "mix up melons with cherries" (apples and pears would be ok). The Chi-squared techniques described in the next section do not require that.

3.2 Approaches based on the Chi-squared technique

In the ANOVA approach, mutual consistency of different results $x_l \pm u(x_l)$ – where the x_l may be single or mean values, and the $u(x_l)$ may be experimental standard deviations or compre-hensive standard uncertainties – was performed using the average difference of two squares, $<(x_k - x_l)^2 - u^2(x_k - x_l)>_{k \neq l}$. Given that the specified uncertainties cover the various differences $(x_k - x_l)$, the bracket term should be ≤ 0. Instead of the average difference, we may consider the average quotient of the squares concerned, $<(x_k - x_l)^2 / u^2(x_k - x_l)>_{k \neq l}$. This step will open the door to another data evaluation technique, based on the χ^2 distribution. Inserting the general expression for the variance of a difference, $u^2(x_k - x_l) = u^2(x_k) + u^2(x_l) - 2u(x_k, x_l)$, the average quotient of squares is obtained as

$$\chi_{std}^2 = \frac{1}{L(L-1)}\sum_k \sum_{l \neq k} \frac{\left(x_k - x_l\right)^2}{u^2\left(x_k\right) + u^2\left(x_l\right) - 2u\left(x_k, x_l\right)} \tag{28}$$

Remark: For the purpose of this paper it is not necessary to be familiar with the Chi-squared distribution. However, like ANOVA, this is a very versatile and powerful tool for data evaluation, and the study of Chi-squared techniques is mandatory for professionals in this field.

Given mutually consistent results $x_l \pm u(x_l)$, the expected value of χ^2_{std} is 1. Therefore $\chi^2_{std} \leq 1$ would be taken to indicate consistency and $\chi^2_{std} > 1$ to indicate inconsistency. A comprehensive treatment of the pair-difference Chi-squared technique for assessing the mutual compatibility of Key Comparison data is given in ref. [12].

As before, an excess variance may be introduced to account for lack of consistency, i.e. lack of coverage of the differences $(x_k - x_l)$ by the specified uncertainties $u(x_k)$ and $u(x_l)$ concerned. This cannot be done by difference as before. Instead, the excess variance is introduced in the expression for χ^2_{std}, and the expression so obtained is equated to 1. This gives a non-linear equation for the excess variance u^2_{ex} as follows:

$$\frac{1}{L(L-1)}\sum_k \sum_{l \neq k} \frac{(x_k - x_l)^2}{u^2(x_k)+u^2(x_l)+2u^2_{ex}-2u(x_k,x_l)} = 1 \tag{29}$$

which may be solved for u^2_{ex} using numerical methods.

To get a feeling for this approach, consider a case where all standard uncertainties $u(x_l)$ are about the same and covariances $u(x_k, x_l)$ can be neglected. For an approximate solution we may replace the sums $u^2(x_k) + u^2(x_l)$ in the denominators by the average $< u^2(x_k) + u^2(x_l) >_{k \neq l}$. Then eq. (29) takes the form $<(x_k - x_l)^2>_{k \neq l} = < u^2(x_k) + u^2(x_l)>_{k \neq l} + 2\,u^2_{ex}$. Now $<(x_k - x_l)^2>_{k \neq l} = 2s^2_x$ (where s_x = standard deviation of the values x_l) and $< u^2(x_k) + u^2(x_l)>_{k \neq l} = 2 <u^2(x)>$ (where $<u^2(x)>$ = average of the variances $u^2(x_l)$). This gives $u^2_{ex} = s^2_x - <u^2(x)>$ as an approximate solution – the ANOVA estimate. For measurement results $x_l \pm u(x_l)$ with widely varying uncertainties the estimate u^2_{ex} of the excess variance obtained by the Chi-squared technique may be quite different from the ANOVA estimate.

Having solved eq. (29), the variance of the mean value $\bar{x} = \sum x_l / L$ is obtained as $u^2_{am}(\bar{x}) = [\sum u^2_{am}(x_l)] / L^2 = \{\sum[u^2(x_l) + u^2_{ex}] / L^2$. This gives

$$u_{am}(\bar{x}) = \frac{1}{L}\sqrt{\sum_l u^2(x_l)+ L \cdot u^2_{ex}} \quad \text{(inconsistent results, } u^2_{ex} > 0) \tag{30}$$

For consistent results the standard uncertainty of the mean value is given by

$$u(\bar{x}) = \frac{1}{L}\sqrt{\sum_l u^2(x_l)} \quad \text{(consistent results, } u^2_{ex} \leq 0) \tag{31}$$

The estimate for the excess variance obtained from eq. (29) could also be used if the common mean $\bar{x}$ is replaced by a weighted mean. In case of major differences between the specified uncertainties $u(x_l)$ the inverse-variance weighted mean according to

$$\bar{x}_w = \frac{\sum_l w_l \cdot x_l}{\sum_l w_l} \quad \text{with weights} \quad w_l = \frac{1}{u^2(x_l)+u^2_{ex}} \tag{32}$$

would be an appropriate choice. This will give a standard uncertainty as follows:

$$u(\bar{x}_w) = \frac{1}{\sqrt{\sum_l 1/u^2(x_l)}} \quad \text{(consistent results, } u^2_{ex} \leq 0)$$

$$\tag{33}$$

$$u_{am}(\bar{x}_w) = \frac{1}{\sqrt{\sum_l 1/(u^2(x_l)+u^2_{ex})}} \quad \text{(inconsistent results, } u^2_{ex} > 0)$$

Here it should be noted that the value of the weighted mean depends on the excess variance used in the weights.

As far as the authors are aware, the approach above to estimate an excess variance accounting for between-results bias using averages over all pairs of measurements has not been used in any published paper. Instead of average coverage of the differences ($x_k - x_l$) between any two measurement results, published approaches consider average coverage of the deviations ($x_l - \bar{x}$) from an appropriate mean value. The starting point for this approach is the comparison of the average squared deviation from the mean, $<(x_l - \bar{x})^2>_l$ with the average variance of these deviations, $<u^2(x_l - \bar{x})>_l$. In this case the variances $u^2(x_l - \bar{x})$ include a covariance to account for the correlation between the mean value $\bar{x}$ and the individual values x_l. Using the unweighted mean, and determining the excess variance from the equation $<(x_l - \bar{x})^2>_l = <u^2_{am}(x_l - \bar{x})>_l$ returns the ANOVA solution. Here correct evaluation of the amended variances $u^2_{am}(x_l - \bar{x})$ is of critical importance. As noted above, the mean value $\bar{x}$ and the individual values x_l are correlated. This fact has to be taken into account, either by eliminating $\bar{x}$ as a separate (but dependent) variable or by including covariances $u(x_l, \bar{x})$. The result so obtained is

$$u^2_{am}(x_l - \bar{x}) = \frac{L-2}{L}u^2(x_l) + \frac{1}{L}\langle u^2(x)\rangle + \frac{L-1}{L}u^2_{ex} \qquad (34)$$

Turning to the Chi-squared technique, quotients $(x_l - \bar{x})^2 / u^2_{am}(x_l - \bar{x})$ are averaged instead of differences $(x_l - \bar{x})^2 - u^2_{am}(x_l - \bar{x})$. This gives an equation for the excess variance as follows:

$$\frac{1}{L}\sum_l \frac{(x_l - \bar{x})^2}{u^2_{am}(x_l - \bar{x})} = 1 \qquad (35)$$

where the denominator is given by eq. (34).

Most often, for practical purposes, the variances and covariances relating to the mean value are neglected, and the amended variances of the deviations from the mean are approximated according to $u^2_{am}(x_l - \bar{x}) \approx u^2(x_l) + u^2_{ex}$. Accordingly the expected value on the right-hand side of eq. (35) is reduced to $(L - 1)/L$. This gives a modified equation for the excess variance as follows:

$$\sum_l \frac{(x_l - \bar{x})^2}{u^2(x_l) + u^2_{ex}} = L - 1 \qquad (36)$$

In case of major differences between the specified uncertainties $u(x_l)$, the inverse-variance weighted mean $\bar{x}_w$ according to eq. (32) may be preferred over the common mean. Inserting this expression into the equation above gives a modified equation for the excess variance u^2_{ex}. Having solved that, the mean value value $\bar{x}_w$ and its standard uncertainty $u(\bar{x}_w)$ are obtained from eqs. (32) and (33).

This approach was introduced by Paule and Mandel in a paper entitled "Consensus values and weighting factors" [13], and has been recognised as one of the leading approaches for the combination of measurement results. It was in particular considered by NIST for in-house characterisation of reference materials [14]. The statistical basis of the Paule/Mandel approach has been a subject of mathematical investigations [15], and its performance was investigated by simulation studies, see e.g. [16]. Ref. [17] compares various approaches for combining interlaboratory data based on random effects models, including ANOVA and the Chi-squared approach of Paule and Mandel, from a generic statistical perspective.

3.3 Other approaches to account for lack of consistency

The approaches described in sections 3.1 and 3.2 operate according to the same basic scheme: measurement results $x_l \pm u(x_l)$ are investigated for consistency, i.e. whether the specified uncertainties cover the differences $(x_k - x_l)$ among the various measurements, or the deviations $(x_l - \bar{x})$ from the mean value. If this is indeed the case, the uncertainty of the mean is given by combination of the individual uncertainties according to

$$u(\bar{x}) = \frac{1}{L}\sqrt{\sum_l u^2(x_l)} \qquad (37)$$

If the results exhibit a lack of consistency, an additional uncertainty contribution (the excess variance u^2_{ex}) is introduced to account for between-results bias. This term is obtained from the requirement that the amended uncertainties $u_{am}(x_l) = [u^2(x_l) + u^2_{ex}]^{½}$ provide for consistency. Having determined the excess variance u^2_{ex}, the uncertainty of the mean is given by

$$u_{am}(\bar{x}) = \frac{1}{L}\sqrt{\sum_l u^2(x_l) + L \cdot u^2_{ex}} \qquad (38)$$

If weighted means are used instead of the common mean, other expressions for the uncertainty of a mean apply, see eq. (33).

However, the use of an additional uncertainty contribution to account for between-results bias is not the only strategy to accommodate lack of consistency.

3.3.1 Birge approach. A alternative approach was introduced by Birge [18] more than 70 years ago to account for discrepancies in interlaboratory studies for the determination of fundamental constants. This approach is based on a very simple assumption: all participants have underestimated their uncertainties by about the same factor. Accordingly a factor is introduced to amend the uncertainties according to $u_{am}(x_l) = f \cdot u(x_l)$, and fixed at the smallest value effecting consistency. If consistency is assessed on the average differences $<(x_k - x_l)^2 - u^2(x_k - x_l)>_{k \neq l}$, the Birge factor is given by $f^2 = <(x_k - x_l)^2>_{k \neq l} / <u^2(x_k - x_l)>_{k \neq l} = s^2_x / <u^2(x)>$. For the common mean, this gives an amended uncertainty of

$$u_{am}(\bar{x}) = \frac{1}{L}\sqrt{\sum_l f^2 \cdot u^2(x_l)} = \frac{f}{L}\sqrt{\sum_l u^2(x_l)} = \frac{s_x}{\sqrt{L}} \qquad (39)$$

This is just the ANOVA result. If consistency is assessed on the average quotients, $<(x_k - x_l)^2 / u^2(x_k - x_l)>_{k \neq l}$, the Birge factor is given by $f^2 = <(x_k - x_l)^2 / u^2(x_k - x_l)>_{k \neq l} = \chi^2_{std}$. This gives an amended uncertainty of the common mean as follows:

$$u_{am}(\bar{x}) = \frac{\sqrt{\chi^2_{std}}}{L}\sqrt{\sum_l u^2(x_l)} \qquad (40)$$

The Birge approach can also be used for weighted means. In this case, the Birge factor f is fixed such that

$$\sum_l \frac{(x_l - \bar{x}_w)^2}{f^2 u^2(x_l)} = L - 1 \qquad (41)$$

and this factor is then used in the amended uncertainty of the weighted mean according to

$$u_{am}(\bar{x}_w) = \frac{f}{\sqrt{\sum_l 1/u^2(x_l)}} \qquad (42)$$

Here $\bar{x}_w$ is the weighted mean obtained using the inverse variances $1/u^2(x_l)$ as weights, and in contrast with the Paule/Mandel approach amendment will not change that.

3.3.2 Approaches based on bias correction. Instead of amending the uncertainties $u^2(x_l)$ such that compatibility is achieved between the differences $(x_k - x_l)$ among the various measurements, or the deviations $(x_l - \bar{x})$ from the mean value, and the uncertainty attributed to these differences/deviations, discrepancies may be utilised to derive corrections, either of the mean value $\bar{x}$ or the individual values x_l. In addition to this, the uncertainty on these corrections is evaluated, and this uncertainty is included in the uncertainty of the mean value obtained, either by direct correction or from the corrected individual values. A procedure for simultaneous correction of individual biases of the values x_l and amendment of individual uncertainties $u(x_l)$ is described in [19]. Ref. [20] specifies a procedure for the correction of bias in mean values, e.g. based on a comparison of the mean value $\bar{x}$ and the midrange value $x_{mid} = \frac{1}{2}(x_{min} + x_{max})$, and reviews other approaches from this perspective.

3.4 Robust statistics and related tools

Combination of different results $x_l \pm u(x_l)$ for the same quantity is a two-step process. In the first step the data x_l are combined into some kind of mean value. In the second step the uncertainty of this mean value is evaluated. So far this paper has only given marginal attention to the first step: the common mean value was used by default, supplemented by the inverse-variance weighted mean to account for major differences among the uncertainty data. This attitude was also motivated by the fact that, e.g., with just two measurement results $a \pm u(a)$ and $b \pm u(b)$ there is only very limited scope for statistical data processing. However, if a major number L of measurement results $x_l \pm u(x_l)$ have to be combined, the common mean value may not be the best choice. This is a good estimate if the data x_l behave like a sample from a normal distribution. But if the data distribution is skewed, other estimates perform better. As another major problem, contamination with outliers will spoil the performance of the common mean. Thus outliers should be removed or, preferably, robust means should be used.

A well-known robust mean is the median. This is obtained by arranging the data in an increasing sequence, $x_1 \leq x_2 \leq .... \leq x_L$. If L is an odd number, $L = 2N + 1$, the median is the middle value x_{N+1}. If L is an even number, $L = 2N$, the median is the mean of x_N and x_{N+1}. The median is obviously insensitive to outliers. On the other hand, for well-behaved data from a normal distribution the median gives similar results like the common mean. These are highly attractive features, but unfortunately this medal has a backside, too: it is impossible to evaluate the uncertainty of the median by propagation of the uncertainties $u(x_l)$ – a problem shared by many other robust means. Thus the uncertainty data $u(x_l)$ cannot be used in the estimation of uncertainty for the median, i.e. a major part of the information contained in the data $x_l \pm u(x_l)$ is thrown away.

If the median is used, the standard uncertainty of the median is estimated from the data x_l alone, discarding the uncertainty data $u(x_l)$. For this purpose the median of the absolute deviations $| x_k - \text{med}\{x_l\} |$ may be used with an appropriate numerical factor (to the effect that for normally distributed data the standard deviation of the mean is recovered).

Ref. [21] gives an introduction to the median and its use for a robust evaluation of interlaboratory comparisons. The contribution by Duewer in this volume [22] provides a comprehensive survey of robust techniques for the determination of interlaboratory consensus values, focussed on Key Comparisons.

3.5 Combined evaluation of measurements

Taking up the issue of section 1.3, if possible, the task of combining measurements should not be deferred to final results $x_l \pm u(x_l)$ according to the demand: take the data or leave them. Instead, combination should already be addressed in the planning phase, be carried out on the complete data obtained by measurement or otherwise, and utilise method performance data from validation studies and quality control.

A proposal how such an approach could be utilised in interlaboratory studies for the certification of reference materials by was described in ref. [23]. It is based on the availability of a comprehensive uncertainty budget for each laboratory result $x_l \pm u(x_l)$. These uncertainty budgets are analysed, e.g. concerning joint uncertainty sources, and the various uncertainty contributions are combined into an uncertainty budget for the mean value. However, this idea is not easily put into practice because interlaboratory studies where all participants provide comprehensive and comparable uncertainty budgets for their results are still rare.

3.6 Handling of correlated results revisited

One of the most popular results from statistical evaluation of measurements is the "$1/\sqrt{N}$ Law": Given data $\{x_1, x_2, ..., x_N\}$ from a set of (independent) replicate measurements, the standard deviation of the mean value is obtained from the standard deviation of the single values according to

$$s_{mean} = \frac{s_{\sin gle}}{\sqrt{N}} \tag{43}$$

This result is only valid if the data are uncorrelated. Otherwise, e.g. in case of a time series with a significant trend, the standard deviation of the mean may be much larger. In general (excluding the rather uncommon case of negative correlation) the relation is as follows:

$$\frac{s_{\sin gle}}{\sqrt{N}} \leq s_{mean} \leq s_{\sin gle} \tag{44}$$

With a large number of measurements, the factor $1/\sqrt{N}$ makes a large difference. For 10 or more measurements eq. (43) should therefore not be used unless appropriate efforts have been taken to ensure that the data are indeed uncorrelated.

With standard deviations replaced by standard uncertainties according to the GUM, similar results apply as follows: Given data $\{x_1, x_2, ..., x_N\}$ from a set of independent replicate measurements, and assuming the same standard uncertainty, $u(x)$, for all replicates, the standard uncertainty of the mean value, $u(\bar{x})$, is obtained from the standard deviation of the single values according to

$$u(\bar{x}) = \frac{u(x)}{\sqrt{N}} \tag{45}$$

However, if the measurements are correlated, eq. (45) is invalid. Instead, assuming uniform correlation with a correlation coefficient of r ($0 < r \leq 1$), the standard uncertainty of the mean value is obtained as

$$u(\bar{x}) = u(x) \cdot \sqrt{r + \frac{1-r}{N}} \tag{46}$$

This equation clearly shows that neglect of correlation, even at moderate levels will lead to substantial underestimation of the standard uncertainty of mean values. For example, a correlation coefficient of $r = 0.5$ gives $u(\bar{x}) \geq u(x) / \sqrt{2}$ instead of $u(\bar{x}) = u(x) / \sqrt{N}$.

If correlation between replicates shows up in the data $\{x_1, x_2, ..., x_N\}$ themselves, e.g. as a trend in a time series, correlation coefficients can be determined by statistical analysis. Often, however, correlations do not show up directly in the data $\{x_1, x_2, ..., x_N\}$ but only in the uncertainty budget for the mean value $\bar{x} = x_1 + x_2 + ... + x_N$. This will happen if the individual measurement results x_1, x_2, ..., x_N have major uncertainty contributions in common. As a pragmatic approach, uncertainty contributions accounting for random effects may be taken as independent, while those accounting for systematic effects are shared in full.

As shown in ref. [24] this approach gives an uncertainty as follows.

$$u(\bar{x}) = \sqrt{\frac{s_{\sin gle}^2}{N} + u_{syst}^2(x)} \tag{47}$$

In this expression s_{single} is the applicable standard deviation, either calculated from the replicates or, preferably, taken from quality control data. The second term, $u_{syst}(x)$, is the combined standard uncertainty restricted to systematic effects, i.e. effects or influence quantities which do not change between replicate measurements (e.g. calibration). This term may either be calculated from a restricted uncertainty budget or be determined by difference according to $u_{syst}^2(x) = u^2(x) - s_{single}^2$.

The essence of the result agrees with plain common sense: Taking averages only reduces data variability. While this entails major reduction of the random-effects contribution to measurement uncertainty, this benefial effect does not apply to the uncertainty arising from systematic effects.

The importance to account for correlations is not restricted to the uncertainty of mean values from a large number of replicate measurements. Therefore the general approach using uncertainty propagation is also outlined here. In this approach the standard uncertainty of a mean is given by

$$u^2(\bar{x}) = \frac{1}{N^2}\left[\sum_i u^2(x_i) + \sum_{j \neq k} u(x_j, x_k)\right] \tag{48}$$

In this uncertainty budget the covariances $u(x_j, x_k)$ account for correlations between errors in the j-th and the k-th measurement. The input for equation (48) is, in turn, obtained using the uncertainty budgets for the individual measurement results $x_i \pm u(x_i)$. The variances $u^2(x_j)$ are obtained from those of the relevant input quantities for x_j while the covariances $u(x_j, x_k)$ are obtained from the variances of the input quantities shared by x_j and x_k according to

$$u(x_i, x_k) = \sum_z c_i(z) \cdot c_k(z) \cdot u^2(z) \tag{49}$$

In the equation above the sum is over all input quantities shared, and the c´s are the sensitivity coefficients concerned. In principle there may also be correlation among input quantities, but there is no need to introduce this additional complication here.

4 EXAMPLE CALCULATIONS

In sections 3.1 – 3.3 various techniques were described how to estimate the uncertainty of a mean value obtained from discrepant results. The main techniques will now be illustrated by example calculations.

4.1 Design and data

The techniques addressed in the calculations are as follows:
- one-way ANOVA (un-weighted mean only)
- χ^2 technique for differences (un-weighted mean and weighted mean)

- χ^2 technique for deviations form the mean (un-weighted mean and weighted mean)
- Birge factor (un-weighted mean and weighted mean)

Five data sets $\{x_l \pm u_l \, ; \, l = 2, \, ..., \, 10\}$ are used to simulate different scenarios for the outcome of an interlaboratory study including 10 laboratories.

- (approximately) consistent data – similar uncertainties (CS)
- (approximately) consistent data – dissimilar uncertainties (CD)
- discrepant data – dissimilar uncertainties (DD)
- (approximately) consistent data – dissimilar uncertainties – particular feature (CDp)
- discrepant data – dissimilar uncertainties – particular feature (DDp)

In each of the five data sets the values $x_1, x_2, ..., x_{10}$ vary about 100 (values in arbitrary units). Uncertainty data specify standard uncertainties, expressed as absolute values (same unit as the measured values). For each set the data $x_l \pm u_l$ are assumed to be independent, i.e. correlations are assumed to be negligible.

Table 4 *Data sets simulating different scenarios for the outcome of an interlaboratory study including 10 laboratories*

CS	88±7	90±7	95±7	96±7	98±7	100±7	101±7	102±7	108±7	110±7
CD	88±10	90±10	95±8	96±5	98±5	100±3	101±5	102±5	108±8	110±10
DD	80±10	85±10	88±8	90±5	95±5	100±3	110±5	115±5	120±8	125±10
CDp	88±10	90±10	95±1	96±5	98±5	100±3	101±5	102±5	108±8	110±10
DDp	80±10	85±10	88±1	90±5	95±5	100±3	110±5	115±5	120±8	125±10

The level of discrepancy is governed by the proportion between deviations from the mean and specified uncertainties. An overall indicator is the proportion between the standard deviation of the values x_l and the root mean square of the standard uncertainties u_l. However, using the same data $\{u_l\}$ and $\{x_l\}$, different levels of discrepancy may be generated, depending on whether large uncertainties are matched to large deviations and small uncertainties to small deviations, or vice versa. The particular feature in the last two data sets is a particularly small uncertainty.

Table 5 contains basic summary statistics of these data: (a) mean value, (b) standard deviation, (c) standard deviation of the mean and (d) root mean square of standard uncertainties.

Table 5 *–Summary statistics of the data in Table 4*

	mean $\bar{x}$	standard deviation s_x	SD of the mean $s_x/\sqrt{10}$	RMS of the u_l $\sqrt{\sum u_l^2/10}$
CS	98.8	7.02	2.22	7.00
CD	98.8	7.02	2.22	7.33
DD	100.8	15.8	4.99	7.33
CDp	98.8	7.02	2.22	6.88
DDp	100.8	15.8	4.99	6.88

4.2 Techniqes examplified

Tables 6 and 7 give a summary of how to calculate the standard uncertainty of a mean value under consideration according to the different techniques.

Table 6   *Summary of calculations for the common (un-weighted) mean*

Common mean (un-weighted)	ANOVA	χ^2 for differences	χ^2 for deviations from the mean	Birge factor
Check of data consistency	check on s^2_{ex} obtained from eq. (26)	check on u^2_{ex} obtained from eq. (29)	check on u^2_{ex} obtained from eq. (36)	check on χ^2_{std} obtained from eq. (28)
Mean value	calculate as usual	calculate as usual	calculate as usual	calculate as usual
Std. uncertainty of mean for consistent data	use eq. (31)	use eq. (31)	use eq. (31)	use eq. (31)
Std. uncertainty of mean for inconsistd. data	use the standard deviation of the mean	use eq. (30) with u^2_{ex} determined from eq. (29)	use eq. (30) with u^2_{ex} determined from eq. (36)	use eq. (40) with χ^2_{std} determined from eq. (28)

Table 7 –*Summary of calculations for the (inverse-variance) weighted mean*

Weighted mean (inv. variances)	ANOVA	χ^2 for differences	χ^2 for deviations from the mean	Birge factor
Check of data consistency	(*not considered, see ref. [10]*)	check on u^2_{ex} determined from eq. (29)	check on u^2_{ex} obtained from eq. (36) adapted to the weighted mean	check on Birge factor f obtained from eq. (41)
Mean value for consistent data	(*not considered, see ref. [10]*)	use eq. (32) with $u^2_{ex} = 0$	use eq. (32) with $u^2_{ex} = 0$	use eq. (32) with $u^2_{ex} = 0$
Std. uncertainty of mean for consistent data	(*not considered, see ref. [10]*)	use eq. (33), upper part	use eq. (33), upper part	use eq. (33), upper part
Mean value for inconsistd. data	(*not considered, see ref. [10]*)	use eq. (32) with u^2_{ex} determined from eq. (29)	use eq. (32) with u^2_{ex} determined from eq. (36)	use eq. (32) with $u^2_{ex} = 0$
Std. uncertainty of mean for inconsistd. data	(*not considered, see ref. [10]*)	use eq. (33), lower part with u^2_{ex} determined from eq. (29)	use eq. (33), lower part with u^2_{ex} determined from eq. (36)	use eq. (42) with Birge factor f obtained from eq. (41)

Note: Data $x_l \pm u_l$ from different laboratories are assumed to be independent, i.e. correlations are assumed to be negligible. Therefore the covariances $u(x_j, x_k)$ in eqs. (28) and (29) are put to zero.

4.3 Results

Tables 8 and 9 present the results of the example calculations, obtained from the data in

Table 4, using the techniques summarised in Tables 6 and 7. For comparative purposes numerical values are given with a much larger number of digits than reasonable otherwise.

The data for the results of consistency checks refer to the characteristics given in Tables 6 and 7, i.e. the values obtained for s^2_{ex}, u^2_{ex}, χ^2_{std} or the Birge factor f, respectively. The letters following the data are used to denote the following cases:

- sd – data are only slightly discrepant;
- fc – data are fully consistent;
- d – data are clearly discrepant.

The various consistency checks utilise different criteria. For borderline data sets (such as CDp) different criteria may give different results.

Data in brackets denote estimates which would be considered invalid under the given circumstances:

- standard uncertainties for consistent data, if the data are in fact inconsistent;
- weighted means for consistent data, if the data are in fact inconsistent;
- standard uncertainties for inconsistent data, if the data are in fact consistent;
- weighted means for inconsistent data, if the data are in fact consistent.

Table 8 *Results of example calculations for the common (un-weighted) mean*

Common mean (un-weighted)	ANOVA	χ^2 for differences	χ^2 for deviations from the mean	Birge factor
Check of data consistency	CS: 0.0206 - *sd*	CS: 0.5375 - *sd*	CS: 0.5375 - *sd*	CS: 1.0059 - *sd*
	CD: 0 - *fc*	CD: 0 - *fc*	CD: 0 - *fc*	CD: 0.7722 - *fc*
	DD: 8.4397 - *d*	DD: 13.6179 – *d*	DD: 13.2542 - *d*	DD: 4.5276 - *d*
	CDp: 0.1358 - *sd*	CDp: 0 - *fc*	CDp: 1.7138 - *sd*	CDp: 0.9313 - *fc*
	DDp: 8.8830 - *d*	DDp: 13.8404 - *d*	DDp: 13.4931 - *d*	DDp: 5.8639 - *d*
Mean value	CS: 98.8000	CS: 98.8000	CS: 98.8000	CS: 98.8000
	CD: 98.8000	CD: 98.8000	CD: 98.8000	CD: 98.8000
	DD: 100.8000	DD: 100.8000	DD: 100.8000	DD: 100.8000
	CDp: 98.8000	CDp: 98.8000	CDp: 98.8000	CDp: 98.8000
	DDp: 100.8000	DDp: 100.8000	DDp: 100.8000	DDp: 100.8000
Std. uncertainty of mean for consistent data	CS: (2.2136)	CS: (2.2136)	CS: (2.2136)	CS: (2.2136)
	CD: 2.3173	CD: 2.3173	CD: 2.3173	CD: 2.3173
	DD: (2.3173)	DD: (2.3173)	DD: (2.3173)	DD: (2.3173)
	CDp: (2.1772)	CDp: 2.1772	CDp: (2.1772)	CDp: 2.1772
	DDp: (2.1772)	DDp: (2.1772)	DDp: (2.1772)	DDp: (2.1772)
Std. uncertainty of mean for inconsistd. data	CS: 2.2201	CS: 2.2201	CS: 2.2201	CS: 2.2201
	CD: (2.2201)	CD: (2.3173)	CD: (2.3173)	CD: (2.0363)
	DD: 4.9862	DD: 4.8903	DD: 4.7893	DD: 4.9308
	CDp: 2.2201	CDp: (2.1172)	CDp: 2.2436	CDp: (2.1011)
	DDp: 4.9862	DDp: 4.8883	DDp: 4.7902	DDp: 5.2721

Table 9 – *Results of example calculations for the (inverse-variance) weighted mean*

Weighted mean (inv. variances)	ANOVA	χ^2 for differences	χ^2 for deviations from the mean	Birge factor
Check of data consistency	(*not calculated*)	CS: 0.5375 - *sd* CD: 0 - *fc* DD: 13.6179 - *d* CDp: 0 - *fc* DDp: 13.8404 - *d*	CS: 0.5375 - *sd* CD: 0 - *fc* DD: 13.2536 - *d* CDp: 1.5250 - *sd* DDp: 16.8822 - *d*	CS: 1.0029 - *sd* CD: 0.7971 - *fc* DD: 2.0671 - *d* CDp: 1.0837 - *sd* DDp: 3.0234 - *d*
Mean value for consistent data	(*not calculated*)	CS: (98.8000) CD: 99.4189 DD: (101.2787) CDp: 96.1154 DDp: (91.3517)	CS: (98.8000) CD: 99.4189 DD: (101.2787) CDp: (96.1154) DDp: (91.3517)	CS: (98.8000) CD: 99.4189 DD: (101.2787) CDp: (96.1154) DDp: (91.3517)
Std. uncertainty of mean for consistent data	(*not calculated*)	CS: (2.2136) CD: 1.7346 DD: (1.7346) CDp: 0.8715 DDp: (0.8715)	CS: (2.2136) CD: 1.7346 DD: (1.7346) CDp: (0.8715) DDp: (0.8715)	CS: (2.2136) CD: 1.7346 DD: (1.7346) CDp: (0.8715) DDp: (0.8715)
Mean value for inconsistd. data	(*not calculated*)	CS: 98.8000 CD: (99.4189) DD: 101.0440 CDp: (96.1154) DDp: 100.6492	CS: 98.8000 CD: (99.4189) DD: 101.0530 CDp: 97.2232 DDp: 100.7102	CS: 98.8000 CD: (99.4189) DD: 101.2787 CDp: 96.1154 DDp: 91.3517
Std. uncertainty of mean for inconsistd. data	(*not calculated*)	CS: 2.2201 CD: (1.7346) DD: 4.8406 CDp: (0.8715) DDp: 4.8307	CS: 2.2201 CD: (1.7346) DD: 4.7363 CDp: 1.3133 DDp: 5.7300	CS: 2.2201 CD: (1.3826) DD: 3.5856 CDp: 0.9444 DDp: 2.6348

Figures 1 – 5 present the results of the example calculations in graphical form, one for each data set.

4.4 Discussion

For homoscedastic data (i.e. all uncertainties are about the same) all approaches give about the same results. For data set CS (which is strictly homoscedastic), both the means and the corresponding uncertainties are independent of the combination technique applied (Fig. 1).

For consistent data there is only the choice between the use of the common or the weighted mean. With heteroscedastic data (i.e. dissimilar uncertainties) the weighted mean deviates from the un-weighted mean, and has a lower uncertainty than the latter. If the uncertainties are deemed credible, there are no objections against using the weighted mean. Within each group (weighted and un-weighted techniques considered separately), and for (merely) heteroscedastic but consistent data all combination techniques give about the same result except the Birge factor approach (see Fig. 2).

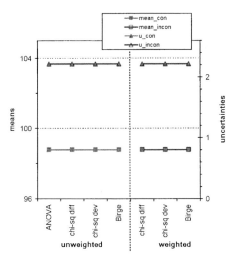

Figure 1: *Results of mean value and uncertainty calculation for data set CS, using the techniques listed inTables 6 and 7. Mean values are calculated for the data treated as consistent (mean_con) and inconsistent (mean_incon), the corresponding uncertainties are u_con and u_incon. Mean values refer to the scale on the left-hand side, uncertainties to the scale on the right-hand side. Categories for unweighted techniques are ANOVA, χ^2 for differences (chi-sq diff), χ^2 for deviations from the mean (chi-sq dev), and the Birge factor approach. For weighted techniques, the same categories except ANOVA apply.*

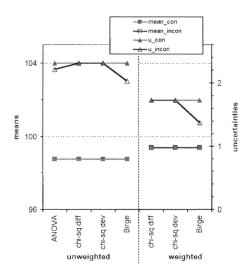

Figure 2: *Results of mean value and uncertainty calculation for data set CD, using the techniques listed in Tables 6 and 7. Symbols and abbreviations as in Fig. 1.*

If the variability (SD) of the values x_l is much larger than the average uncertainty specified (expressed as root mean square of the u_l), ANOVA and the two χ^2 approaches give approximately the same uncertainties for an un-weighted approach, while the result obtained using the weighted approaches may differ quite substantially (see Fig. 3 for the inconsistent data set DD).

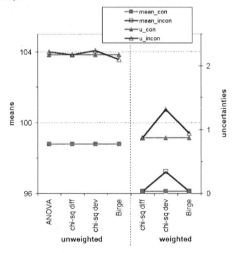

Figure 3: *Results of mean value and uncertainty calculation for data set DD, using the techniques listed in Tables 6 and 7. Symbols and abbreviations as in Fig. 1.*

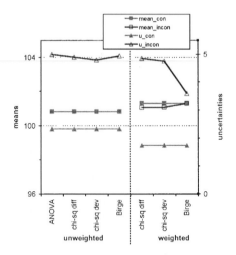

Figure 4 *Results of mean value and uncertainty calculation for data set CDp, using the techniques listed inTables 6 and 7. Symbols and abbreviations as in Fig. 1.*

The differences between the un-weighted and the weighted approach, but also between the different combination techniques become evident when particular features (like a particular uncertainty much smaller than the rest) of the data sets come in. Results between the groups, and also within each group may vary widely, especially the Birge factor results may be quite different from the results of the other techniques. This applies already to consistent data for the weighted approaches (see Fig. 4).

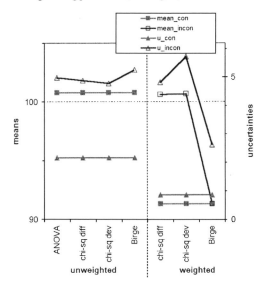

Figure 5 *Results of mean value and uncertainty calculation for data set DDp, using the techniques listed in Tables 6 and 7. Symbols and abbreviations as in Fig. 1.*

Finally, for inconsistent data with some exceptionally small uncertainties (see Fig. 5) the weighted approach is highly problematic. With a skewed distribution of results with small uncertainties the weighted mean may be shifted far away from the common (un-weighted) mean. Unless the small uncertainties are deemed credible, the use of the weighted mean is doubtful. In the first place, this concerns the use of the weighted mean with uncertainties as specified (Birge factor), in particular when the data are incompatible.

With one exception the consistency checks for the different approaches give the same results. Only data set CDp is partly classified as slightly discrepant and partly as fully consistent. For the common mean the uncertainties are nevertheless almost the same, irrespectively of whether the data are treated as consistent or inconsistent. For the weighted mean differences are appreciable, e.g. 96.1 ± 0.9 vs. 97.2 ± 1.3 for the two Chi-squared approaches. As it should, for any approach the applicable uncertainty estimate is always the larger one of the uncertainties obtained treating the data as consistent or inconsistent, respectively.

CONCLUSIONS

For the main issues addressed in this paper, discrepancy between, and correlation among measurement results subject to combination, practical relevance is largely complementary. The more measurements have in common, the higher the level of correlation. The less measurements have in common, the higher the level of discrepancy. Considering a common hierarchical scheme for multiple measurements (see ref. [7]), this may be summarised as follows.

Table 10 *Relevance of discrepancy and correlation for combination of measurement results*

Origin of results	Discrepancy	Correlation
different replicates	low	high
different runs	moderate	high
different laboratories	high	moderate
different methods	high	low

As another, rather obvious conclusion from this review, for the problem at hand – combination of different measurement results – there is no such thing like the master approach that would give the definitive solution. This is true for the vast majority of statistical data evaluation problems: There are different approaches available, and agreement of results is rather the exception than the rule. Among others, this is because statistical data evaluation procedures are based on assumptions (models) concerning the data. Thus different statistical data evaluation procedures for the same task behave much the same as different measurement procedures for the same measurand: they give different answers. For a rational measurand different measurement procedures can be compared with respect to the true-value recovery – at least in principle, but also in practice through traceability to the SI. Unfortunately nothing like that is available for statistical data evaluation procedures: the results are largely method-dependent.

References

1   *Guide to the Expression of Uncertainty in Measurement*
     1st corr. Edition, ISO, Geneva 1995, ISBN 92-67-10188-9
2   *Metrologia* (2000 ff.), e.g., contains many papers dedicated to the determination of Key Comparison Reference Values (KCRV).
3   L.J. Lyons, *J Phys A Math Gen*, 1992, **25**, 1967-1979
4   M.S. Levenson et al, *J Res NIST*, 2000, **105**, 571-579
5   S.L.R. Ellison et al, *Accred Qual Assur*, 2001, **6**, 274-277
6   ISO 5725 (6 parts), *Accuracy (trueness and precision) of measurement methods and results*
7   M. Thompson, *Analyst*, 2000, **125**, 2020-2025
8   ISO/TS 21749, *Measurement uncertainty for metrological applications - Repeated measurements and nested experiments* (1st Edition 2005)
9   ISO Guide 35, *Reference materials – General and statistical principles for certification*   (3rd Edition 2006)
10   W.G. Cochran, *Biometrics*, 1954, **10**, 101-129
11   P. Dehouk et al, *Analyt Chim Acta*, 2003, **481**, 261-272
12   R.J. Douglas and A.G. Steele, *Metrologia*, 2006, **43**, 89-97
13   C.R. Paule and J. Mandel, *J Res Natl Bur Stand*, 1982, **87**, 377-385
14   S.B. Schiller and K.R. Eberhardt, *Spectrochimica Acta*, 1991, **46B**, 1607-1613
15   A.L. Rukhin and M.G. Vangel, *J Am Stat Assoc*, 1998, **93**, 303-308
16   W. Hässelbarth, W. Bremser and R. Pradel, *Fresenius J Anal Chem*, 1998, **360**, 317-321
17   R. Kacker, *Metrologia*, 2004, **41**, 132-136
18   R.T. Birge, *Phys Rev*, 1932, **40**, 207-227
19   W. Weise and W. Wöger, *Meas Sci Technol*, 2000, **11**, 1649-1658
20   R. Kacker, R. Datla and A. Parr, *Metrologia*, 2002, **39**, 279-293

21   J.W. Müller, *J Res NIST*, 2000, **105**, 551-555
22   D.L. Duever, article in this volume
23   J. Pauwels, A. Lamberty and H. Schimmel, *Accred Qual Assur*, 1998, **3**, 180-184
24   W. Hässelbarth and W. Bremser, *Accred Qual Assur*, 2004, **9**, 597-600

METROLOGICAL CHARACTERISTICS OF THE CONVENTIONAL MEASUREMENT SCALES FOR HYDROGEN AND OXYGEN STABLE ISOTOPE AMOUNT RATIOS: THE δ-SCALES

Manfred Gröning, Michael van Duren, Liliana Andreescu

International Atomic Energy Agency (IAEA), Agency's Laboratories Seibersdorf and Vienna, Isotope Hydrology Laboratory, P.O.Box 100, A-1400 Vienna, Austria

1 INTRODUCTION

Natural variations of stable isotope amount ratios for the elements hydrogen and oxygen are used in diverse scientific fields to delineate for example the origin of water masses, past climatic conditions, biological processes or the authenticity of food products.

Here, only a brief outline is provided on measurement methods. The focus lays on discussion on properties of the measurement scale, which can be described as a conventional measurement scale based on isotope ratio properties of arbitrarily chosen artefact materials (calibrators). The traceability chain for such measurements is thus limited back to the used artifacts. Any change of those calibrators potentially induces problems in scale consistency and in any case increases the combined uncertainty of measurements. The recent efforts are described to limit the increase of uncertainty due to the necessary change of calibrators (exhaustion of used artefact material).

The internationally used calibrators are pure water samples and therefore the discussion below focuses on hydrogen and oxygen stable isotopes in water. Similar arguments are valid for other substances containing hydrogen and oxygen isotopes. Furthermore similar concepts hold for measurements of isotope amount ratios of other elements like carbon, nitrogen and sulfur.

2 HYDROGEN / OXYGEN STABLE ISOTOPE AMOUNT RATIO MEASUREMENTS

The naturally occurring isotopic fractionation processes for hydrogen and oxygen are rather small and result in a rather limited variability of isotopic amount ratios of these elements on earth. For example, the total range for the oxygen $^{18}O/^{16}O$ ratio in water and ice on earth is only nine percent, and increases only to 17 percent, if all oxygen bearing materials are considered.[1] Even for hydrogen and its large relative isotope mass difference, variations on earth reach only a factor of two. Therefore, such isotope amount ratio measurements for any local study have to be carried out with high precision and very small uncertainties in the range of tenth of per mill in order to provide useful data.

So far, for more than five decades, mass spectrometry has been the ultimate analytical tool for the analysis of stable isotope ratios at environmental levels[2]. Other methods like nuclear magnetic resonance or optical infrared absorption spectroscopy yet cannot achieve the necessary precision and accuracy for routine application and are mostly used for analysis of isotopically enriched samples[3]. Especially for the light elements H, C, N, O and S, until recently most measurements have been carried out using gas isotope ratio mass spectrometers using the dual-inlet configuration after conversion of samples into suitable gases like H_2, CO_2, CO, N_2, O_2, SO_2 or SF_6.[4,5,6]

That involves the handling of large sample amounts in the range of 50 mg to 5 g, conversion into suitable gases as mentioned above and transfer into variable steel bellow volumes for subsequent dynamic inlet into the mass spectrometer. In the mass spectrometer, under high vacuum conditions, the gas is ionized by electron impact, accelerated by high voltage and the resulting ion beam is mass separated by a magnetic field. The resulting separate ion beams hit highly sensitive faraday cup detectors and the tiny ion currents are recorded simultaneously.

The dual inlet approach uses two parallel inlet systems with the unknown sample gas in one and a known gas reference (called transfer gas or working standard) in the other inlet branch. During the measurement, the gas stream into the mass spectrometer is repeatedly switched between sample and reference. As great advantage this provides a virtually direct comparison of each sample with a gas reference. This enables to keep very stable measurement conditions over a full measurement day and to calibrate measurements precisely by use of laboratory standards treated similar to the unknown samples, while canceling out completely most systematic biases.

During the last decade, a new development in mass spectrometry has emerged allowing the analysis of very tiny samples at the microgram level. These continuous flow techniques allow an on-line sample preparation by combustion in elemental analyzers, separation of gases by chromatographic columns and gas transfer via helium carrier gas stream into the mass spectrometer. The use of very tiny sample sizes is very demanding for the proper use of the reference materials and for the calibration. Recently major breakthroughs have considerably improved the precision and accuracy of water stable isotope analyses using high temperature pyrolysis systems and continuous flow mass spectrometry.[7]

3 INVENTION OF δ-MEASUREMENT SCALES

For reporting an isotope amount ratio R for two isotopes of an element (e.g. $R^{18/16} = {}^{18}O/{}^{16}O$), the ion currents of gas molecules representing the interesting isotopes are measured (here mass 44: ${}^{12}C^{16}O_2$; mass 46: ${}^{12}C^{16}O\,{}^{18}O$). The ratio $R^{46/44}$ is calculated by dividing the ion current values obtained for the respective masses. This ratio value has then to be corrected, e.g. for isobaric interferences induced by the presence of ${}^{13}C$ and ${}^{17}O$.[8]

Due to the low abundance of the minor isotopes ${}^{2}H$ and ${}^{18}O$ in respect to the major ones (${}^{1}H$ and ${}^{16}O$), this results in very small ratio values with only very tiny natural variations. In most applications of stable isotopes in earth sciences, it is of much more interest to know the relative differences in isotopic ratios between samples than to know the real (or absolute) isotopic amount ratios of the samples.

In the early fifties of last century such a relative measurement scale was proposed and is still used throughout the measurement community. Here in the so-called δ-scale isotope amount ratios are reported as deviation from the isotope ratio of an artificially selected reference. In this approach the knowledge of the mole fraction of an isotope of a given reference material is not necessary, since these isotope ratio δ-scales are defined completely arbitrarily relative to the isotope ratio of a selected primary reference material. This primary reference material serves as the end of the traceability chain. This concept as described is the realization of a "conventional" scale.[9] It is not traceable back to the S.I. system, since it is not based on fundamental constants, but on arbitrarily selected properties, e.g. on the stable isotope amount ratio in a sample of a chosen primary reference material (the pH scale being another example for a conventional scale).

The general definition of δ scales is given by the following formula:

$$\delta = \frac{R_{sample} - R_{reference}}{R_{reference}} \tag{1}$$

with δ (e.g. δ^2H, $\delta^{13}C$, $\delta^{15}N$, $\delta^{18}O$, $\delta^{34}S$) being the difference of the isotope concentration ratios R ($^2H/^1H$, $^{13}C/^{12}C$, $^{15}N/^{14}N$, $^{18}O/^{16}O$, $^{34}S/^{32}S$) of the sample and of the reference (primary reference material, calibrator). δ-values are therefore unitless numbers, like the isotope ratios itself. As the differences between sample and reference are normally very small, the δ-values are usually expressed as per mill difference (parts per thousand).

Thus, we have:

$$\delta = \frac{R_{sample} - R_{reference}}{R_{reference}} \times 1000\,\%_0 \tag{2}$$

This modified equation 2 results in more convenient numbers being reported, without several leading zeros. The values can be positive or negative, with e.g. negative numbers indicating a lower abundance of the minor isotope in the sample than in the reference. Note that the per mill sign ($\%_0$) represents the factor 10^{-3} and therefore is part of the notation used by all laboratories and may not be omitted. Detailed discussions of the features of the δ notation as realization of a conventional scale can be found elsewhere.[8,10] An important, but often forgotten feature of δ-scales is their non-linearity. As extreme example a sample without a single atom of the minor isotope left would be indicated by a δ-value of $-1000\%_0$. On the other hand, a sample without a single atom of the major isotope would request the equation to go to infinity.

In case of hydrogen and oxygen, a second reference material (ref2) is used to normalize results, which modifies the formula according to:

$$\delta = \frac{R_{sample} - R_{reference}}{R_{reference}} \cdot \left(\delta_{ref2} \middle/ \frac{R_{ref2} - R_{reference}}{R_{reference}} \right) \times 1000\,\%_0, \tag{3}$$

with δ referring to δ^2H or $\delta^{18}O$, R being the corresponding $^2H/^1H$ or $^{18}O/^{16}O$ ratios and the additional term in the bracket providing the normalization of the respective δ-scale using a pre-defined and calibrated δ_{ref2} value for the second reference material (ref2).[11]

4 REFERENCE MATERIALS USED

No obvious anchor point for natural isotope amount ratios is available. Therefore it was proposed in the 1950's for both hydrogen and oxygen isotope amount ratios to use the average isotopic composition of seawater as arbitrary anchor point. The oceans as major reservoirs of hydrogen and oxygen could serve as ideal standard and therefore the virtual isotopic composition of a hypothetical water mixing all oceans together was compiled from available ocean isotope data. This virtual material was called Standard Mean Ocean Water (SMOW).[12] In fact it was defined as a certain isotopic δ-value difference from an existing river water sample called NBS 1, which was used as calibrator. Due to problems with proper storage of NBS1 and in order to improve the intercalibration of laboratories, a new calibrator was produced in 1967 from distilled sea water with small additions of other water samples with a resulting isotopic composition very close to SMOW. Its $\delta^{18}O$ value was reported as practically identical and the δ^2H value lower by $-0.2‰$, which was within the measurement uncertainty for all laboratories. This material was subsequently named VSMOW (Vienna-SMOW, previously also abbreviated V-SMOW), to distinguish it from the virtual material SMOW. A quantity of 70 liters of VSMOW was produced. After extensive measurements an IAEA expert panel decided in 1976 to consider the isotopic composition of VSMOW as identical to that of SMOW and to adopt VSMOW as zero point for the δ^2H and $\delta^{18}O$ scales for natural compounds.[13]

In addition a second natural sample was obtained from Antarctic firn in 1967 and was introduced as second standard called SLAP (Standard Light Antarctic Precipitation).[13] This second standard, which was much more depleted in the heavier isotopes of both hydrogen and oxygen and was close to their lowest limits observed in natural water, should enable a two-point calibration (normalization) to improve the data consistency between laboratories. Its isotopic composition was fixed by convention using the available data measured by numerous laboratories (Table 1).

Later on a third sample was added with an isotopic composition intermediate to VSMOW and SLAP. This sample obtained from Greenland firn was called GISP (Greenland Ice Sheet Precipitation) and was used as quality control material to check the calibration/normalization performed by VSMOW/SLAP.

Table 1: *Isotopic composition of international water reference materials*

	$\delta^{18}O$	δ^2H	Remarks
VSMOW	0 ‰	0 ‰	per definition
GISP	-24.78 ± 0.09 ‰	-189.5 ± 1.2 ‰	(± 1σ)
SLAP	-55.5 ‰	-428 ‰	per convention

All materials were stored in sealed 10 liter bulk glass containers. For distribution the contents of one container at a time was transferred into flame sealed glass ampoules in 20 ml quantities.

Concerning the traceability of measurements, it is clearly limited to VSMOW and SLAP as basic calibrators. In principle the traceability chain could be extended back to SMOW, but since less information is available on those early measurements as desirable, it is preferable to trace all measurements back to VSMOW.

The actual isotopic amount ratios for $^{18}O/^{16}O$, $^{17}O/^{16}O$ and $^2H/^1H$ on VSMOW were measured by isotope dilution methods. A data compilation is provided elsewhere.[11] It would be therefore possible to compare any future material to VSMOW without directly measuring both materials at the same time. The complexity of such direct isotope amount measurements has so far limited their application; only five studies were published in the last three decades measuring one of the three isotopic amount ratios listed above.

The supply of both VSMOW and SLAP is close to exhaustion in the year 2006. A rigid distribution policy was applied to supply only one unit of each material to a laboratory within a three year's period. However this could stretch the availability of stocks only until now.

To keep the comparability and consistency of data reporting for long time scales, it was decided at IAEA a few years ago to establish two new standards as replacement for VSMOW and SLAP with isotopic compositions for hydrogen and oxygen as close as possible to those of the original standards.

4.1 Production of VSMOW2 as Replacement for VSMOW:

In order to reproduce the original VSMOW $^2H/^1H$ isotope amount ratio as well as those of $^{18}O/^{16}O$ and $^{17}O/^{16}O$ in the successor material, it was necessary to mix only natural raw water samples (isotopically enriched materials do not preserve the ^{18}O-^{17}O ratio). Therefore three carefully selected water samples were isotopically well characterized by five experienced laboratories. One of the water samples had to be slightly adjusted in its deuterium concentration in order to obtain the necessary amount of the successor material. This was performed by use of platinum catalyst and hydrogen gas isotopic exchange between the original sample and highly deuteriated water. Both water samples were separated by a semi-permeable membrane, which was only open for hydrogen gas diffusion, but hindered any transfer of water vapor. This ensured that the ^{18}O-^{17}O-^{16}O ratios were not altered by use of the highly deuteriated water. Control measurements revealed no change in the ^{18}O composition during the deuterium enrichment process. The three water samples were subsequently mixed in proportions to reproduce the isotopic composition of VSMOW (Fig.1).

The final product VSMOW2 was then isotopically compared directly with aliquots of VSMOW. More than 100 measurements were performed in three laboratories and revealed only very small isotopic differences of the two materials: about 0.002‰ for $\delta^{18}O$ and 0.10‰ for δ^2H. Both differences were insignificant compared to the standard errors of the mean calculated for the measurement of VSMOW2 and VSMOW (±0.007‰ for $\delta^{18}O$ and ±0.09‰ for δ^2H) (Fig.1, right diagram, small inner circle).

In view of routine uncertainties of about 0.1‰ and 1‰ or higher for $\delta^{18}O$ and δ^2H measurements in most laboratories (Fig.1, large circle in right diagram), such small isotopic differences do not contribute significantly to the combined measurement uncertainties.

So far as an initial batch 5000 glass ampoules with 20 ml VSMOW2 each were flame sealed and will be checked soon for isotopic homogeneity. Details on the production will be published elsewhere.

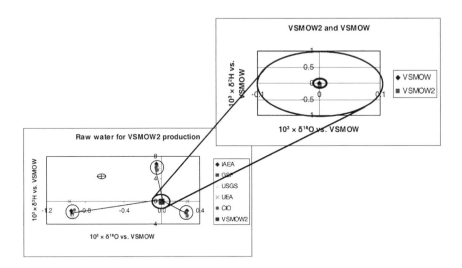

Figure 1: Calibration of isotopic compositions of three raw waters by five laboratories to mix VSMOW2 (lower left figure). Precision of VSMOW2 characterization (upper right figure).

4.2 Production of SLAP2 as replacement for SLAP:

Similarly, the preparation of a replacement for SLAP is ongoing with the isotopic characterization of four Antarctic raw water samples being completed. Those samples were obtained from the Southpole, the Vostok ice core drilling and from Dome C and Dome F sites in Antarctica. One sample (ANT-Abe) had to be isotopically enriched in deuterium in order to span the triangle in Fig.2 and to enable the mixing of a sufficiently large amount of SLAP2. The isotopic enrichment was done quantitatively by use of isotopically enriched hydrogen gas and water-H_2 equilibration using platinum catalyst. Recently approximately 300 liters of SLAP2 were prepared by appropriate mixing of the raw waters.

Upon completion of the SLAP2 preparation by finishing its bottling into 20 ml breakseal glass ampoules, the final isotopic assessment for both new materials will be performed. It will include tests of isotopic homogeneity in the produced ampoules and direct comparisons of the isotopic compositions of the new materials with VSMOW and SLAP.

The amount of standards produced is about 300 liters each and should be sufficient for the next fifty years. The large amounts were produced to limit necessary calibrator transitions in future and to lower the restriction on distribution policy of once per three years per laboratory. A possibility of more frequent calibrations of internal laboratory standards will certainly improve the achievable uncertainty of calibrations. Strategies for proper long-term storage are available to ensure the stability of the isotopic compositions of the calibrators and to be able to detect even small isotopic drifts well in time.

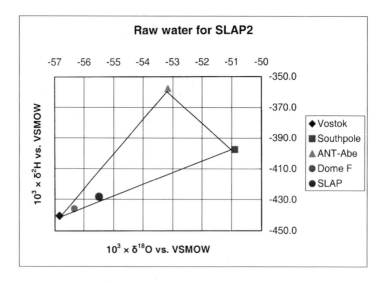

Figure 2: Isotopic composition of raw waters designated for mixing of SLAP2.

5 TRACEABILITY

The traceability chain for stable isotope amount ratio measurements stops at the chosen calibrator material, e.g. VSMOW and SLAP in case of hydrogen and oxygen (Fig.3). The arbitrarily defined VSMOW/SLAP δ-scale is therefore an example of a conventional calibration scale, as in the case of pH measurements or of temperature measurements using the Celsius scale.

Even in the case that measurements are considered to be traceable with reference to VSMOW, a potentially severe problem exists in proper storage conditions for the used internal laboratory standards. Any evaporation out of storage containers or isotopic exchange with laboratory air moisture can cause significant isotopic offsets. The same applies (to the strongly discouraged) storage of unused portions of primary reference materials or calibrators. Such potentially altered materials should not anymore be used for calibration purposes.

Absolute isotope ratio determinations of the calibrators exist partially (however, SLAP was never characterized in this way for its oxygen isotope amount ratios), and can be considered as additional information to anchor the calibrator materials to an absolute isotope abundance scale.

The process used for production and characterization of VSMOW2 and SLAP2 should ultimately ensure that the VSMOW/SLAP scale can be retained without any significant change despite complete change of used calibrators. This is achieved by the careful calibration of the new materials versus the existing calibrators. The chosen procedure to produce isotopically indistinguishable materials is considered as inherently safe concerning traceability for all existing analytical techniques of stable isotope measurements. That means, that no problem occurs, as long as no existing analytical technique is able to detect in routine mode a difference between former and new calibrators.

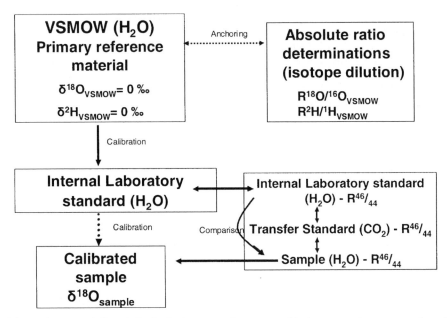

Figure 3: *Traceability chain for hydrogen and oxygen stable isotope ratios measured on the VSMOW/SLAP δ-scale (SLAP omitted here for simplification).*

6 UNCERTAINTY

The VSMOW/SLAP scale can be retained and the traceability chain be preserved due to well measured and very small isotopic differences to the successor calibrators. Due to the intrinsic properties of the δ-scale measurements using the dual inlet scheme, most of the possible systematic offsets both affect sample and reference thus cancel out. Offsets and reproducibility can be assessed by a measurement scheme of direct comparisons of old and new calibrators under identical conditions. Huge efforts were made during calibrator preparation to ensure a very small resulting isotopic offset, which in turn ensures a small uncertainty contribution to the combined uncertainty of routine measurements.

The major uncertainty components of stable isotope amount ratio measurements using the dual inlet techniques can be summarized as follows:

1. Repeatability of mass spectrometric measurements (stability of ion current measurements including H_3^+-factor and other mass spectrometric corrections)[8]
2. Reproducibility of sample measurements versus transfer gas (also often called incorrectly working standard)[9]
3. Daily determination of the mean isotopic composition of samples of internal laboratory standards versus transfer gas
4. Daily measurements of the isotopic composition of quality control samples
5. Estimated magnitude of possible residual offsets for linearity corrections
6. Calibration uncertainty of internal laboratory standards versus the international VSMOW/SLAP scale (depending on the number of such calibrations performed)
7. Uncertainty of measurements of samples of the calibrator material during a calibration
8. Uncertainty of the isotopic amount ratio of the calibrator itself (zero in case of VSMOW and SLAP, not zero for any successor material).

Other possible non-statistical effects due to possible laboratory bias, incomplete linearization or other calibration offsets depend on the individual laboratory and cannot be easily quantified.

The largest individual component is the long-term reproducibility of measurements. At the IAEA laboratory, over many years all $\delta^{18}O$ measurements were performed in duplicate, at different dates and in most cases by different operators and with two completely independent mass spectrometers. A proxy of the annual reproducibility of routine measurements was derived from the average range of duplicate measurements from the annually up to 2800 samples (Fig.4).

Similarly, the other uncertainty components were compiled from long-term quality control charts on daily measurements, mass spectrometer raw data records and internal laboratory standard calibrations with reference to VSMOW.

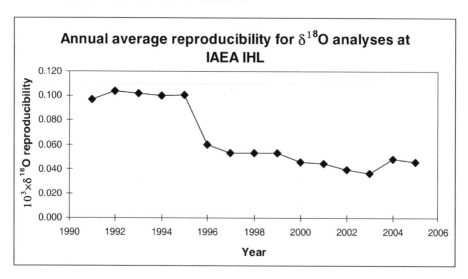

Figure 4: *Annual average reproducibility for sample measurements at the IAEA Isotope Hydrology Laboratory. Each yearly average value is compiled from all range values of independent duplicate sample measurements for at least 400 up to 2800 measurement pairs. The reproducibility is calculated from the average range as its standard deviation (range divided by $d_2=1.128$ for 2 measurements to convert it into a standard deviation)[14].*

An example is given for the assessment of the individual uncertainty components at the IAEA Isotope Hydrology Laboratory by long term data evaluations. For more than ten years all measurements and calibrations performed annually in the laboratory were analyzed to provide proxies for the uncertainty components.[15] The data are provided in Table 2.

Table 2: *Long-term averaged uncertainty components for hydrogen and oxygen stable isotope measurements on water samples analyzed at the IAEA Isotope Hydrology Laboratory*

	Uncertainty for $\delta^{18}O$ in per mill w.r.t. VSMOW/SLAP	Uncertainty for δ^2H in "per mill" w.r.t. VSMOW/SLAP
1. Sample measurement repeatability	0.010	0.1
2. Sample measurement reproducibility	0.048	0.76
3. Daily laboratory standard measurements	0.030	0.6
4. Daily measurements of control samples	0.018	0.3
5. Linearity correction offset	0.020	0.3
6. Calibration of internal laboratory standards	0.015	0.3
7. Uncertainty of measurements of international calibrators	0.020	0.3
8. Uncertainty of international calibrator	0 (for existing calibrators)	0 (for existing calibrators)
Combined standard uncertainty:	**0.07**	**1.1**

This results in a statement at the IAEA Isotope Hydrology Laboratory on the combined standard uncertainty of measurements performed to be in general close to ±0.1‰ for $\delta^{18}O$ and ±1‰ for δ^2H. Comparing the magnitude of the uncertainty components with the estimated uncertainty of the new calibrators proves the suitability of the calibration: no significant increase of routine measurement uncertainty should be caused by the transition to the new materials.

REFERENCES

1. T.B. Coplen, J.A. Hopple, J.K. Böhlke, H.S. Peiser, S.E. Rieder, H.R. Krouse, K.J.R. Rosman, T. Ding, R.D. Vocke Jr., K.M. Révész, A. Lamberty, P. Taylor, P. De Bièvre (2002): Compilation of minimum and maximum isotope ratios of selected elements in

naturally occurring terrestrial materials and reagents. U.S. Geological Survey Water-Resources Investigation Report 01-4222 (rev. August 2002), 98p. (also available under url: http://pubs.water.usgs.gov/wri014222).

2. W. Brand (2004): Mass spectrometer hardware for analyzing stable isotope ratios. In: 'Handbook of Stable Isotope Analytical Techniques', Elsevier, Vol.1, Chapter 38, 835-856.

3. E. Kerstel (2004): Isotope ratio infrared spectrometry. In: 'Handbook of Stable Isotope Analytical Techniques', Elsevier, Vol.1, Chapter 34, 759-787.

4. A.O. Nier (1947): A mass spectrometer for isotope and gas analysis. Rev. Sci. Instrum. 18, 398-411.

5. C.R. McKinney, J.M. McCrea, S. Epstein, H.A. Allen, H.C. Urey (1950): Improvements in mass spectrometers for the measurement of small differences in isotope abundance ratios. Rev. Sci. Instrum. 21, 1950, 724 –730.

6. J. Horita, C. Kendall (2004): Stable isotope analysis of water and aqueous solutions by conventional dual-inlet mass spectrometry. In: 'Handbook of Stable Isotope Analytical Techniques', Elsevier, Vol.1, Chapter 1, 1-37.

7. M. Gehre, H. Geilmann, J. Richter, R.A. Werner, and W.A. Brand (2004): Continuous flow $^2H/^1H$ and $^{18}O/^{16}O$ analysis of water samples with dual inlet precision. Rapid Comm. Mass Spectrom. 18: 2650 - 2660

8. R. Gonfiantini (1981): The δ-notation and the mass-spectrometric measurement techniques. In: 'Stable Isotope Hydrology, deuterium and oxygen-18 in the water cycle' (J.R. Gat and R. Gonfiantini, Ed.), Technical Report Series No. 210, International Atomic Energy Agency, Vienna, Chapter 4, 35-84.

9. M. Gröning, K. Fröhlich, P.P. De Regge, P.R. Danesi (1999): Intended use of the IAEA reference materials Part II: Examples on reference materials for stable isotope composition, in: 'The use of matrix reference materials in environmental analytical processes' (Eds.: A. Fajgelj and M. Parkany), Special Publication No. 238, Royal Society of Chemistry, Cambridge, 81-92.

10. T.B. Coplen (submitted for publication): Expression of relative amounts of isotopes and gases: the δ value. IUPAC Technical Report, International Union for Pure and Applied Chemistry.

11. M. Gröning (2004): International stable isotope reference materials. In: 'Handbook of Stable Isotope Analytical Techniques', Elsevier, Vol.1, Chapter 40, 874-906.

12. H. Craig (1961): Standard for reporting concentrations of deuterium and oxygen-18 in natural waters. Science, 133, No. 3467, 1833-1834.

13. R. Gonfiantini (1978): Standards for stable isotope measurements in natural compounds. Nature, 271: p. 534-536.

14. J. Mandel (1964): The statistical analysis of experimental data. Dover Publications, New York, p.110.

15. M. Gröning , M. van Duren, L. Andreescu, A. Tanweer, M. Jaklitsch, M. Gattin (submitted): Quality assurance for stable isotope analysis of water samples by mass spectrometry. Journal of Accreditation and Quality Assurance.

EXPERIENCE WITH METROLOGICAL TRACEABILITY AND MEASUREMENT UNCERTAINTY IN CLINICAL CHEMISTRY

Anders Kallner

Dept Clinical Chemistry Karolinska University hospital Stockholm Sweden

1 SETTING THE SCENE

The biochemistry of life has been a mystery for centuries but already in medieval times physicians of the time began to study the body fluids in health and disease. This was reflected early in the four colours of medicine red for blood, white for mucus and yellow and black for the yellow and black bile. Urine and stool were the first systems to be studied whereas blood and other fluids like lymph and cerebrospinal fluid were much more difficult to come by. Probably clinical chemistry can be traced back to the days of Berzelius in the beginning of the 19th century. The birth of the discipline is differently dated in all countries with a scientific history that includes that century.

The first observations were purely qualitative and thus expressed on the nominal scale, e.g. the colour and appearance of urine. The presence or absence of precipitates and casts became quite a science when the microscope became a common tool, and still is in a number of countries and laboratories. The familiar disease diabetes or *Diabetes mellius* got its name because the urine attracted bees and wasps and was found to have a sweet taste. As late as in the mid 90-ties a comprehensive book on how urine shall be collected and analysed was published[1].

Modern clinical chemistry was probably evolved from the days when Otto Folin at Harvard University in Boston developed what was then understood as micro methods to estimate the concentration of reducing substances in blood[2]. Without the development of these methods the work of Banting and Best in Toronto who in 1921/22 discovered the cause of diabetes and the active principle, insulin, that largely regulates the blood concentration of glucose would have been much more demanding.

The development of clinical chemistry is very much linked to diabetes. In 1947 Helen Free at Ames laboratories in Indianapolis, invented the rapid test for glucose or rather reducing substances in urine that was later applied also to blood. This was a remarkable invention that has had an enormous impact on the diagnosis and treatment of diabetes. These methods were named semi-quantitative with reference to a undefined accuracy (trueness and precision), a name that has withstood the wear and tear of decades and has the unspoken flavour of something that is not too good. The techniques have been developed and are definitely fit for purpose. We prefer to refer to them as "ordinal"methods, indicating that the results are reported on an ordinal scale. Methods to estimate and report the uncertainty of ordinal results are still not fully agreed.

A further even more sophisticated step was taken in the wake of polio that prompted the development of techniques to measure the blood gas and acid-base status of patients.

The initial instruments were designed by Paul Astrup at the University of Copenhagen in Denmark. The instruments have developed considerably to almost fool proof designs, automatic calibrations and cleansing and "one-use" techniques have paved the way for these complex measurement procedures to intensive care units and patient bedside.

1.1 Automation

The discipline experienced several leaps forward in the 1950-ies. A driving force was the ever increasing demand from the clinical physicians to get the concentration of all sorts of biochemical markers measured rapidly, 24 hours per day, 7 days per week. This prompted the development of automated instruments and probably the first developed for routine use was invented by Skeggs[3] and subsequently marketed by Technicon in New York from 1957 and onwards. It was an ingenious system that was based on removing aliquots of the sample from a primary tube, separating it from other samples by air bubbles, dialysis of the sample and retrieving the small molecules for further mixing with appropriate reagents. The products were finally measured by spectrometers and reported analogously. This not only increased the productivity of the laboratories but also continued the increasing dependence of the medical profession on prompt and exhaustive biochemical measurements. The repeatability and reproducibility of the measurements increased thanks to the automation.

The huge number of samples that are processed daily, under severe time pressure, has motivated investment in robotics, made possible by computer technology, that identify the samples, the quantities to be measured, aliquots the sample if necessary, separates the serum or plasma from the formed elements of blood and transports the sample to the assay system. Manual pipetting is almost gone and substituted by various dispensing instruments. This and the development of minute detectors have led to a reduction of the sample and reagent volumes to μL levels. This can even jeopardize the homogeneity of the sample!

1.2 New Methods

Simultaneously a rapid development took place in methodology e.g. it became possible to measure enzyme activities in the blood. Enzymatic kinetic and endpoint methods were introduced to specifically measure the concentration or activity of a variety components e.g. glucose, cholesterol and critical enzymes. This was manly made possible by the work of Professor Hans-Ulrich Bergmeyer at Boehringer-Mannheim[4]. A great spin-off of this was that laboratories to a largely could eliminate the use of dangerous solvents e.g. sulphuric acid and acetic acid anhydride for cholesterol concentration measurements or organic solvents like ether and chloroform for measurement of the concentration of lipid soluble components.

The concentration of all desirable components could still not be measured and in the sixties new principles were introduced; the use of antigen-antibody reactions and the use of radioactive or enzyme markers. Not Folin, nor Skeggs or Bergmeyer were awarded the Nobel Prize for their inventions whereas Yarlow[5] who was named the inventor of immunoassays was. Many other well known scientists like Berson, Ekins, Engvall and Perlman contributed enormously to make these techniques practical and useful in the laboratories. A new era began in which the number of measurable quantities in the blood was increased manifold. A large hospital laboratory routinely measures the concentration of 300 - 500 components that occur in concentrations in pmol/L to mmol/L ranges with an impressive speed and accuracy. Many methods of measurement are used, e.g. GC-ID-MS, chromatographic techniques, immunoassays and traditional spectrometric methods.

1.3 Preanalytical Sources of Uncertainty

A major concern in clinical chemistry is the preanalytical treatment of the patient and the sample. The problem is to obtain a representative, homogenous sample of from the patient. Usually samples are obtained by venipuncture or a fingerpick. In the first case a tourniquet is applied to facilitate identification of veins. If this is applied too long or too much interfering substances may be released from the tissues. If the bleeding from a fingerpick is not sufficient, interstitial fluid may contaminate the sample. The fasting state of the patient is important for measurement of some quantities and the posture may influence the transport of liquid and its components out of the blood into the interstitial space. In many cases the preanalytical factors may influence the value of the investigation more than the uncertainty in the measurement.

1.4 The Clinical Laboratory is Different

Laboratory medicine as a discipline includes many specialities but in this presentation those that are based on chemical methods and procedures are considered. These laboratories handle huge numbers of samples, they often work under a time constrain, have a large number of measurement procedures and use techniques of all kinds. The samples may originate from healthy individuals and patients suffering from different diseases that might influence the matrix of the samples. The component of the quantity is not always known, there may be multiple species of the molecules, each reacting differently and having different values in the diagnosis of diseases. Too often the laboratory is not informed about the condition of the patient. Patient preparation and other preanalytical conditions may have large impact on the measurements and on the concentrations of the quantity that is studied. Automation of the measurement procedures is common but still expensive and therefore require a certain workload to be justified – and considerable financial strengths. Laboratories of all sizes exist, small at physicians' offices that operate without specially trained staff to huge laboratories with hundreds of staff and sophisticated equipment. Laboratories have to cope with all these variables and still produce reliable and transferable results.

2 UNCERTAINTY OF MEASUREMENTS

2.1 Quality Monitoring in the Clinical Laboratory

Uncertainty of measurement or "measurement uncertainty", MU, as the profession prefers, is a relatively new concept in laboratory medicine. Nevertheless quality management has been a major issue in laboratory medicine for more than 50 years. The major tools to monitor the performance have been Internal Quality Control (IQC) and External Quality Assessment (EQA) The latter is also known as proficiency testing (PT), and recently the topic of an IUPAC technical report[6]. Both routines are based on measuring the concentration of unknown samples. The IQC is totally managed by the laboratory by including control material in the runs, evaluating the results and taking the necessary corrective and preventive actions. Certain rules that have the capacity to reveal changes in precision as well as trueness have been developed (Westgard rules)[7] and are widely and generally used. The EQA is generally designed as interlaboratory comparisons[8] and thus focus on the trueness.

2.2 Accreditation

Many laboratories are nowadays accredited according to EN/ISO 15189[9], a stand alone accreditation standard that has been developed from ISO 9000:2000[10] and ISO/IEC 17025[11] to provide for the special competence requirements of medical laboratories. Regarding uncertainties, the standards state "The laboratory shall determine the uncertainty of results, where relevant and possible. Uncertainty components which are of importance shall be taken into account. Sources that contribute to uncertainty may include sampling, sample preparation, sample portion selection, calibrators, reference materials, input quantities, equipment used, environmental conditions, condition for the sample and changes or operator." (EN/ISO 15189) and "Testing laboratories shall have and shall apply procedures for estimating uncertainty of measurement. In certain cases the nature of the test method may preclude rigorous, metrologically and statistically valid calculation of uncertainty of measurement. In these cases the laboratory shall at least attempt to identify all the components of uncertainty and make a reasonable estimation, and shall ensure that the form of reporting of the result does not give wrong impression of the uncertainty. Reasonable estimation shall be based on knowledge of the performance of the method and on the measurement scope and shall make use of, for example previous experience and validation data" (ISO/IEC 17025).

2.3 Guide to the Expression of Uncertainty in Measurements

Laboratories, at least those in Europe and other countries where accreditation has been using the above ISO standards, have been requested to establish "uncertainty budgets" according to the "Guide to the Expression of Uncertainty in Measurements" (GUM)[12] or the applied text published by Eurachem/CITAG[13]. Some countries have developed guides how the uncertainty should be derived[14]. ISO TC 212 and CLSI (Clinical Laboratory Standards Institute) have work items with a view to produce a Technical Specification or recommendations on estimation of uncertainty.

The procedures described in the standards and other documents have not been particularly well received by the profession. The main arguments seem to be that the GUM presents an intellectually stimulating model that – as all models – needs to be validated in practical life and the validation procedure directly is preferred. Since laboratories agree that they have already a substantial wealth of data from IQC and EQA the model does not contribute much to quality improvement. The only exception is when the imprecision that has been estimated from empirical data is not sufficient; then the uncertainty budget can advice a systematic analysis of the problem. Another argument is that the profession does not really experience a problem with the precision but with bias and the guiding principle of GUM is to eliminate or disregard the bias. Metrologically, the bias is controlled by calibration of the measurement procedure and by calibration of all steps in the procedure e.g. volumes, wavelengths, temperatures. In laboratory medicine the weak point is the calibration of the measurement procedure and this problem will be addressed below.

Much effort and thought has been spent on the education and training of staff in preparation for abiding by the requirements of the accreditation bodies. It has not been very productive and the interpretation of the texts in the standards does not unequivocally require that the laboratories establish uncertainty budgets and estimate the measurement uncertainty according to GUM. Accordingly, most laboratories seem to have denounced the GUM in routine or first line estimation of MU their quality management.

2.4 Use of Results of Measurements

There are two main uses of laboratory data in the clinic, one is to underpin or reach diagnosis of a disease, the other is to monitor treatment. In the first case the physician should decide if the result is above or below certain setpoints, i.e. limits of a biological reference interval. In the second case the physician shall decide if a given treatment has had any effect on the quantity that has been measured. It goes without saying that the farther from a setpoint a result is the more diagnostic it is and the less important is the uncertainty of the result. However, in the second case judging the effect of a treatment requires a sufficient precision at the very level that is applicable. Therefore the laboratory needs to establish imprecision profiles that cover the entire biologically possible concentration interval. Practical solutions to this problem have been advised by CLSI[15,16] based on repeated measurements of patient or reference materials.

The issue is not necessarily to reduce the uncertainty as much as possible but to ascertain that the clinician is provided with results that are fit for purpose.

So far very few physicians request, or are interested in, the measurement uncertainty. Most likely the reason for this lack of interest in – or value of considering – the uncertainty in clinical practice is probably the diagnostic procedure. The physician throws out a net of requests to cover the suspicions that other signs and symptoms may induce. Only in a few cases will the result of a single measurement verify or overthrow a clinically based suspicion. The result or results will, together with other observations form a pattern that might make sense. It seems that the laboratories have not been successful in informing the users of their services with respect to the limitations and potencies of the results.

2.5 Other Uses of Uncertainty Propagation Principle

If the clinician is not particularly concerned about the uncertainty of the results the laboratory should take on itself to make the uncertainty suitable for the use of the result.

The laboratories recognise a random and a systematic or constant contribution to the uncertainty of the result. It has been claimed that the analytical specification can based on the biological variation, i.e. within- (CV_w) and between-(CV_b) person variation of the quantity[17] Thus, the imprecision (I) and the bias (B) are estimated

$$I < 0.5 \times CV_w$$
$$B < 0.25\sqrt{CV_w^2 + CV_b^2}$$

1

and the total allowable error (TE)
$$TE < 1.65 \times I + B$$

2

A list of TE for more than 300 of the most common quantities has been published[17] and is updated on the Westgard homepage[6]. The laboratories can normally and routinely estimate the imprecision of their measurements from the IQC results and make that available to the physician. That would essentially correspond to the imprecision and it should thus be less than half the within individual variation. A major task remains: to agree on how the uncertainty should be estimated to be comparable to the TE. It is reasonable to suggest that the procedures described by the CLSI[15,16] will serve this purpose.

In a number of cases combinations of two or more quantities may give a new quantity that is a better marker for certain conditions. Anle example is the combination of

Apolipoprotein A1 (ApoA1) and Apolipoprotein B (ApoB). These molecules carry the otherwise insoluble cholesterol and triglycerides in plasma and are frequently monitored in lipid disorders. Large epidemiological studies,[18,19,20] studies have shown that the ratio ApoB/ApoA1 is directly correlated to the risk for developing cardiovascular diseases. The choice of target risk will not be discussed but it is argued if the target ratio should be reduced from 1.0 for men and 0.9 for women to 0.8 and 0.7, respectively. Do measurements allow to differentiate between 0.9 and 1.0? In ISO and IUPAC terminology[21] this would be to assess the "analytical sensitivity"(**sensitivity** quotient of the change in the indication of a measuring system and the corresponding change in the value of the quantity being measured) that needs to consider the uncertainty of the result. The difference should be larger than the combined uncertainty of the results. Thus, if D is the difference between results and u the uncertainty of the result,

$$D \geq \sqrt{u_1^2 + u_2^2}$$
$$\textit{if } u_1 = u_2, \textit{ then } D \geq u \times \sqrt{2}$$

3

considering the usually applied 95 % confidence interval a useful rule of thumb is that the difference should be larger than $3\ x\ u$ for a significant difference between results.

We can usually measure the concentration of both apolipoproteins with the same uncertainty. It remains to calculate the combined uncertainty of the ratio. Since the relative variance of a ratio is the sum of the relative variances of the nominator and denominator the maximum allowable uncertainty of the ApoA1 and ApoB would be about 3.0 % (figure 1) to allow differentiation of the results at the desirable level. Provided the laboratory can deliver results with this uncertainty the clinician could then assume that a change in the ratio with 0.1 would be analytically significant.

Similar calculations can be made for other quantities that are computed in the laboratory, based on single quantities, e.g. clearance, anion- or osmotic gap. Such calculations may change the technical specifications and thus the allowable total error of the measurement may be based on other input data.

2.6 Conglomerates of Laboratories and Instruments

Hospitals and laboratories are consolidated with a view to improve the financial outcome. Laboratories have a need for a certain production volume to afford automation and thus optimize the staff. Even if the laboratory exceeds the threshold where profitability begins consolidation continues and huge laboratory conglomerates are formed. They need not – in the beginning – operate the same instruments or methods. Samples are distributed among the laboratories but the clinician rarely knows which instrument has been used. It may seem a matter of trueness to align the results of the instruments and this will be discussed in the next section. However, the laboratory needs to report or be able to report the measurement uncertainty in an understandable form. This can be achieved by setting up an experiment in which the same sample is measured repeatedly with the participating instruments or methods. A suitable method is advised in[16] and described in a recent publication[22].

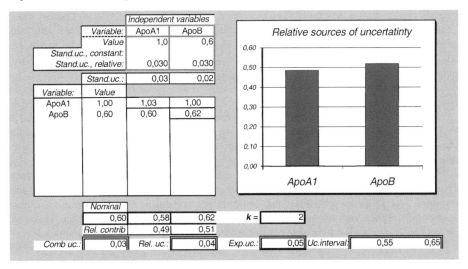

		Independent variables	
	Variable:	ApoA1	ApoB
	Value	1,0	0,6
Stand.uc., constant:			
Stand.uc., relative:		0,030	0,030
	Stand.uc.:	0,03	0,02
Variable:	Value		
ApoA1	1,00	1,03	1,00
ApoB	0,60	0,60	0,62

	Nominal						
	0,60	0,58	0,62	k =	2		
	Rel. contrib	0,49	0,51				
Comb uc.:	0,03	Rel. uc.:	0,04	Exp.uc.:	0,05	Uc.interval:	0,55	0,65

Figure 1. *Uncertainty budget applied to calculation of ApoB/ApoA1 ratio according to Kragten[23]. Nominal = ApoB/ApoA1, k= coverage factor.*

3 BIAS ESTIMATION

3.1 Definitions

In the draft VIM 2006 (unpublished), trueness is defined as "closeness of agreement between the average of an infinite number of replicate measured quantity values and a true value of the measurand"

Three notes are attached:

1. Measurement trueness cannot be expressed numerically.
2. Measurement trueness is inversely related to only systematic measurement error.
3. The term 'measurement trueness' should not be used for measurement accuracy.

Bias is then defined as "estimate of a systematic measurement error"

Thus, trueness when estimated is expressed numerically as bias.

In many cases laboratory medicine is unable to estimate the bias according to VIM because the true value of the measurand cannot be established.

3.2 Precision and Trueness

As long as most patients' samples were measured by one and the same laboratory and evaluated by the same physician against the same reference interval the precision appeared far more important than the trueness. Accordingly IQC procedures and rules primarily aimed at monitoring the intermediary reproducibility i.e. the performance from one day to another. Globalisation and movement of patients and physicians have changed this problem and quality improvement now focuses on trueness. This does not mean that the imprecision is not important but it seems that in most cases the imprecision is under control whereas the bias is not. The problem with the VIM definition is recognised in other disciplines also. Thus, O'Donnal and Hibbert[24] wrote "Many sectors of analytical chemistry use empirical methods to give comparability of analytical results when "trueness" cannot be achieved by any practical means. The results of such analyses are dependent on the method used and are not related to the true value. The results are traceable to that method only." Many procedures used in laboratory medicine would qualify for the empirical method. But that may not be acceptable because diagnosis are made in relation to biological reference intervals and dangerous misclassifications may be a result if the bias were not coordinated with the reference interval.

Different ways to resolve this have been tried. Laboratories can for instance establish their own reference intervals. Many do but it is by far too expensive and cumbersome for most laboratories. The transferability of results and experience would also be seriously reduced.

Other attempts have been to establish reference methods using isotope dilution –mass spectrometry[25]. These early attempts focus on small molecules and steroid hormones. The intention was to develop methods that not only could measure the concentration of pure compounds but also of components when present in patient serum or plasma. There are only a few such methods available but this would satisfy the definition of trueness

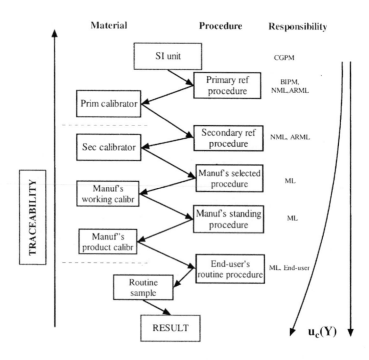

Figure 2. *A complete traceability chain, from the realisation of the SI unit to the patient results. Adapted from* [25]. *Materials and procedures specified between dashed lines are the responsibilities of manufacturers. ML= Manufacturer's laboratory, NML=National Metrological Institute*

Table 1. *Instruments participating in the comparison. The right columns show the deviation (%) of the results form the lowest value.*

Instrument	Manufacturer	ApoA1	ApoB
Advia 1650	Bayer	14,2	21,3
Advia 1650	Bayer	9,1	19,8
Advia 1650	Bayer	2,6	23,0
BN II	Dade-Behring	20,4	24,7
Immage	Beckman	**0,0**	16,8
Modular	Roche	1,6	16,3
Modular	Roche	2,4	11,3
Modular	Roche	11,3	**0,0**

3.3 Traceable Calibrators

Another approach to control the bias is to define the calibrator. This has been described and tried in laboratory medicine in two ISO standards; the most important is the EN/ISO 17511[26]. Professional groups, industry and government organisations, led by the IFCC (International Federation of Clinical Chemistry and Laboratory Medicine), have produced important calibrators for many proteins, Haemoglobin A1c that is an important marker for diabetes, and lipoproteins. A special committee, Joint Committee on Traceability in Laboratory Medicine has been established to coordinate the development of traceability in laboratory medicine[27]. (Figure 2)

Theoretically this approach would result in true values with known uncertainty. The availability of some calibrators that have become generally available and acceptable has improved the situation but it still does not solve the problem.

For the measurement of catalytic activity concentration of enzymes that play an important role in laboratory medicine a hierarchy of calibrators, very similar to that in EN/ISO 17511, is described in the EN/ISO 18153[28]. The top of the hierarchy is the derived coherent unit katal per cubic meter, followed by a series of reference procedures and materials.

3.4 Results of a Pilot Experiment

In a recent experiment we compared the results of measuring the concentration of ApoA1 and ApoB in 40 different patient samples in eleven different laboratories during routine conditions. All laboratories participated in EQA schemes and their performances were claimed to be within specifications. Also the performance had bee stable (within specifications) judged by the IQC. The assay methods are based on antigen-antibody reactions and the laboratories used their routine methods with the original kits from the instrument manufacturer. The kits contain all necessary reagents and calibrators. The calibrators were traceable to the IFCC reference calibrator[29]. It has not been possible to obtain the stated uncertainty of the calibrators. The participating instruments are listed in table 1. There is no obvious pattern, i.e. systematic variation with respect to the various instruments but it seems that the variation of results is larger in the measurements of Apo B than ApoA1.

The components are stable under the conditions used and the samples were pooled patient samples to obtain a sufficient spread of the concentrations, thus about 20 % of the samples were above or below the middle 40 percentile of the samples. The combined regressions are shown in 'Figure 3. The average difference between the maximum and minimum value that was recorded for each patient sample was 21 % for ApoA1 and 22 % for ApoB. Simultaneously one sample with high concentration and one with a low concentration were measured six times each by the laboratories. The within- and between-laboratory variation and the total variation were calculated for each concentration. The results are shown in table 2. The regression functions, in relation to the mean of all the results that were obtained from the two-sample part of the experiment, were then used to recalculate the patient samples. The mean of the maximum difference as above was then reduced to 12 % and 10 % for ApoA1 and ApoB, respectively.

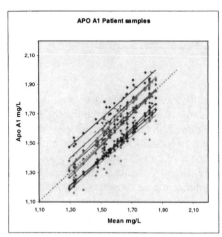
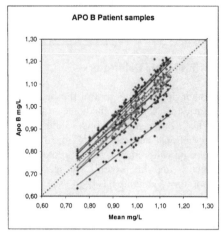

Figure 3. *Summary regression of results of measurement of patient samples against the mean of all observations. Dotted line is the 'equal line'*

As illustrated in table 2 the within laboratory uncertainty, disregarding the bias, is adequate with reference to the estimated analytical goal that was estimated in section 2.5, above whereas the interlaboratory variation is about twice as large as would be acceptable. It is likely that techniques could be advised whereby the interlaboratory variation could be minimized, at least regionally, by active interlaboratory comparisons and recalibrations using patient material. It will not remove the insufficient knowledge about the bias but it may reduce the likelihood for severe misdiagnosis and maltreatment.

Table 2. *Imprecision of apolipoprotein measurements in eleven different laboratories using four different instruments and calibrators traceable to the IFCC calibrator*

Apo A1						
Mean mg/L:	Within SD	Within CV%	Betw SD	Betw CV%	Tot SD	Tot CV%
2,02	0,05	2,7	0,13	6,5	0,14	7,0
1,20	0,0	1,9	0,1	5,4	0,1	5,7
Apo B						
1,45	0,02	1,3	0,09	6,4	0,10	6,6
0,61	0,01	1,7	0,04	6,9	0,04	7,1

3.5 Heterogeneous Components or Components with Unstable Species Distribution

The immunotechnique has indeed revolutionized the assay of components that occur in minute concentrations in blood. They are all based on the ability of the immunosystems of certain mammals, usually rabbit, goat or horse, to produce antibodies against specific molecules. However, the antibodies that are formed may react differently against the complex molecules they are raised against i.e. the epitopes in the naturally present antigens may vary from patient to patient and in health and disease. A normal or common distribution of the molecular species can usually be found but in disease the frequency of the species may vary and thus the trueness will be changed. In addition, proteins may be partially metabolised or degraded and yet retain the reactive epitope. As a consequence the results will be depending on the reagents that are used and the user cannot always be sure that the reagents from one manufacturer will really measure the same measurand as the other although specificity test with pure compounds may indicate something else. Frequently it is found that although the calibrators are traceable to a recognised metrological level they can only be used in dedicated systems. Thus the calibrators themselves give different results when measured by different measurement procedures. The measurand is therefore defined by the combination reagent-sample-calibrator.

3.6 Commutability

The term commutability describes how well a calibrator resembles the behaviour of the component in natural systems. When the commutability is not good enough then it is no surprise that the trueness will be suffering. However, it seems as a similar term needs to be defined to describe the ability of a reagent set to give the same results with natural and purified compounds. A present dilemma is that the component whose properties are assayed is defined by the interactions between at least the component, the antibody and the calibrator. The true value cannot be assessed and the laboratories have to refer to a conventional value or a conventional reference method. The procedure used needs to be indicated in the name of the measurand, as is possible and recommended in the IFCC/IUPAC nomenclature[21].

CONCLUSIONS

The importance of declaring and using the measurement uncertainty in clinical work is slowly becoming understood. This puts pressure on the laboratories to agree on a common method to estimate and describe the measurement uncertainty. The GUM approach does not seem to meet the requirements of the profession that rather prefers using empirical data and only systematically establish and use uncertainty budgets when the uncertainty does not seem fit for purpose. Special problems will occur during the consolidation of laboratories and the establishment of conglomerates of laboratories and instruments. Although the trueness of results is of great importance, the VIM definition that makes reference to a true value does not always make sense in the clinical laboratory because the truth is illusive and cannot be determined. Traceable calibrators may improve the transferability of results but does generally not provide an acceptable trueness within a group of laboratories. Therefore, recalibration using patient materials in interlaboratory comparisons in relation to a conventional procedure seems to be a way forward to reduce the interlaboratory variation.

References

1   T Kouri, G Fogazzi, V Gant, H Hallander, W Hofman, WG Guder European Guidelines. *Scand J Clin Lab Invest* 2000,60(Suppl 231) 1-96.

2   S Meites, *Otto Folin: America's first clinical biochemist.* AACC, Washington DC USA ISBN 0-915274-48-5 1989

3   L T Skeggs, *Clin chem.*, 2000,46,1425.

4   HU Bergmeyer, Methods in Enzymatic analysis, 2nd Ed Academic press NY 1984

5   RS Yarlow, SA Berson, *J. Clin. Invest.* 1960, 39, 1157

6   IUPAC Technical Report. The international harmonized protocol for the proficiency testing of analytical chemistry laboratories. *Pure Appl Chem* 2006,78,145.

7   J Westgard, in depth references available at www.Westgard.com (visited 060307)

8   ISO guide 43. Proficiency testing by interlaboratory comparisons -- Part 1: Development and operation of proficiency testing schemes. Geneva 1994.

9   EN/ISO 15189:2003 Medical laboratories -- Particular requirements for quality and competence, Geneva 2003.

10 ISO 9000 2000 Quality management Geneva 2000.

11 ISO/IEC 17025:2005 General requirements for the competence of testing and calibration laboratories Geneva 2005.

12 BIPM, IEC, IFCC, ISO, IUPAC, IUPAP, OIML *Guide to the expression of uncertainty in measurement* Geneva 1995.

13 EURACHEM/CITAC *Quantifying Uncertainty in Analytical Measurements* 2 ed 2000 www.eurachem.ul.pt/ (visited 060306).

14 GH White, I Farrance, *Uncertainty of measurement* www.aacb.asn.au (visited 060306)

15 CLSI/NCCLS EP9 Method comparison and bias Estimation Using Patient Samples. Approved Guidline -2nd ED ISBN 1-56238-472-4. Wayne 2002.

16 CLSI EP 15 User Verification of Performance for Precision and Trueness; Approved Guideline – 2nd ed ISBN ISBN 1-56238-574-7. Wayne 2005.

17 C Ricos, V Alvarez, F Cava, JV Carcia-Larioi, A Hernandez, CV Jimenez et al. *Scand J Clin Lab Invest* 1999,59,491.

18 I Jungner, AD Sniderman, C Furberg, AH Aistveit, I Holme; G Walldius *Am J Cardiol* 2006, 97,xxx. Available online 060213.

19 G Walldius, AH Aastveit, I Jungner. *J Intern Med* 2006,259,259.

20 BJ Barter, CM Ballantyne, R Carmena, M Castro Cabezas, MJ Chapman, P Couture et al. *J Intern Med* 2006,259,247.

21 http://www.iupac.org/publications/epub/index.html. (visited 060306).

22 A Kallner, L Khorovskaya, T Pettersson, *Scand J Clin Lab Invest* 2005,65,551.

23 J Kragten *Analyst* 1994,119,2161.

24 GE O'Donnell, DB Hibbert *The Analyst* 2005,130,721.

25 I Björkhem. In Proceedings of the fourth international conference on stable isotopes. Elsevier Sci Publ., Amsterdam (1982). pp 593-604.

26 EN/ISO 17511 In vitro diagnostic medical devises – Measurement of quantities in biological samples – Metrological traceability of values assigned to calibrators and control materials. Geneva 2002.

27 Activities of the JCLM http://www.bipm.fr/utils/en/pdf/jctlm_activities.pdf (visited 060306).

28 EN/ISO 18153 In vitro diagnostic medical devises – Measurement of quantities in biological samples – Metrological traceability of values for catalytic concentration of enzymes assigned to calibrators and control materials. Geneva 2002.

29 SM Marcovina, JJ Albers;F Dati, TB Ledue, RF Richie *Clin Chem* 1991, 37, 1676.

TRACEABILITY OF pH IN A METROLOGICAL CONTEXT

G. Meinrath[1], M.F. Camões[2], P. Spitzer[3], H. Bühler[4], M. Máriássy[5], K. Pratt[6], C. Rivier[7]:
The IUPAC Working Party "Comparable pH Measurement by Metrological Traceability"

[1] RER Consultants, Passau, Germany
[2] Faculdade de Ciências, Universidade de Lisboa, Portugal
[3] Physikalisch-Technische Bundesanstalt (PTB), Sect. 3.1, Braunschweig, Germany
[4] Hamilton Bonaduz AG, Bonaduz, Switzerland
[5] Slovak Institute of Metrology (SMÚ), Bratislava, Slovakia
[6] National Institute of Standards and Technology (NIST), Gaithersburg, MD, USA
[7] Laboratoire National de Métrologie et d'Essais (LNE), Paris, France

1 INTRODUCTION

Undoubtedly, pH is one of the most often measured chemical quantities. Since its introduction in 1909 it has been recognised as an important quantity in analytical chemistry, biosciences, clinical chemistry, food and pharmaceutical industries, cosmetic industries and environmental sciences. Despite its important role, pH is still reported in the vast majority of cases without a valid statement of an associated measurement uncertainty. This fact is the more surprising as the requirement for such an uncertainty statement was issued almost a decade ago [1].

By issuing the 2002 IUPAC recommendation on pH measurement [2], IUPAC has paved the way for introducing metrological concepts such as the complete measurement uncertainty budget in the measurement of pH values.

Table 1: *A comparison between the (now obsolete) 1985 'compromise' recommendation [4] and the current 2002 pH recommendation [2]. Note that the mutually exclusive operational definitions ('single standard' and 'bracketing' procedures) have been replaced by a statistically founded approach relying on traceability.*

IUPAC Recommendation 1985	*IUPAC Recommendation 2002*
Two coexisting definitions of pH	Thermodynamic definition of pH
Two mutually exclusive protocols for primary pH standards	Single primary measurement system (Harned cell)
No uncertainty information for either protocol	Detailed uncertainty analysis presented for primary technique
No mention of traceability	Traceability to worldwide accepted references
Secondary pH metrology glossed over	Secondary techniques described

The different approaches are contrasted in Table 1. The 2002 IUPAC recommendation [2,3] illustrates the almost revolutionary change in the attitude towards chemical measurements: while the 1983 'compromise' recommendation [4] almost exclusively focused on the operational procedure to perform a measurement the 2002 Recommendation places great emphasis on uncertainty and traceability issues. In great part, this change originated with the increased requirements for traceability in pH metrology on the part of institutions that perform pH measurements. In relying on comparability of measurement values the 2002 recommendation requires an extension of the international metrological structure to this field of chemistry. The introduction of ISO's Guide to the Expression of Uncertainty in Measurement [1] in 1993 was the prerequisite to indicate the trustworthiness of a measurement result.

Because of its wide applications, the quantity pH has received significant attention from metrology in chemistry divisions at the NMI (National Metrology Institutes). International key Comparison between the NMI under the auspices of the BIPM [5,6], and national proficiency testing schemes have promoted the establishment of a traceability chain from the field application level to the highest metrological accuracy at the NMI. Appropriate protocols are currently introduced into national norms of pH measurement on basis of EURACHEM/CITAC's bottom-up approach [7] (e.g. [8,9]). Protocols have been developed for two-point and multi-point calibrations.

These increased requirements for traceability have enhanced the need for further development of the metrological foundations for pH metrology, specifically in areas of secondary and practical pH metrology that apply to real-world applications.

2 CAUSE-AND-EFFECT ANALYSIS: BOTTOM-UP APPROACH

A cause-and-effect diagram for the two-point calibration procedure is shown in Figure 2.

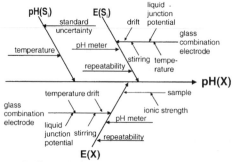

Figure 2 *Cause-and-effect diagram for a two-point calibration using calibration standards S_1 and S_2 with potential readings $E(S_1)$ and $E(S_2)$ and assigned values $pH(S_1)$ and $pH(S_2)$ with appropriate measurement uncertainties. The unknown sample X has a potential $E(X)$.*

The two-point calibration assumes that the uncertainty in the calibration itself is adequately represented by the measurement uncertainties given for the standards $pH(S_1)$ and $pH(S_2)$. Multi-point calibration assesses the uncertainty in the calibration step from a small sample ($n \geq 3$) of standards using the equation (scheme 1), where t gives the appropriate Student's t value for (n-2) degrees of freedom and confidence interval ($1-\alpha$). Application of Scheme 1 implicitly assumes that the major contribution to measurement uncertainty of an measurand pH(X) is associated with calibration. At the present level of understanding, this assumption is valid using state-of-the-art calibration standards pH(S$_i$) [10].

$$\frac{s(pH(X))_o}{s(pH(X))_u} = (pH(X) - pH_m)g \pm \frac{\dfrac{t\, s_e}{\beta}\left\{ \dfrac{(pH(X) - pH_m)^2}{\sum_i (pH(i) - pH_m)^2} + \dfrac{1-g}{n} \right\}^{\frac{1}{2}}}{1-g}$$

with

$$g = \frac{t^2}{\left[\beta / \sqrt{\dfrac{s_e^2}{\sum_i (pH(i) - pH_m)^2}} \right]^2}$$

$$\beta = \frac{n\sum_i (pH(i) * E(i)) - \sum_i pH(i)\sum_i E(i)}{n\sum_i pH(i)^2 - \left(\sum_i pH(i)\right)^2}$$

$$s_e = \sqrt{\frac{\sum_i (E(i))^2 - \alpha\sum_i E(i) - \beta\sum_i (pH(i)E(i))}{n-2}}$$

$$pH_m = \frac{1}{n}\sum_i pH(i)$$

Scheme 1

Scheme 1 is statistically correct. Manual evaluation of the equations in Scheme 1 is not recommended. However, this evaluation is conveniently implemented into a computer code. The quantities s(pH)$_o$ and s(pH)$_u$ represent the upper (subscript o) and lower (subscript u) confidences (depending on the choice of (1-α)). Strictly speaking, both values will differ somewhat. However, under most circumstances the difference is negligible for metrological purposes. Multi-point calibration is applied preferentially in situations where the demand for reliability is high and in situations where information on the response over a wide range of pH values is required..

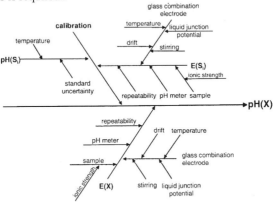

Figure 3 *Cause-and-effect diagram for multi-point calibration (MPC). While the majority of influence factors are the same as for 2-point calibration, in fact the calibration process is the central element in multi-point calibration.*

Figure 3 shows the corresponding cause-and-effect diagram, differing from Figure 2 by its focus on the calibration process. Scheme 1 yields a mean-value based regression line with upper and lower confidence regions. The regression line is obtained assuming that the abscissa values (pH(S$_i$)) are not affected by uncertainty, while all uncertainty is due to the ordinate quantities (E(S$_i$) and E(X)). The estimates s(pH)$_o$ and s(pH)$_u$ give the abscissa positions of the intercepts of the upper and lower measurement uncertainty limits of E(X) with confidence bands of a regression line. This approach is shown schematically in Fig. 4.

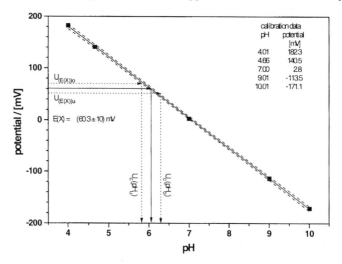

Figure 4   *Example of pH measurement by five-point calibration on basis of the cause-and-effect diagram Figure 3.*

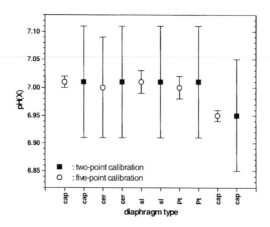

Figure 5   *pH measurement results by five-point calibration (MPC) and two-point calibration (TPC). Nominal pH value of pH(X): 7.00. (cap: capillary, cer: ceramic; sl: sleeve; Pt: platinum)*

In Figure 5, results with both techniques are given. The pH measurements are carried out at 25 °C in the laboratory [14] under controlled conditions. Differences in temperature and drift conditions for calibration and measurement are negligible. The resolution of the meter was 0.01 mV. pH standards with an expanded uncertainty (k = 2) of 0.005 are used. The type of diaphragm of the electrodes covers ceramic, capillary, sleeve and platinum. The standard uncertainties of the potential measured in the pH standard solution and in the unknown solution are estimated to 2 mV.

3 PROFICIENCY TESTING SCHEMES

It is interesting to contrast the typical measurement uncertainty estimates obtained by two-point and multi-point calibration (bottom-up approach) with results of round-robin studies. Figure 6 shows the result of a round-robin study performed in water with 30 participants following a standardized procedure. The reference value (nominal true value) is given as pH 4.450. The tolerable limits (|z| score < 3.000) are pH 4.152 and pH 4.752 resulting in a standard deviation of about 0.1 pH. The scatter within the data is certainly surprising but not uncommon in interlaboratory studies (cf. e.g. IMEP-6 [12]).

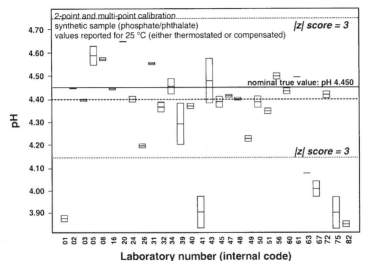

Figure 6 *Results of a proficiency test on pH measurement with 30 participating laboratories (BAM 2003) following DIN 38402-A45 [11]. The boxes given the estimated standard uncertainty as estimated by the participating laboratories.The central bar in each box represents the reported mean value while the height of the boxes represents the reported standard uncertainty. Sample was a synthetic phthalate/phosphate material.*

Figure 7 presents results of a round-robin study on pH in a drinking water (BIPEA 2005) with 95 participating laboratories. The interpretation of the reported mean pH values by a Normal distribution on basis of a Kolmogorov-Smirnov test results in the dotted cumulative distribution curve with a standard deviation in pH of 0.18.

4 CONCLUSIONS

Previous efforts in pH metrology [2] have provided a traceability chain for pH calibration materials from the highest metrological level down to the field laboratory level. The current efforts concentrate on appropriate protocols for the application of these materials in the evaluation of measurement uncertainty budgets. Cause-and-effect diagrams are shown for two-point calibration and multi-point calibration schemes. From a statistical point of view both calibration procedures do have their deficiencies which, however, appear to be of limited practical influence. Uncertainty budgets for both pH calibration procedures yield pH measurement uncertainties in the range of 0.01 to 0.1.

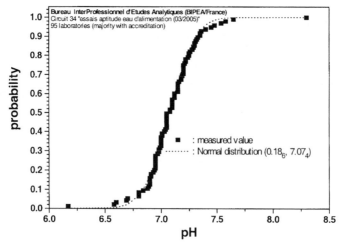

Figure 7 *Measurement of pH in drinking water with 95 participating laboratories (BIPEA/France) [13]. The reported mean value results are given as empirical probability distribution and interpreted by its closest-fitting cumulative Normal distribution with a standard deviation of 0.18 pH units.*

Two available recent round-robin studies on pH measurement with standard deviations of 0.1 to 0.2 in the mean value indecate that interlaboratory comparison results are somewhat larger than uncertainties obtained by bottom-up approaches. Hence, interlaboratory comparisons yield wider uncertainty margins compared to bottom-up approaches. These differences call for a more precise standardisation of calibration procedures in case of interlaboratory comparisons and request additional efforts for quantification of uncertainty contributions, especially with respect to temperature effects and liquid junction potentials. The IUPAC Working Group is currently preparing the metrological basis for respective research activities.

The present discussion needs to be very abbreviated. A full account on this work will be given elsewhere.

References
1   ISO Guide to the Expression of Uncertainty in Measurement, ISO, Geneva/CH 1993
2   R.P. Buck, S. Rondinini, F.G.K. Baucke, M.F. Camões, A.K. Covington, M. J. T. Milton, T. Mussini, R. Naumann, K. W. Pratt, P. Spitzer, G. S. Wilson *Pure Appl. Chem.* 2002, **74**, 2169
3   F. G. K. Baucke, *Anal Bioanal Chem* 2002, **374**, 772
4   A. K. Covington, R. G. Bates, R. A. Durst, *Pure Appl. Chem.* 1985, **57**, 531

5   Spitzer P., Hongyu X., Dazhou C., Famnin M., Kristensen H.B., Hjelmer B., Nakamura S., Hwang E., Lee H., Castro E., Mendoza M.M., Kozlowski W., Wyszynska J., Mateuszuk A., Pawlina M., Karpov O.V., Zdorikov N., Seyku E., Maximov I., Vyskocil L., Máriássy M., Pratt K.W., Vospelova A., Giera J., Eberhardt R. Metrologia, **40**, 2003, *Tech. Suppl.*, 08006; Final Report of Key Comparison CCQM-K17, February 2003. Available online at: http://kcdb.bipm.org/AppendixB/appbresults/ccqm-k17/ccqm-k17_final_report.pdf

6   Final Report of Key Comparison CCQM-K9, 12 December 2001. Available online at: http://kcdb.bipm.org/AppendixB/appbresults/ccqm-k9/ccqm-k9_final_report.pdf; Final Report of Key Comparison CCQM-K17, February 2003. Available online at http://kcdb.bipm.org/AppendixB/appbresults/ccqm-k17/ccqm-k17_final_report.pdf

7   EURACHEM/CITAC Quantifying Uncertainty in Chemical Masurement. EURACHEM/CITAC 2000

8   DIN 19268 "pH-Messung – pH-Messung von wässrigen Lösungen mit pH-Messketten mit pH-Glaselektroden und Abschätzung der Messunsicherheit" Deutsches Institut für Normung

9   DIN 19261 "Terminology for measuring techniques using potentiometric cells" Deutsches Institut für Normung

10 G. Meinrath, P. Spitzer, Mikrochim. Acta 2000, **135**, 155

11 BAM, 25. Ringversuch Wasseranalytik, Bundesanstalt für Materialforschung und – prüfung (BAM), Berlin, 2003.

12 IRMM IMEP-9, Institute of Reference Materials and Measurements, CEC Geel/Belgium

13 BIPEA, 2005

14 R.Naumann, Ch. Alexander-Weber, R. Eberhardt, J. Giera, P. Spitzer, Anal. Bioanl. Chem. 2002, **374**, 778

DETERMINATION OF PCBs IN ORGANIC SOLUTIONS: AN EXAMPLE OF TRACEABILITY CHAIN

Michela Sega, Elena Amico di Meane

Istituto Nazionale di Ricerca Metrologica (I.N.RI.M), Strada delle Cacce 91, I-10135 Torino

1 INTRODUCTION

Persistent Organic Pollutants (POPs) are carbon-containing chemical compounds that, to a varying degree, resist photochemical, biological and chemical degradation. POPs are often halogenated and characterised by low water solubility and high lipid solubility, leading to bioaccumulation in fatty tissues. This fact, together with their persistence in the environment, causes a potential health impact on a wide variety of biota, including human beings.[1, 2, 3] Although many different chemicals, both natural and anthropogenic may be defined as POPs, twelve classes of them, all chlorine-containing compounds, have been chosen as priority pollutants by the United Nations Environment Programme (UNEP) under the Stockholm Convention in 1997 for their impact on human health and environment.[4] Polychlorinated byphenils (PCBs) are included among these twelve POPs.

The European Community developed a strategy for dioxins, furans and PCBs[5] with the purpose of developing monitoring programmes to control compliance with existing legislation and to monitor the effects of this strategy, the state of the environment and the trends.

In the metrological framework various international comparisons on PCBs determination are carried out in order to improve the quality measurements and to obtain comparable, consistent, reliable and high quality measurement results. I.N.RI.M. (previously as Istituto di Metrologia "G. Colonnetti" - IMGC) took part in the first stage of the international comparison EUROMET/METCHEM Project n° 833 "PCB congeners in organic solution".

In order to obtain high quality reliable results and to compare results of measurements carried out in different places and by different laboratories, traceability is needed. Traceability is the property of the result of a measurement or the value of a standard whereby it can be related to stated references, usually national or international standards, through an unbroken chain of comparisons all having stated uncertainties. The unbroken chain of comparisons is called traceability chain.[6] A correct traceability chain needs to be established. Many methods already exist for the analysis of PCBs.[7, 8, 9, 10, 11, 12] The present work deals with the establishment of a traceability chain for the analysis of various PCBs congeners in organic solutions.

2 EXPERIMENTAL SETUP

The analytes chosen are the PCBs shown in Table 1.

Table 1 *Name and corresponding IUPAC number of the analytes*

PCB	IUPAC number
2,2',5,5'- tetrachlorobiphenyl	52
2,2',4,5,5'- pentachlorobiphenyl	101
2,3',4,4',5- pentachlorobiphenyl	118
2,2',3,4,4',5'- hexachlorobiphenyl	138
2,2',3,4,4',5,5'- heptachlorobiphenyl	180

The measurements were carried out on three different solutions of the same batch containing the five PCBs congeners in the range from 500 to 1000 ng/g.

The analyses were carried out by means of a gas-chromatograph coupled with an ion-trap mass spectrometer (GC-MS).

For the establishment of a correct traceability chain, depicted in Figure 1, three standards were prepared by weighing, diluting in *iso*-octane the certified reference material BCR 365 (an *iso*-octane solution of ten PCBs congeners with a standard uncertainty of about 2,5%). The certificate states that each certified value is the unweighted mean of the means of four sets of results provided by different laboratories and different GC methods. Furthermore the certified values with their uncertainties are traceable either to a SI unit or an internationally accepted reference.

The advantage of the weighing method compared to volumetric dilution is the lower uncertainty that can be achieved. Calibrated mass standards were used for the weighings following the ABBA scheme (double substitution weighings). In addition a correction to take into account the buoyancy effect was applied. In such a way three calibration solutions containing the five analytes in the ranges from 500 to 1000 ng/g were prepared with standard uncertainties of about 3-3,5 %.

For the analytes identification a series of analyses were carried out using the MS detector in scan mode from 30 to 450 *m/z*. The obtained mass spectra were interpreted by matching them with the library ones. In Figure 2 the mass spectrum experimentally determined for the PCB congener 52 is presented.

The analytes quantification was done by Selected Ion Monitoring (SIM).

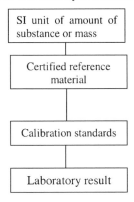

Figure 1 *Traceability chain*

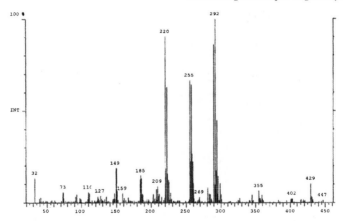

Figure 2 *Mass spectrum of PCB 52*

3 RESULTS AND DISCUSSION

For all the analytes the calibration curves were determined by means of an Excel worksheet,[13] developed at I.N.RI.M., based on the Weighted Least Squares method, which calculates a linear correction to be applied to the instrument readings according to the following equation:

$$x = y + d(y) = y + \alpha_0 + \alpha_1 y \qquad (1)$$

where x is the concentration of the analyte in the standard solutions, y is the instrument output and $d(y) = \alpha_0 + \alpha_1 y$ is the correction. The measurands are the polynomial coefficients α_0 and α_1. The estimation algorithm takes care of different sources of uncertainty: the standard solutions uncertainty, the repeatability of the instrument, the lack of fit, the instrument resolution. Being the standard solutions prepared from the same certified reference material, a correlation coefficient of 0,9 was adopted in the calculation.

For each congener a concentration value with its combined standard uncertainty was estimated in the three ampoules in duplicate, for a total of six estimations.

After the calibration process α_0 and α_1 being known, if a set of n_r instrument readings, arranged in a vector r, are to be corrected by the calibration algorithm, the matrix R can be defined, whose columns are the first two powers of r:

$$R = (r^0 \; r) \qquad (2)$$

The correction vector $d(r)$ can be computed from $d(r) = R\,a$, where a is the vector of the coefficients α_0 and α_1. The corrected readings are:

$$q = d(r) + r \qquad (3)$$

The covariance matrix of the readings is $\psi_r = s^2 I$, where s is the repeatability standard uncertainty of the instrument and I an identity matrix. The covariance matrix ψ_d of d can be estimated starting from the law of propagation of uncertainty:

$$\psi_d = \nabla_\alpha(d)\,\psi_\alpha\,\nabla_\alpha(d)^\mathrm{T} + \nabla_r(d)\,\psi_r\,\nabla_r(d)^\mathrm{T} \qquad (4)$$

where the symbol $\nabla_z(w)$ means the Jacobian matrix, i.e. the matrix derivative, of the vector w with respect to the vector z and ψ_α is the variance-covariance matrix of the coefficients α_0 and α_1.

Table 2 *Estimated concentration value and its combined standard uncertainty for each analyte*

PCB congener	Estimated concentration ng/g	Combined standard uncertainty, ng/g
PCB 52	705,83	41,50
PCB 101	671,97	35,44
PCB 118	858,09	40,30
PCB 138	462,56	34,51
PCB 180	672,70	35,80

From equation 3 it follows that the combined standard uncertainty of a result derives from a term due to the correction obtained by the calibration curve and from a term due to instrument repeatability:

$$u^2_c(q) = u^2(d(r)) + u^2(r) \tag{5}$$

Since the six results and their uncertainties are non-negligibly scattered and systematic effects are known to be negligible, it was chosen to express the final result as a weighted mean, calculating the corresponding combined standard uncertainty[14] of about 5%. The estimated concentrations and their combined standard uncertainties are reported in table 2.

4 CONCLUSION

The aim of the present work was to establish a correct traceability chain for the determination of PCBs in organic solutions. The above goal was fulfilled by using a certified reference material to prepare calibration standards in traceable conditions, by following a proper weighing procedure and by taking into account all significant contributions to uncertainty at each step. The main contributions to the combined standard uncertainty of the results are due to the calibration curve and to repeatability of readings of sample measurements.

Future developments are foreseen to improve the analitical performances in order to achieve better repeatability and to ensure traceability for the PCBs determination also in complex matrices.

References
1 S. B. Singh and G. Kulshrestha, *J. Chromatogr. A.*, 1997, **774**, 97.
2 A. Fernandes, S. White, K. D'Silva and M. Rose, *Talanta*, 2004, **63**, 1147.
3 C. Linding, *Chemosphere*, 1998, **37**, 405.
4 http://www.unep.org
5 COM(2001) 593 final.
6 VIM - *International Vocabulary of Basic and General Terms in Metrology*; Geneva: ISO, 1993.
7 G. Font, J. Mañes, J. C. Moltó and Y. Picó, *J. Chromatogr. A*, 1996, **733**, 449.
8 J. W. Cochran and G. M. Frame, *J. Chromatogr. A*, 1999, **843**, 323.
9 J.-F. Focant, A. Sjödin and D. G. Patterson Jr., *J. Chromatogr. A*, 2004, **1040**, 227.
10 W. Vetter, B. Luckas and J. Buijten, *J. Chromatogr. A*, 1998, **799**, 249.
11 A. K. Djien Liem, *Trends Anal. Chem.*, 1999, **18**, 499.
12 R. M. Smith, *J. Chromatogr. A*, 2002, **975**, 31.
13 M. Plassa, M. Mosca and M. Sega, *Proc. 16th Int. Conf. IMEKO TC3/APMF '98*, Myung Sai Chung Ed.; Taejon, Korea, **1998**, 183.
14 M. G. Cox, *Metrologia*, 2002, **39**, 589.

QUALITY CONTROL OF pH MEASUREMENTS CONSIDERING ACTIVITY AND CONCENTRATION SCALES: UNCERTAINTY BUDGET OF PRIMARY AND SECONDARY APPARATUSES

P. Fisicaro[a], E. Ferrara[a], E. Prenesti[b] and S. Berto[b]

[a] Istituto Nazionale di Ricerca Metrologica, Strada delle Cacce 91, 10135 Torino, Italy
[b] Dipartimento di Chimica Analitica dell'Università, via Pietro Giuria 5, 10125 Torino, Italy

1 INTRODUCTION

The concept of pH is very special in the field of physicochemical quantities, because of its widespread use as an important control property and process descriptor and because of the difficulty of defining its theoretical meaning. Efforts have been devised to assess both the theoretical rationale and the experimental strategy for assignment of primary pH values, however these have not yet been accomplished satisfactorily, because the definition of pH involves the activity of a single ion[1], which, according to thermodynamics, cannot be inferred alone in practice. To bypass this point, several technical details for the measurement of pH have been developed, along with theoretical considerations. Currently, primary measurements are available with the Harned cell, although traceability and comparability are achievable only within the limitations stated by international conventions that cannot account for all possible situations of pH measurements. Under these circumstances, it has been shown[2] that comparison with primary results can be contradictory when secondary measurements (e.g. with glass electrodes) are obtained by considering the concentration scale of the hydrogen ion (pcH), for example when acid–base titration, aimed at formation constant refinement, is appreciated through potentiometric data.[3,4] To harmonise results of different methods of pH measurement, we previously proposed[2] the analysis of a series of comparative exercises supported by modifications of the theoretical approach adopted to calculate activity coefficients. Primary measurements (Harned cell) of the pH of a phosphate buffer have been considered and the results have been compared with secondary (glass electrode) measurements, executed according to either activity (paH) or concentration (pcH) scales. In the current paper we aim at performing a careful inspection of all those parameters significantly affecting the overall uncertainty on both paH and pcH. Uncertainty budgets have been evaluated considering both primary and secondary apparatuses and also including in the study the management of experimental and theoretical aspects of pH in concentration scale.

2 EXPERIMENTAL

Ref. 2 reports any detail related to the materials and methods adopted for the primary and secondary (paH) apparatuses. Further information on secondary measurement (pcH), related to the development performed for the current paper, is given below.

Symbols: paH = $-\log a_H$, pcH = $-\log [H^+]$.

2.1 Chemicals

The buffer under measurement is composed of potassium dihydrogen phosphate and disodium hydrogen phosphate, approximately 0.025 mol·kg^{-1} each, $I = 0.1$ mol·kg^{-1}.

The electrode calibration was performed at 25°C using the certified pH-metric buffer at paH = 4.005, U(paH) = 0.006, and paH = 6.98, U(paH) = 0.02 by Merck.

The concentration of the HCl solution prepared to measure pcH daily by the glass electrode was refined during the equilibrium calculation of the acid-base titration data (HCl solution, at a stated ionic strength controlled by adding KCl (Merck, 99.995%)). KOH solutions were prepared by diluting Fluka concentrate products and standardised against potassium hydrogenphthalate (Fluka, puriss.) as primary acid-base standard substance.

2.2 Volumetric and gravimetric measurements

Micropipettes were from Raining, U.S.A., (models 1000 and 5000), while burette of 25 ml capacity was from Hirschmann EM Techcolor, Germany. Table 1 reports details on volumetric dispensers.

The calibration of the technical balance Sartorius (model BP211D), was performed using reference certified masses by Kern. The balance calibration uncertainty was estimated as $u = 8.5 \cdot 10^{-5}$ g, while repeatability uncertainty resulted $u = 2 \cdot 10^{-5}$ g.

Table 1 *Features of volumetric dispensers.*

Dispenser	Capacity	Resolution	Accuracy
Micropipette 1000	1 ml	2.0 µl	0.8 % (relative)
Micropipette 5000	5 ml	5.0 µl	0.6 % (relative)
Automatic Burette	1 ml	1.0 µl	0.3 % (relative)
Burette	25 ml	0.05 ml	0.03 ml (absolute)
Flask	1000 ml	–	0.4 ml (absolute)

2.3 Data analysis and calculations

The non linear least squares computer program ESAB2M[5] was used to refine all the parameters related to the pcH measurement. User is able to select the parameter to be refined as a function of experimental procedure. The software can account for titration points at different ionic strengths considering a Debye-Huckel type equation[6,7] for the dependence of all thermodynamic quantities on ionic strength.

The uncertainty budgets are evaluated in accordance with recommended procedures[1,8,9] using commercial spreadsheets.

3 RESULTS AND DISCUSSION

3.1 Uncertainty budget of primary apparatus

According to IUPAC recommendations[1], the combined standard uncertainty of the primary measurement of paH is:

$$u_c(paH) = u_c(pa_H\gamma_{Cl}°) \quad (1)$$

and

$$u_c(pa_H\gamma_{Cl}°) = [u_c{}^2(pa_H\gamma_{Cl\,m=0.005}) + u_c{}^2(intercept)]^{1/2} \quad (2)$$

where: i) $u(pa_H\gamma_{Cl\,m=0.005})$ is the uncertainty of the acidity function for the corresponding amount of added chloride (taking into account the acidity function related to the smallest concentration of added KCl, usually corresponding to $m_{Cl} = 0.005$ mol·kg^{-1}, is the recommended procedure), ii) $u(intercept)$ is the uncertainty of the acidity function, $pa_H\gamma_{Cl}°$, extrapolated to zero chloride molality, obtained by a linear least squares fit of the $pa_H\gamma_{Cl}$ values at the different chloride molalities (if the scatter around the regression line is large, the uncertainty of the intercept can become the major contribution to the overall uncertainty).

As a first step the uncertainty in the determination of the acidity function is estimated according to eq. (3):

$$p(a_H\,\gamma_{Cl}) = -\log(a_H\,\gamma_{Cl}) = (E - E°) / [(RT/F)\,\ln10] + \log m_{Cl} - \tfrac{1}{2}\log(p°/p_{H2}) \quad (3)$$

The standard uncertainty $u(p(a_H\,\gamma_{Cl}))$ is determined from the estimate x_i of the input quantities $(E, E°, T, m_{Cl}, p_{H2})$ and their own associated standard uncertainties, $u(x_i)$. Since estimation of the uncertainty of the acidity function requires the knowledge of the uncertainty of $E°$, whose value derives from an independent measurement based on eq. (4) ($p° = $ standard pressure, $p_{H2} = $ partial pressure of H_2, $\gamma_{\pm HCl} = $ mean ionic activity coeff.), it requires an independent budget:

$$E° = E - 2\,[(RT/F)\,\ln10]\log m_{HCl} + 2\,[(RT/F)\,\ln10]\log\gamma_{\pm HCl} - \tfrac{1}{2}[(RT/F)\,\ln10]\log(p°/p_{H2}) \quad (4)$$

The uncertainty budget of the phosphate buffer at 25°C, is given in Table 2. Table 3 reports the budget for the standard potential $E°$.

Table 2 *Data for uncertainty budget of primary measurement of paH.*[a]

Quantity	Unit	Estimate x_i	Standard uncertainty u_i	Sensitivity coefficient c_i	Contribution $u_i(y)$
$E°$	V	0.22213	8.76E-05	16.90	1.48E-03
E	V	0.76964	2.20E-05	16.90	3.71E-04
T	K	298.15	1.00E-03	0.031	3.10E-05
m_{KCl}	mol·kg^{-1}	0.005	2.63E-05	86.86	2.28E-03
p_{H2}	Pa	98872.3	19.5	2.20E-06	4.28E-05
Combined standard uncertainty of $p(a_H\gamma_{Cl})$					**2.75E-03**

[a] The small uncertainties of F and R do not contribute significantly to the budget and are neglected (the same is for calculations in Table 3).

The measurement procedure is described in details in our previous paper[2]. The resulting acidity function at zero chloride molality is 6.957 (paH resulted 6.847 using $\gamma_{Cl} = 0.777$) with the corresponding standard uncertainty of $2.75·10^{-3}$.

The $u(intercept)$ was equal to $1.10·10^{-3}$ (fitting obtained by six measurements). Combining $u(pa_H\gamma_{Cl})_{0.005}$ and $u(intercept)$, the combined standard uncertainty of the paH value is:

$$u_c(paH) = u_c(pa_H\gamma_{Cl}°) = [(2.75·10^{-3})^2 + (1.10·10^{-3})^2]^{1/2} = 0.003,$$
and $U(paH) = 0.006$ ($k = 2$).

Table 3 *Data for uncertainty budget of primary measurement of paH.*

Quantity	Unit	Estimate x_i	Standard uncertainty u_i	Sensitivity coefficient c_i	Contribution $u_i(y)$
E	V	0.46298	5.07E-05	1.0	5.066E-05
T	K	298.15	1.00E-03	-8.11E-04	-8.110E-07
m_{HCl}	$mol \cdot kg^{-1}$	0.01	9.36E-06	5.14	4.810E-05
p_{H2}	Pa	98872.3	19.5	1.30E-07	2.534E-06
$\gamma_{\pm HCl}$		0.9042	9.30E-04	0.0568	5.285E-05
Combined standard uncertainty of $E°$					**8.765E-05**

3.2 Secondary apparatus: the junction potential estimation.

As well known, a glass electrode operates according to the general equation:

$$E = K + s\,pH + E_J \qquad (5)$$

where the term E_J can be expressed as:

$$E_J = J_A[H^+] + J_B[OH^-] \qquad (6)$$

being i) E the instrumental reading (e.m.f.), ii) K the formal potential difference, iii) s the slope of the Nernst equation (59.16 mV at 25°C), and iv) J the parameter related to the junction potential (E_J) acting in the acidic (J_A) or in the basic (J_B) field. The linear relationship occurring between E and pH is hence distorted owing to effects mainly related to the salt bridge, which allows the glass electrode response to be dependent upon the activity of the hydrogen ion alone (on the other side, primary apparatus responds to both H^+ and Cl^- activities). The parameter J_A can be estimated during the refinement process of the acid-base titration data leading to the contemporary optimization of J_A, C_{HCl} and K. The process requires, as input, titrimetric data (electrode potential and titrant volume), recorded during the alkalimetric titration of HCl with KOH, from pcH 1.5 to 3.5. In the very acidic field the effect of J is emphasized and the refinement is more accurate. Considering titration data until pcH nearly 12 the calculation can include the refinement of J_B and pK_w. The effect of J_A on pcH value can usually be neglected for pcH > 2.5, but this strictly depends on each glass electrode characteristic (e.g. membrane composition, nature of the liquid junction, filling electrolyte). The corresponding uncertainty contribution is included in the uncertainty budget (see below).

In ref. 10, IUPAC proposed for J_A an estimation equal to 67 ± 6 mV·mol^{-1}·l. J depends on ionic strength value of the testing solution, as confirmed by the following experimental results. We have refined, for I equal to 3.0, 1.0, 0.10 or 0.030 M the following values of J_A: 75, –20, –32 and –55 mV·mol^{-1}·l, respectively (standard deviation ranged between $\pm$ 1 and $\pm$ 6). The parameter J is related to the gap of overall ion concentration between internal and external solution with respect to the glass membrane. Unfortunately, overlapped effects of either acidic or basic field of pH (including acid error and alkaline error) cause further difficulties in the rationalization of this matter. Moreover, the composition of the glass membrane, the nature and concentration of the filling electrolyte (usually KCl 3 M) and the material used to build the liquid junction (ceramic junction, capillary junction, and so on) affect the behaviour of the electrode inducing phenomena difficult to be predicted and quantified. Nevertheless, we can observe that the value of 67 mV·mol^{-1}·l proposed in ref. 10 is probably obtained considering for the evaluation of K a fluid exhibiting very high ionic strength: deviation from nernstian linearity following the opposite trend has been found handling testing solution at low values of ionic strength.

If attention is paid with respect to the ionic strength of the commercial paH-metric buffers, the estimation of E_J of a glass electrode can be tried according to the shown procedure, which is typical of pH measurement according to the concentration scale. On the other hand, although an activity scale is considered in a secondary measurement, the effect of E_J can be fairly estimated taking care of the mentioned features. The uncertainty budget of pH, whatever the scale, will hence be improved.

3.3 Uncertainty budget of secondary apparatus in activity scale

The expanded standard uncertainty associated to the paH value of the same solution above described, measured in activity scale by the secondary apparatus (paH = 6.85 [2]), is $U(paH) = 0.02$ (k = 2). The evaluation has been performed considering a two point calibration (paH(S$_1$) = 4.005 and paH(S$_2$) = 6.98). We report below the algorithms employed in the measurement with paH-metric buffers:

$$s = \frac{E(S_1) - E(S_2) + E_J(S_1) - E_J(S_2)}{paH(S_2) - paH(S_1)} \qquad (7)$$

$$paH(X) = paH(S_1) - [E(X) - E(S_1)]/s - (E_J(X) - E_J(S_1)]/s \qquad (8)$$

where: i) s is the practical slope of the glass electrode, ii) $E(S)$ and $E(X)$ are the measured potentials of the standard buffers and of the sample, iii) $E_J(S)$ and $E_J(X)$ are the liquid junction potentials, and iv) paH(S) and paH(X) are the paH values of the standard buffers and of the sample.

Table 4 reports the uncertainty budget based on eq. (8). Uncertainty contribution of s deriving from eq. (7) is $1.689 \cdot 10^{-4}$ V (calculation not reported).

Uncertainty on E_J has been evaluated considering a J_A value of -54 mV·mol^{-1}·l (from solution at low ionic strength as the buffer at paH 4) according to the procedure described in the following paragraph.

Table 4 *Data for uncertainty budget of secondary measurement of paH with paH-metric buffers.*

Quantity	Unit	Estimate x_i	Standard uncertainty u_i	Sensitivity coefficient c_i	Contribution $u_i(y)$
pH(S1)		4.00	5.774E-03	1.0	5.774E-03
ΔE	V	-0.1661	2.160E-04	-17.07	-3.687E-03
ΔE_J	V	5.46E-03	1.500E-04	-17.07	-2.560E-03
s	V	0.0586	1.689E-04	46.78	7.901E-03
T	K	298.15	2.887E-02	9.193E-03	2.654E-04
δpH_{ris}		–	2.887E-04	1.0	2.887E-04
Combined standard uncertainty					**1.077E-02**
Expanded uncertainty U **(k = 2)**					**2.154E-02**

3.4 Uncertainty budget of secondary apparatus in concentration scale

Rearranging eq. (5), pcH can be expressed as follows:

$$pcH = K/s - E/s + E_J/s \qquad (9)$$

For pcH uncertainty budget, we must consider the contributions acting on K, E and E_J. The standard uncertainty is:

$$u_c(\text{pcH}) = \{[u(K)\cdot c(K)]^2 + [u(E)\,c(E)]^2 + [u(E_J)\,c(E_J)]^2\}^{1/2} \qquad (10)$$

where the terms $c(x_i) = \partial(\text{pcH})/\partial x_i$ are the sensitivity coefficients of E, K and E_J, all equal to $1/s$.

Further details on instruments employed are provided, to assess the uncertainty budget related to the secondary measurement of pcH, which requires an articulated procedure of solution preparation involving various calibrated liquid dispensers (see data in Table 1).

Five quantities contribute to the overall uncertainty of K. Evaluation of K is run by means of a refinement procedure based on the treatment of titration data; similarly, the uncertainty estimation must follow the same path. In particular, we have to consider the following scheme.

1) The preparation of the HCl/KCl solution step (total volume 25 ml, $I = 0.1$ M) to evaluate K, which brings uncertainty related to the volume dispensers and to the preparation of 1 M KCl stock solution (flask volume and weight of KCl). The estimation of $u(V)$ (relative standard uncertainty on volume) has been assessed on 0.41%. As for ionic strength, an overall uncertainty contribution of 0.016% has been estimated.

2) All what concerns the titrant (including the weight of potassium hydrogenphtalate step to standardise KOH titrating solution). The estimation of $u(C_{KOH})$ (relative standard uncertainty on KOH concentration) has been assessed on 0.27%. The influence of both $u(V)$ and $u(C_{KOH})$ has been tested running ESAB2M[5] to quantify the corresponding effect on the refinement of K.

3) The resolution of the automatic burette. The estimation of $u(V_{tit})$ (relative uncertainty on volume dispensed by the automatic burette) has been assessed on 0.12%, assuming a triangular distribution[9] on the stated accuracy of the burette.

4) The resolution of the potentiometer. The estimation of $u(E)$ (relative uncertainty on potential reading) has been assessed on 0.03%, assuming a rectangular distribution on the instrument resolution[9] and considering a value – the last point of the titration considered in the K refinement - of 193.8 mV.

5) The standard deviation on K from the refinement process. The value of standard deviation on K has been evaluated by ESAB2M[5] as 0.04 mV, average value of three replicates. It accounts for the accordance degree between experimental and calculated titration curves taking into account in the calculation the covariance(s) among the parameters under refinement.

Combining the aforementioned points we obtained the relative standard uncertainty of K equal to 0.14%.

The uncertainty contribution on E has been evaluated considering the potential resolution and the repeatability.

As for E_J, at $I = 0.1$ M we obtained $J_A = -32$ and $J_B = 20$ mV·mol^{-1}·l. The uncertainty on J_A comes from the same quantities and experimental/calculation procedures affecting K. In this connection, from ESAB2M[5] refinement we obtained a standard deviation of 1.2 (average value of three replicates); moreover, the volume produces a contribution of 0.84% while the titrant (C_{KOH}) gives a contribution of 0.55%, each of them estimated thanks to apposite runs of ESAB2M.[5] The relative standard uncertainty $u_c(J_A)$ is then 3.9%. Very similar value is obtained for $u_c(J_B)$. Applying $u_c(J_A)$ and $u_c(J_B)$ to E_J, for $[H^+] = 10^{-6.80}$ M ($pK_w = 13.78$ at $I = 0.1$ M), $u(E_J) = 2.158\cdot 10^{-7}$ mV.

In Table 5 the contributions of each quantity on the overall uncertainty of pcH are shown. For pcH = 6.80 we obtained the extended $U(\text{pcH}) = 0.02$, using a coverage factor k = 2.

Table 5 *Data for uncertainty budget of secondary measurement of pcH.*

Quantity	Unit	Estimate x_i	Standard uncertainty u_i	Sensitivity coefficient c_i	Contribution $u_i(y)$
K	mV	407.6	5.710E-01	1.69E-02	9.65E-03
E	mV	5.3	4.670E-02	1.69E-02	7.89E-04
E_J	mV	-2.98E-06	2.158E-07	1.69E-02	3.65E-09
Combined standard uncertainty					**9.68E-03**
Expanded uncertainty U (k = 2)					**1.94E-02**

4 CONCLUSIONS

The accuracy of the secondary measurement of paH, using the widely diffused two-point calibration, suffers for different junction potentials - each of them being a function of the ionic strength - affecting the slope of the calibrating function.[4] Moreover, in this case the uncertainty budget is hard to be satisfactorily estimated, since the ionic strength of commercial paH-metric buffers is usually undeclared, so as the relative standard uncertainty on E_J. As to IUPAC recommendations,[1] the relative standard uncertainty on E_J is arbitrarily assumed equal to 100% with respect to the quantity, apparently missing a sharp linking with individual experimental features.

The pcH measurement allows an estimation of both E_J and the corresponding relative standard uncertainty. The amount of substance involved in the calculation of the formal potential K is well controlled, since an absolute method of analysis (titrimetry) is employed. Moreover, the method allows to ignore the activity coefficient while forces the operator to strictly monitor the ionic strength of the fluids under investigation. It results in two advantages: the junction potential can be fairly estimated and the relevant uncertainty of the activity coefficient of mono-charged ions (γ_H) ignored. On the other side, the primary method assumes γ_{Cl} as an axiomatic quantity to obtain paH by the acidity function: the secondary measurement of paH deriving from a calibration by paH-metric buffers includes the consequences of such a choice. As for the value of γ_i, we have already underlined[2] how it can be critical in order to harmonise the dissemination of a pH value obtained according to either activity or concentration scales.

Finally, the uncertainty budget estimated for the primary method informs us regarding possible ways of improvement: major contributions are ascribed to both standard potential and chloride molality. A focus on uncertainty contributions affecting these two quantities can be foreseen for future enhancements.

References
1   R.P. Buck, S. Rondinini, A.K. Covington, F.G.K. Baucke, C.M.A. Brett, M.F. Camoes, M.J.T. Milton, T. Mussini, R. Naumann, K.W. Pratt, P. Spitzer and G.S. Wilson, IUPAC recommendations, *Pure Appl. Chem.*, 2002, **74**, 2169.
2   P. Fisicaro, E. Ferrara, E. Prenesti and S. Berto, *Anal. Bioanal. Chem.*, 2005, **383**, 341.
3   I. Brandariz, T. Vilarino, P. Alonso, R. Herrero, S. Fiol and M.E. Sastre de Vicente, *Talanta*, 1998, **46**, 1469.
4   I. Brandariz, J.L. Barriada, T. Vilarino and M.E. Sastre de Vicente, *Monatsh. Chem.*, 2004, **135**, 1475.
5   C. De Stefano, P. Princi, C. Rigano, and S. Sammartano, *Ann. Chim.*, 1987, **77**, 643.

6   P.G. Daniele, A. De Robertis, C. De Stefano, S. Sammartano and C. Rigano, *J. Chem. Soc. Dalton Trans.*, 1985, 2353.
7   P.G. Daniele, C. Rigano and S. Sammartano, *Anal. Chem.*, 1985, **57**, 2956.
8   Guide to the expression of uncertainty (GUM) BIPM, IEC, IFCC, ISO, IUPAC, IUPAP, OIML, 1993.
9   EURACHEM/CITAC Guide: Quantifying Uncertainty in Analytical Measurement, second edition, 2000.
10   A. Braibanti, G. Ostacoli, P. Paoletti, L.D. Pettit and S. Sammartano, 1987, *Pure Appl. Chem.*, **59**, 1721.

METROLOGY IN COMPLEX SITUATIONS: EXPERIENCES WITH THERMO-DYNAMIC DATA

G. Meinrath[1,4], S. Lis[2], A. Kufelnicki[3]

[1] RER Consultants, Passau, Germany
[2] Adam Mickiewicz University, Poznan, Poland
[3] Medical University of Łódź, Łódź,, Poland
[4] Technische Universität Bergakademie Freiberg, Freiberg/Germany

1 INTRODUCTION

Thermodynamic data are typical examples for measurands which cannot be expressed in a closed mathematical equation. The measurand, commonly a formation quotient, is generally evaluated from a series of multivariate masurements. Such situation will be termed 'complex situations' in the following. A priori assumptions on the distribution of measurement uncertainty may be misleading, thereby introducing bias into the measurand's value. An alternative approach is the evaluation of the empirical distribution of the measurement uncertainty [1]. In fact measurement values of thermodynamic data are notorious for being poorly reproducible and poorly comparable even under highly comparable conditions [2-4]. Nevertheless, determination of formation quotients is a widespread activity. The forwarded values are routinely applied in environmental and technical simulation and modeling. Thermodynamic data are considered as constants of nature which motivates their use in a variety of applications.

A larger number of critical eviews of various thermodynamic data are available; commonly with different selection criteria and widely varying recommendations. In almost all cases, mean values or –occasionally- repeatabilities are reported with published with thermodynamic data rendering any claim of objectivity in the assessment of values for thermodynamic data void. Data which are not reported with detailed information on their quality remain questionable – independent of the good will of whatsoever review panel.

2 COMPLETE MEASUREMENT UNCERTAINTY BUDGET

The recognition that a common, internationally accepted convention on the communication and documentation of uncertainty in measurement is lacking has lead to the publication of the ISO Guide to the Expression of Uncertainty in Measurement (GUM) [5]. This guide documents an intenational convention which explicitly states that 'if all quantities on which the result of a measurement depends are varied, its uncertainty can be evaluated by statistical means' [5]. Hence, the predominance of the Normal distribution in discussion of measurement uncertainty, which is for instance evident by the recommendation of classical progression-of-uncertainty approaches in the evaluation of standard uncertainties from the individual uncertainty contributions of relevant influence factors in the EURACHEM/ CITAC Guide [6], is caused by its (relative) simplicity; it is not a requirement of the GUM. In fact, concepts like the mean value, the standard deviation, the variance-covariance matrix, the Pearson correlation coefficient, the progression-of-error rules and others are intimately related to the Normal distribution.

The complexity of operations required in the evaluation of thermodynamic data also requires statistically valid numerical approaches independent of the limitations of the Normal distribution, especially correlation, non-linearity and non-Normally distributed residuals. The complete measurement uncertainty budget results from two groups of operations. The first group affects the individual data points. The second group affects the parameters of the model function interpreting the data points in terms of a (usually multiparameteric) model function.

The interpretation of data points by multiparametric models is commonly describes as 'fitting curves to data'. A variety of methods have been described to obtain parameter estimates usually using the least-sum-of-squared-residuals criterion as quality measure [7]. Least-square fitting methods, however, do not provide an estimate for the likely doubt to be associated with the forwarded parameter estimate(s). Approximations, e.g. estimates of the variance-covariance matrix in the minimum of the response surface described by the mathematical model and the data points, rely on a quadratic form of the response surface [8]. Non-linearity, correlation and non-Normality are only spuriously considered. For the application of more adequate statistical models, the underlying numerical and mathematical structures are often overly complex. There is, for example, no application of Bates' and Watts' curvature measures [9] to thermodynamic data to the knowledge of the authors. The classical least-squares approaches may be rather biased [10,11]

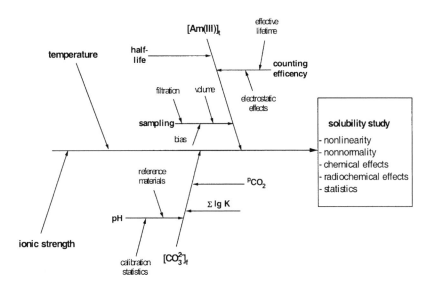

Figure 1: *Cause-and-effect diagram for the formation constants of carbonate with Am(III) in aqueous solutions. The cause-and-effect diagram is separated into two parts. The classical, fishbone, part represents the uncertainties obtained by Type B evaluation, the right-hand-side box summarizes those effects which can be accounted for by bootstrap resampling (Type A evaluation) [1].*

The advent of fast and cheap computing power allows the routine application of computer-intensive resampling algorithms to these problems, especially bootstrap methods [12,13]. Bootstrap methods are an efficient statistical approach to obtain uncertainty estimates from complex models, whereby complicated and often unavailable analytical analysis is replaced by brute computing power. The central element of bootstrapping is the resampled bootstrap distribution F*(x), which is the best unbiased estimate of the true distribution F(x), from which the samples (measurement values of data points) have been obtained. The bootstrap distribution F*(x) is obtaind from the samples by random resampling with replacement. Since its introduction in 1979, the bootstrap is widely applied in many filds of science, humanics and technology. Thus, the application of bootstrap methods to thermodynamic data evaluation is a feasible way to obtain the empirical probability distribution of the parameters [1,10].

Limited information is commonly available to estimate the influence factors acting upon the measurement process of the individual data points. The state-of-the-art in metrology in chemistry is the expression of influence factor contributions by standard deviations of supposed Normal distributions. In fact, this approach in most situations seem to forward satisfactory results. Classical cause-and-effect diagrams can be derived to account for influence factors following the guidelines of EURACHEM/CITAC [6].

Figure 1 shows a typical cause-and-effect diagram for a complex situation, here the evaluation of formation constants of Am(III) carbonato species from solubility measurements using the α-decay of ^{241}Am for concentration determination by liquid scintillation counting [1].

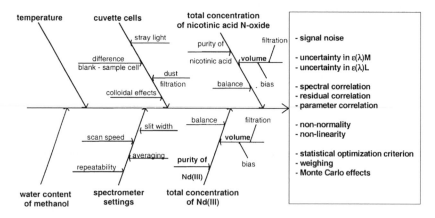

Figure 2: *Cause-and-effect diagram for the evaluation of formation constants of nicotinic acid N-oxide with lanthanides by UV-Vis absorption spectroscopy using TB-CAT [14].*

For complex situations, classical progresion-of-error approaches can be applied to the fishbone part of the cause-and-effect diagram. The fishbone part thus summarizes those influence factors which can be quantified via ISO Type B evaluation. These factors influence the determination of the individual data points. The final interpretation of the measurement data by curve fitting procedures has to be modelled numerically within a bootstrap shell. The bootstrap algorithms randomly resamples the measurement data between 1000 to 2000 times, thereby varying the data within the uncertainty ranges set by the Type B evaluation.

Coverage considerations may play a role. To assure with at least 99% probability that at least one sample is included into the analysis coming simultaneous from the 95% percentiles of the distribution, 2000 samples have to be drawn if there are two influence factors, 36840 if there are three influence factors and 736825 sampls if there a four influence factors. These figures leave no doubt that the classical Monte Carlo strategy used here for the inclusion of Type B uncertainties is not fully satisfactory to estimate tail probabilities with high accuracy. The empirical distribution of the mean, however, can be satisfactorily estimated from the empirical distribution F*(x).

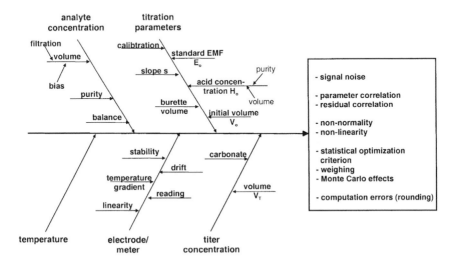

Figure 3 *Cause-and-effect diagram for the evaluation of protonation constants of phosphoric acid by glass electrode potentiometry [15].*

Figures 2 and 3 present cause-and-effect diagrams for other relevant method for determination of thermodynamic data: UV-Vis absorption spectroscopy (Figure 2) and glass electrode potentiometry (figure 3). Despite the fact that the right-hand-side box gives about the same influence factors, the relative magnitude of their influence requires different numerical approaches. Especially UV-Vis spectroscopy differs from the two other methods from the amount of data (commonly more than 20000 data points are treated simultaneously) and the importance of residual correlation.

UV-Vis spectroscopy can be interpreted by the Bouguer-Lambert-Beer Law, given in Eq. (1) in matrix form

$$A = E C + N \tag{1}$$

where A is the matrix of measured absorptions, E is the column matrix of single component spectra (molar absorptions) and C gives the concentration of species in the sample. N is the matrix of residuals, commonly assumed to be Normally distributed, uncorrelated and independently distributed. If E and C are known, A can be computed. The analyst's task, however, is the opposite: A is known from experiment (hence affected by noise and, possibly, bias) while estimats for E and C are required. This task is commonly addressed by factor analysis [16]. UV-Vis spectroscopy is an important method in the analysis of aqueous chemical systems, because it allows to obtain an independent estimate on the number of species present in the system under study. There is no oher method with this feature.

In the analysis of the measurement uncertainty of thermodynamic data derived from UV-Vis spectroscopic data, the correlation of residuals strongly interferes and needs to be considered by appropriate strategies, e.g. threshold bootstrap methods [16]. Past experience shows that residual correlation may widely vary from study to study.

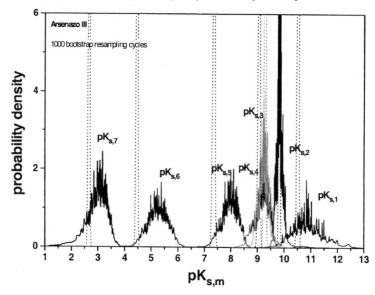

Figure 5 *Empirical complete measurement uncertainty budget distributions for the protolysis constants of Arsenazo III. The dashed curves represent distributions obtain by neglecting uncertanties from Type B evaluation [15].*

Figure 5 gives an example of the results obtained by a bootstrap analysis of a complex systems, using potentiometric titration of arsenazo III as an example [15]. Two different distributions are given for each protonation constant: the complete measurement uncertainty budget (solid curves) and the measurement uncertainty ignoring Type B contributions (dashed lines). In fact, ignoring Type B uncertainties results in very narrow distributions indicating a highly precise method. In general, potentiometric titration has long been assumed to be of high accuracy in the study of aqueous chemical systems provided the necessary precautions are taken (e.g. [2]). A recent analysis, however, it turned out that despite the apparent accuracy of results obtained in one laboratory the between-lab comparability was rather poor [13]. Figure 5 gives an analysis of this observation: as a consequence of the neglect of Type B uncertainties, the observed repeatability of glass electrode potentiometry is high because Type B uncertainties only affect influence quantities becoming apparent in interlaboratory comparisons [3].

However, overinterpretation of thermodynamic data is a common phenomenon and illustrated in Figure 6. Eight mostly discepant (too narrow) formation constants for the

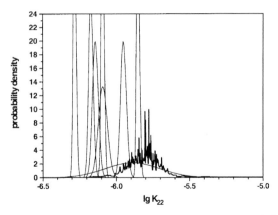

Figure 6 *A comparison of eight formation constants reported for the U(VI) species $(UO_2)_2(OH)_2^{2+}$ with the complete measurement uncertainty distribution (thick line) obtair from UV-Vis specroscopic analysis [4].*

U(VI) species $(UO_2)_2(OH)_2^{2+}$ are compared with the complete measurement uncertainty distribution obtained from a UV-Vis spectroscopic analysis [4]. In fact, the complete measurement uncertainty budget covers most reported values indicating that the literature values are overinterpreted, not inconsistent.

References

1   G. Meinrath, Fresenius J. Anal. Chem. 2001, **369**, 690.
2   M. Filella, P. M. May, Talanta 2005, **65**, 1221.
3   G. Meinrath, A. Kufelnicki, M. Świątek, Accred. Qual. Assur. 2006, **10**, 494.
4   G. Meinrath, S. Lis, Z. Piskula, Z. Glatty, J. Chem. Thermodyn., J. Chem. Thermodyn. accepted for publication (2006).
5   ISO, Guide to the Expression of Uncertainty in Measurement. 2nd ed. 1995. ISC Geneva/CH.
6   EURACHEM/CITAC Quantifying Uncertainty in Chemical Masurement EURACHEM/CITAC 2000.
7   G. Meinrath, Ch. Ekberg, A. Landgren, J.O. Liljenzin, Talanta, 2000, **51,** 231.
8   S. Brumby, Anal. Chem. 1989, **61**, 1783.
9   D. M. Bates, D. G. Watts, J. Royal Stat. Soc. 1980, **42**, 1.
10 G. Meinrath, Chemometrics Intell. Lab. Syst. 2000, **51**, 175.
11 G. Meinrath, Fresenius J. Anal. Chem., 2000, **368**, 574.
12 B. Efron, Ann. Stat. 1979, **7**, 1.
13 B. Efron, R. J. Tibshirani, An Introduction to the Bootstrap, Monographs on Statistic and Probability, Vol. 57, Chapman & Hall, London/UK 1993.
14 G. Meinrath, S. Lis, U. Böhme, J. Alloy Comp. 2006, **408-412**, 962.
15 A. Kufelnicki, S. Lis, G. Meinrath, Anal. Bioanal. Chem. 2005, **382**, 1652.
16 G. Meinrath, S. Lis, Fresenius J. Anal. Chem. 2001, **369**, 124.

SOME TRACEABILITY PROBLEMS IN ANALYTICAL ASSAYS OF INTEREST IN THERMAL METROLOGY

F. Pavese

Istituto Nazionale di Ricerca Metrologica (INRIM), strada delle Cacce 91, 10135 Torino, Italy

1 INTRODUCTION

The variability of the isotopic composition is a well-known issue and is regularly monitored and reviewed by bodies such as the IUPAC. [1] However, these data cover the whole spectrum of the variability observed on the earth. The actual variability that can be observed when buying commercial substances could be smaller to such an extent to alleviate or eliminate practical problems for the temperature standards, or could instead remain relevant.

In recent years it has been recognized that the uncertainty budgets of the fixed-point realizations in temperature standards should include a component for the isotopic composition and that, for some substances, the sample-to-sample isotopic variability is an obstacle to further improvements of the realized uncertainty.

In fact, the recent improvements in thermal techniques and in the understanding and mastering of the underlying phenomena, substantially obtained thanks to an international collaboration Project, [2] allowed a reduction in the total uncertainty of the best temperature standards, based on the realization of the triple point, to nearly 50 μK. [3]

Evidence of the real extent of an isotopic problem for hydrogen became known in metrology only after 1999. [4-6]

For neon, Furukawa [7] had already pointed out in 1972 the problem arising from the sensitivity of the triple point temperature to isotopic composition and suggested the use of the pure isotope ^{20}Ne. However, the first evidence of the extent and significance of the isotopic variability in samples of commercial origin became known only since 2003. [8-9]

For water, recent studies [10-12] have shown that isotopic composition should be verified for each sample used for the realisation of the triple point of water, in order to ensure that it is not a limiting factor of the relevant realisations.

In a recent paper, [13] the author has summarized the implications for the definition of the ITS-90, which, in fact, is only loosely prescribing, except for ^{3}He, a "natural isotopic composition".

This paper, first briefly introduces the modern techniques that make it possible to reduce the overall uncertainty in the realization of low temperature standards at the indicated level, and that also make possible more accurate studies of the isotopic effect on the transition temperature values. Then, it summarizes the new findings, which mainly concern hydrogen (for deuterium, not used today as a temperature standard and its isotopic problems (protium content), see Ref 14-15), water and neon and concludes reporting recent relevant decisions taken by the Comité Consultatif de Thermométrie (CCT).

2 MODERN TECHNIQUES FOR ACCURATE TRIPLE-POINT REALISATIONS

The modern techniques are the last evolution of the technique of the realization of the condensed-state phase transitions of gases in miniature "sealed cells", introduced by the author in 1976. [16] The first cell, sealed in 1975, is still working properly and reproducing the same transition temperature; with the IMGC preparation technique of these devices, no drift in time was ever observed for all devices produced since then, [3,17] except for D_2, where a decrease of a few tenths of a millikelvin in ten years was observed, attributed to a progressive protium concentration growth inside the cell.

However, the overall quality of the melting that can be obtained is also affected by the commercial purity of the substances used, which makes the quality of the calorimeter used for the measurements more or less demanding.

Since 2000, the new generation of these devices developed at IMGC and in Europe, [2] resulted in improved thermal characteristics and much smaller size.

With these elements, it was possible to achieve the following best results. [18] *Argon*: reproducibility (for measurements performed by up to five different laboratories, from 10% to 90% melted fraction), 26 μK (1 σ,), melting slope $(T_{100\%} - T_{50\%})$ <20 μK. *Oxygen*: reproducibility 24 μK, melting slope $(T_{100\%} - T_{50\%})$ <5 μK. *Neon*: reproducibility 26 μK, melting slope $(T_{100\%} - T_{50\%})$ <5 μK. *e-Hydrogen*: reproducibility 44 μK, melting slope $(T_{100\%} - T_{50\%})$ <10 μK.

The improvement obtained with respect to the previous state-of-the-art is reflected in Table 1. This improvement makes it possible today to determine with a corresponding lower uncertainty the effect of the isotopic composition on the triple point temperature.

Table 1 *Typical and aimed uncertainty budget for cryogenic fixed point measurements* [7]

#	Item	Typical 1975–2000 (μK)	Presently available (μK)	Aimed in future (μK)
1	Non-isotopic impurities	$50 \Rightarrow 2000$ [4]	40	10
2	Isotopic composition	up to 700 [5]	30 [6]	20
3	Induced by R_{cs}	50–200	30	10
4	Induced by τ	<100	20	10
5	Cryostat other effects [1]	20–300	10	10
6	Resistance measurement [2]	30–200	30	10 [8]
7	T_{tp} definition [3]	20–300	20	0
	TOTAL	150–1000	70	30

Legend: R_{cs} = static thermal resistance; τ = cell dynamic time constant; T_{tp} = triple point temperature.
[1] For meltings lasting less than ≈12 h. [2] Except with *e*-H_2 when measuring a CSPRT. [3] For $\Delta T_{melt(20-80\%)} < ≈0,1$ mK. [4] For Ar in O_2 or HD in D_2. [5] For D in H. [6] For the best certification uncertainty only. [7] See Ref.17.
[8] Equivalent to ≈1 μΩ for a SPRT above 40 K.

3   ISOTOPIC VARIABILITY OF SOME SUBSTANCES USED FOR LOW
TEMPERATURE FIXED POINTS AND ITS EFFECT ON TEMPERATURE

3.1 Hydrogen

A comprehensive study recently finalized [6] represents the current state-of-the-art, including
both the most recent thermal measurement techniques for the realization of temperature
fixed points and the analytical assays of the isotopic composition of the samples used.
These studies have shown that gas samples from commercial sources used all around the
world exhibit a broad variability in their isotopic composition ($28-155 \ 10^{-6}$), close to the
maximum spread ($26-184 \ 10^{-6}$) indicated by the IUPAC.[1]

This variability in composition results in a range of about 700 μK for the spread of the
fixed-point temperatures, which is significantly greater than the target uncertainty limit of
20 μK indicated in Table 1.

In Ref.6 a new determination of the slope for dilute D/H mixtures dT_{tp}/dx was
performed, being previously available with low accuracy only from a single study on the
whole phase diagram of D-H mixtures. [19] It was found to be ($5.4_2 \pm 0.3_1$) μK per μmol
D/mol H. The uncertainty of the experimental data for compositions x close to 53 μmol
D/mol H was limited to ± 17 μK (2 σ), tripling for the highest deuterium contents, while
the present total uncertainty budget is about 70 μK.
Progress is also dependent on the availability of isotopic assays of gaseous hydrogen. All
the samples measured in Ref.6 have shown a deuterium concentration lower or equal than
the "Vienna Standard Mean Ocean Water" (VSMOW).

The ITS-90 also makes use of two points of the H_2 vapour pressure line, near 17 K
(33 kPa) and 20 K (101 kPa). Also the vapour pressure is affected by the isotopic
composition. An estimate has been computed of the error arising from the isotopic
variability: [5] the observed isotopic variability corresponds to a range of values on the
condensation isobars of ≈700 μK at 33 kPa and ≈650 μK at 101 kPa; on the evapouration
isobars, these spreads are ≈300 μK and ≈350 μK, respectively. The difference between the
evapouration and condensation lines would increase from ≈0.1 mK to ≈0.4 mK from the
lowest D/H concentration to the VSMOW one. No experimental data are presently
available.

3.2 Neon

A recent preliminary international study [9] reports the first evidence that the variability of
the isotopic composition of neon samples from commercial sources is large enough to limit
the uncertainty in the realization of the temperature fixed point.
The European Union Institute for Reference Materials and Measurements (IRMM)
performed the analytical assays. The uncertainty, for the ratio of amount of isotopes
$n(^{22}Ne)/n(^{20}Ne)$, was 0.05%: this turns into a contribution to the temperature measurement
uncertainty of only 2—7% of the evaluated ΔT_{tp}. No other laboratory was found in the
world able to provide a similar uncertainty: usually others are providing, at best, only ten
time larger ones, unsuitable for thermometric use.
These results showed that the isotopic variability range of ^{22}Ne is considerable and is offset
(one-sided) with respect to "natural" (standard air) composition: in fact, it was found to be
$-(0.07-0.27)$ % ^{22}Ne, corresponding to a difference in T_{tp} of about $-(100-400)$ μK.
These data are unusual with respect to known data (dating back 30 years at least), in that
they lie on the mass fractionation line of neon obtained from air, but for higher ^{22}Ne

content with respect to the air standard composition. Again, this should be compared to the target uncertainty of 20 μK, in a total uncertainty budget of about 70 μK. [17]

The effect of the variability of ^{21}Ne is irrelevant for temperature standards.

More work is going on within a EUROMET Project [8] to establish both a variability range based on a more comprehensive and worldwide set of data, and a correction uncertainty. Also a determination of the actual T_{tp} differences for the relevant mixtures will be necessary taking advantage of the decreased uncertainty of thermal measurements. The old determinations of the triple point temperatures of ^{22}Ne and ^{20}Ne [7] have an uncertainty of about 1 mK and no experimental determinations of the actual T_{tp} differences on the liquidus line for the relevant mixtures exist.

3.3 Oxygen

No experimental determination of the triple point temperatures of pure ^{17}O and ^{18}O exists. However, an estimate of the effect on T_{tp} of a variability corresponding to the full range reported for ^{17}O by the IUPAC (about 4 %), indicates that it is expected to be irrelevant (<20 μK). [20]

3.4 Argon

An experimental determination exists of the T_{tp} difference between ^{40}Ar and ^{36}Ar: 59 mK. Considering again the full range of isotopic variability reported by the IUPAC, this would correspond to a difference of only 1.5 μK, totally irrelevant.

3.5 Water

Water cells were found, [10-12] mainly due to the variability of the sample-to-sample deuterium content, realizing temperatures up to 80 μK lower than the definition, compared again to the limit, 20 μK, aimed for this component of the uncertainty budget, while a typical total uncertainty budget is as low as 50 μK. The available experimental data are now thought to reduce the uncertainty in the triple-point temperature due to isotopic composition from 40 μK to better than 10 μK. The recent key comparison CCT K7, in fact, addressed this issue as well. Should it be ascertained that, also for water, the range of the actual variation of isotopic composition of the sample used is not centred on the present ITS-90 definition composition ("substantially ... ocean water" –VSMOW), a situation similar to that reported for neon would occur, but with more profound implications, considering that this fixed point realises the kelvin definition, which appears now to be ambiguous by referring simply to "water".

4 CONSEQUENCES ON THE CHEMICAL TRACEABILITY NEEDED BY THE ITS-90

Information about the large variability of the D/H ratio and the evidence of neon isotopic variability, in commercially available samples, was not available to the thermometry community at the time of the "Mutual Recognition Arrangement" (MRA) "Key Comparison on realizations of the ITS-90, 0.65 K to 24.5561 K, using rhodium-iron resistance thermometers" (CCT K1) [21] and "Key comparison of capsule-type standard platinum resistance thermometers (CSPRT) from 13.8 K to 273.16 K" (CCT K2) [22].

Therefore, these key comparison protocols did not require the capsule SPRT calibrations at the triple point of e-H_2 and Ne to take isotopic variability into account. Consequently, the published data, including those in the key comparison database of the Bureau International des Poids et Mesures (BIPM), do not presently include this effect, to which most of the observed variability can be attributed, of ≈ 0.7 mK for *e-*H_2 [6] and of ≈ 0.4 mK for Ne [9] (see also the comparison database on www.bipm.org).

Since hydrogen and oxygen are the components of water, also water triple-point temperature (which is also the definition of the kelvin) is heavily affected by the variability in their isotopic composition. [12, 13]

The one-sided variability with respect to neon composition in air, generally taken as the "natural" isotopic composition for neon, is particularly problematic, if confirmed. All the correct $R(^{nat}Ne, tp)$ values of thermometers calibrated at this fixed point would become higher by up to more than 200 μK after the data are corrected to the "natural" composition so far required by the ITS-90, therefore "offsetting" these values of these realizations of the ITS-90 with respect to the old un-corrected realizations. Similarly will happen with water, since CCT decided in its 2005 meeting to use, as the reference isotopic composition, the same presently defined for VSMOW, which is close to the upper limit of the water observed isotopic variability, with corrections amounting up to ≈ 100 μK, a lot especially for the kelvin definition. [13]

The urgency to eliminate the ambiguity, for the best standard realizations, carried by these isotopic effects into the ITS-90 definition, was evident.

The preliminary necessary requirement is the availability to the thermal metrology community of chemical assays concerning the isotopic composition of the samples used.

These assays first-of-all require to have a sufficiently low uncertainty for these metrological purposes: this is presently achieved for hydrogen in the gas and in water, thanks to the capabilities of the analytical chemists involved in geochemistry. It is not, at present, for neon, where IRMM is the only laboratory available in the world providing the required uncertainty.

A second requirement, often not necessary in other fields, is the need that the isotopic fractions are determined in an absolute way and not simply relative to another sample. The reason is that the temperature standards are using not reference materials but are referring to the ideal chemical substances (e.g., neon). This problem again is solved in the case of hydrogen but, at present, it is not in the case of neon.

Among the several possibilities analysed in Ref.13, the CCT, in its 2005 meeting, was in the position to make a choice to fix the problem of the value of the temperature standards for hydrogen, based on Ref.5 for the vapour pressure points and on Ref.6 for the triple point, and for water, based on Ref.12. Contrary to water, for which the VSMOW composition was used, the SLAP composition (89.02 μmol D/mol H) was preferred as the reference isotopic composition for hydrogen, being mid-way between the limits of the isotopic variability range, so minimizing the uncertainty increase toward the limit compositions (max < 20 μK).

In addition, a linear correction function was defined to be applied to the measured T_{tp} values of samples with a different and certified isotopic composition, using the slope value indicated above –and the uncertainties of the analytical assay and of the correction have to be added to the uncertainty budget.

Should no isotopic correction be applied (e.g., when no analytical assay is available for a specific sample), the uncertainty to be used in the budget must be the one resulting for the total isotopic variability range and an associated rectangular probability density function: for hydrogen, ≈ 0.2 mK, ten times more.

For the hydrogen vapour-pressure reference points, now "the ITS-90 relations for the vapour pressure of equilibrium hydrogen are referenced to an isotopic ratio of 0.000 089 02 mole D per mole H, the isotopic ratio determined for SLAP".

The above modifications of the ITS-90 for hydrogen and water were approved by the Comité International of Poids et Mesures in October 2005 and are now effective.

Similarly, neon problems will be solved as soon as the EUROMET Project is finalized and the results published.

5 CONCLUSIONS

With the improvement in the accuracy of temperature standard realizations, the effect of the variability of the isotopic composition of some substances used in the definition of these standards became relevant and should today be taken into account for the best realizations. An increasing amount of new data in this respect is becoming available, allowing for the proper corrections to be performed. It also required an adjustment of the ITS-90 definition, effective since October 2005, where the previous definition was creating an ambiguity by not specifying a reference isotopic composition: in fact, the concept of "natural isotopic composition" is today recognized to be ill defined or, at best, empiric –so needing a definition.

To further progress in the accuracy of the temperature standards, a stricter interaction with the analytical chemistry community is becoming essential (this also concerns the chemical impurities) [23], in order to obtain the necessary analytical assays, and with a sufficiently low uncertainty: this was found easier with H_2, but is presently extremely difficult with Ne.

References

1. J. R. de Laeter, J. K. Böhlke, P. De Bièvre, H. Hidaka, H. S. Peiser, K. J. R. Rosman and P. D. P. Taylor, *Pure Appl. Chem.*, 2003, **75**, 683.
2. F. Pavese, B. Fellmuth, D. Head, Y. Hermier, A. Peruzzi, A. Szmyrka Grzebyk and L. Zanin, *Temperature, Its Measurement and Control in Science and Industry* **8**, ed. D.C. Ripple, AIP, New York, 2003, 161.
3. F. Pavese, *Temperature, Its Measurement and Control in Science and Industry* **8**, ed. D.C. Ripple, AIP, New York, 2003, 167.
4. F. Pavese, Communication to CCT Members, August 1999; F. Pavese and W.L. Tew, 2000 Comité Consultatif de Thermométrie (CCT), Bureau International des Poids et Mesures, Sèvres, Doc. CCT/00–9 (www.bipm.org).
5. F. Pavese, W.L. Tew and A.G. Steele, *Proceedings TEMPMEKO 2001 (8th International Symposium on Temperature and Thermal Measurements in Industry and Science)*, ed. B. Fellmuth *et al*, VDE Verlag, Berlin, 2002, 429.
6. B. Fellmuth, L. Wölber, Y. Hermier, F. Pavese, P.P.M. Steur, I. Peroni, A. Szmyrka-Grzebyk, L. Lipinski, W.L. Tew, T. Nakano, H. Sakurai, O. Tamura, D. Head, K.D. Hill and A.G. Steele, *Metrologia*, 2005, **42**, 171.
7. G.T. Furukawa, *Metrologia*, 1972, **8**, 11.
8. F. Pavese, Communication to CCT Members, October 2003; EUROMET Project N.770, April 2004 (www.euromet.org).
9. F. Pavese, B. Fellmuth, D. Head, Y. Hermier, K.D. Hill and S. Valkiers, *Analytical Chemistry*, 2005, **77**, 5076.

10. J.V. Nicholas, D.R. White and T.D. Dransfield, *Proceedings TEMPMEKO 1996*, ed.P. Marcarino, 1997, 9.
11. J.V. Nicholas, T.D. Dransfield, B.W. Mangum, G.F. Strouse, R.L. Rusby and J. Gray, *Proceedings TEMPMEKO 1999*, 2000, 66.
12. D.R. White, T.D. Dransfield, G.F. Strouse, W.L. Tew, R.L. Rusby, J. Gray, *TMCSI*, 2003, **8**, *AIP Conf. Proc.*, **684**, 221.
13. F. Pavese, *Metrologia*, 2005, **42**, 194.
14. F. Pavese and G.T. McConville, *Metrologia*, 1984, **24**, pp. 107.
15. G.T. McConville and F. Pavese, *J. Chem.Thermodynamics*, 1988, **20**, pp. 337.
16. F. Pavese, G. Cagna, and D. Ferri, *Proc. 6th Int. Cryo. Engin. Conf. (ICEC6)*, IPC Science and Tech. Press, Guildford, 1976, 205.
17. F. Pavese, D. Ferri, D. Giraudi and P.P.M. Steur, *Temperature, Its Measurement and Control in Science and Industry* **6**, ed. J. F. Schooley, AIP, New York, 1992, 251.
18. F. Pavese, D. Ferri, I. Peroni, A. Pugliese, P.P.M. Steur, B. Fellmuth, D. Head, L. Lipinski, A. Peruzzi, A. Szmyrka Grzebyk and L. Wölber, *Temperature, Its Measurement and Control in Science and Industry* **8**, ed. D. C. Ripple, AIP, New York, 2003, 173.
19. N. Bereznyak, I. Bogoyavlenskii, L. Karnatsevich and V. Logan, Sov. Phys. JETP **30**, 1970, 1048.
20. R. Gonfiantini, private communication, 2002.
21. R.L. Rusby *et al.*, to be published.
22. A.G. Steele, B.Fellmuth, D. Head, Y. Hermier, K.H. Kang, P.P.M. Steur and W.L. Tew, *Metrologia*, 2002, **39,** pp. 551.
23. K.D. Hill and S. Rudtsch, *Metrologia*, 2005, **42**, L1.

DEVELOPMENTS IN UNCERTAINTY EVALUATION: THE ACTIVITY OF JCGM/WG1

W. Bich[1] and F. Pennecchi[1]

[1] Istituto nazionale di ricerca metrologica INRIM, Torino, Italy, w.bich@inrim.it

1 INTRODUCTION

The Guide to the Expression of Uncertainty in Measurement (GUM) was published in 1993 by seven international organizations in the field of measurement, i.e., BIPM, IEC, IFCC, ISO, IUPAC, IUPAP and OIML.[1] It was reprinted in a slightly revised version in 1995 and since then it has been established as the authoritative document in the field of the uncertainty of measurement. Many Countries and several organizations have adopted it as a standard, or a law, or as a basis for other specific standards.

In 1997 a Joint Committee for Guides in Metrology (JCGM) was created by the same seven organizations that had prepared the GUM (and the International Vocabulary of Basic and General Terms in Metrology, VIM). A further organization joined these seven international organizations, namely, the International Laboratory Accreditation Cooperation (ILAC). This Joint Committee had the tasks *"to promote the use of ... the GUM; to prepare supplemental guides for its broad application; and to revise and promote the use of ... the VIM. The JCGM has taken over responsibility for these two documents from ISO TAG 4, who originally published them."*[2]

JCGM has two Working Groups. Working Group 1, "Expression of Uncertainty in Measurement", has the task to promote the use of the GUM and to prepare supplements for its broad application. Working Group 2, "Working Group on International Vocabulary of Basic and General Terms in Metrology (VIM)", has the task to revise and promote the use of the VIM.

In this paper, the activity of JCGM-WG1 is outlined and the documents in preparation are described.

Keywords: probability density functions, coverage intervals, uncertainty matrix.

2 BACKGROUND

The GUM has been in use for more than a decade. During this period, its merits and drawbacks have been clearly identified.

The main merit is to have proposed a unified method for treating in a comprehensive and logically sound framework both systematic and random effects, thus obviating the confusion existing at the time of its publication. The confusion was due to the lack of a universally accepted method for combining uncertainties, which resulted in uncertainty statements which could hardly be compared.

The main drawbacks are:

a) The assumptions implicit in the method, although sufficiently weak, are not fulfilled in a number of practical cases. This is especially true for the procedure concerning the expanded uncertainty at a prescribed coverage probability.

b) The case, frequent in metrology, in which more than one measurand are estimated, is only addressed marginally and not covered to sufficient detail.

In addition, a certain inconsistency exists, inherent in the use in the same document of frequentist and Bayesian views of probability in the treatment of random and systematic effects, respectively.

To obviate these drawbacks, the JCGM/WG1 is preparing two specific supplements to the GUM addressing two specific cases:

a) When a coverage interval is required at a stipulated coverage probability.

b) When more than a measurand are involved in the measurement.

The first of the two documents is now at an advanced stage and should be issued by mid 2006. The second is at an earlier stage and should appear in 2007. This paper, while describing the two documents, outlines the techniques proposed to address the mentioned issues.

3 THE GUM: A BRIEF REMAINDER

3.1 Combined Standard Uncertainty

The GUM framework is based on the existence of a measurement model

$$Y = f(X_1, X_2, ... X_N) \tag{1}$$

relating the *measurand Y* to *N input quantities* X_i on which it depends. This framework implies that the measurand is not directly observed, but is indirectly obtained from model (1) through the available knowledge on the input quantities. In this respect, it is worth remarking that, in the GUM, each symbol (X_i or Y) in model (1) is used with two different meanings: the first refers to the "natural" quantities (physical, or chemical, or biological), which are considered as having unique, although unknown, values [GUM, 4.1.1]. The second refers to random variables used to represent knowledge on such quantities. In fact, random variables are characterized by probability distributions, or probability density functions (PDFs), which embody the available knowledge on the possible quantity values. In the GUM framework the available knowledge is summarized by input *estimates* (also called best estimates) x_i and their associated standard uncertainties $u(x_i)$. In terms of PDFs, the former are the best available values for the input quantities, i.e., the expectations of the PDFs, the latter are their standard deviations, evaluated by either statistical means (sample standard deviations, i.e., Type A evaluations) or other means (i.e., Type B evaluations).

The establishment of a model and the assignment of estimates and uncertainties to the input quantities can be viewed as the *formulation* stage. The propagation of the estimates and their uncertainties through the model to obtain the measurand estimate and its standard uncertainty is viewed as the *calculation* stage.

The measurand estimate y is thus obtained from

$$y = f(x_1, x_2, ... x_N), \tag{2}$$

that is, by evaluating model (1) at the input estimates or, with a different wording, by *propagating* the input estimates through the model. Accordingly, the estimate uncertainty $u(y)$ is obtained by propagating *in a suitable way* the input uncertainties through the model. To this purpose, the measurement model is approximated by the first-order term of a power series expansion. This approximation works because the uncertainties are usually small compared to the estimates.

Therefore, the combined standard uncertainty $u_c(y)^6$ for the measurand estimate y, given the standard uncertainties $u(x_i)$ of the input estimates x_i for the input quantities X_i is obtained from the well-known formula

$$u(y) = \sqrt{\sum_{i=1}^{N} \left(\frac{\partial f}{\partial X_i} \right)\Bigg|_{x_1, x_2, \ldots, x_n}^{2} u^2(x_i)} \quad , \tag{3}$$

valid for independent input quantities and for models affected by "reasonable" non-linearities in the input quantities. If (some of) these are non-independent, the relevant *covariance* (or *mutual uncertainty*) terms are to be taken into account in eq. (3) [GUM, 5.2.2]. If model (1) is significantly non-linear, higher-order terms are to be included [GUM, 5.1.2, NOTE]. However, these terms are not always easy to calculate, and anyway the involved input quantities must have Gaussian distributions and be independent. Even the first partial derivatives may be difficult to calculate, for example when the model is complicated.

However, in the case that the appropriate conditions are met, the GUM method provides a meaningful standard uncertainty $u(y)$ for the measurand estimate y.

3.2 Expanded Uncertainty

The standard uncertainty $u(y)$ obtained from eq. (3) is a meaningful measure of the amount of information on the measurand value. However, to many (perhaps most) end-users of the measurement process, it has little practical usefulness. For example, it is worth remembering that, although Recommendation 1 (CI-1986) of the CIPM requests that the standard uncertainty *"be used by all participants in giving the results of all international comparisons or other work done under the auspices of the CIPM and Comités Consultatifs"*, the Mutual Recognition Arrangement states that the degrees of equivalence be given at the 95% level of confidence.[3]

In the GUM itself it is recognized that *"...in some commercial, industrial, and regulatory applications, and when health and safety are concerned, it is often necessary to give a measure of uncertainty that may be expected to encompass a large fraction of the distribution of values that could reasonably be attributed to the measurand."* [GUM, 6.1.2].

This further measure of uncertainty for the measurand estimate y suggested in the GUM to fulfill these different needs [GUM, 2.3.5] is the well-known *expanded uncertainty* U. It can be obtained by multiplying the standard uncertainty $u(y)$ by an appropriate coverage factor k, typically such that $2 \leq k \leq 3$. However, there is no increase in knowledge unless the qualifier "large" is quantified, that is, unless the *coverage probability* p corresponding to U is known [GUM, 6.2.3].

An expanded uncertainty having a known coverage probability p is indicated by U_p. Its evaluation implies some knowledge about the PDF of the measurand. In the GUM it is suggested that in most cases, by virtue of the Central Limit Theorem, this can be approximated by a Gaussian, or, for finite degrees of freedom, by a scaled and shifted Student-t distribution. In this case, the Welch-Satterthwaite formula [GUM, eq. (G.2 b)] can be used to evaluate the effective degrees of freedom v_{eff} necessary to select the appropriate k factor. This formula is a sort of weighted mean of the degrees of freedom of

---

[6] Incidentally, the JCGM has decided that the qualifier "combined" and therefore the subscript "c" are superfluous and can be omitted.

the input contributions to uncertainty. Therefore, it is necessary to attach a degrees of freedom not only to Type A, but also to Type B evaluations. However, the notion of degrees of freedom is quite hard to associate to a subjective evaluation, despite the interpretation given in the GUM [GUM, G.4.2]. This is only one of the drawbacks inherent in the procedure. As a further issue, the Welch-Satterthwaite formula has been questioned.[4,5]

Furthermore, the conditions for the output distribution to be a Gaussian or a scaled and shifted Student-*t* distribution in practice are often not fulfilled, which limits the applicability of the procedure. For example, the input estimates may be correlated, or a non-Gaussian input component (say, a uniform distribution arising from a Type B evaluation) may be dominant. In this case the output distribution is trapezoidal-like, and therefore in general $k < 2$.

Last but not least, very little guidance is provided in the GUM on how to obtain an expanded uncertainty with asymmetric distributions [GUM, G.5.3].

4 THE GUM SUPPLEMENT 1: PROPAGATION OF DISTRIBUTIONS USING A MONTE CARLO METHOD

4.1 Formulation Stage

To obviate these difficulties, both practical and conceptual, the JCGM is preparing a Supplement to the GUM,[6] in which a method based on numerical simulation is used to construct an interval of confidence (or, better, a *coverage interval*). This method is more general than the GUM procedure and avoids the internal inconsistencies of the GUM.

In this framework, the probability distributions of the input variables, rather than the estimates and their uncertainties, play the central role. Specifically, the PDFs are assigned by the experimenter to describe his knowledge. In this respect, a common misunderstanding is that "true", unknown PDFs exist and that the experimenter has the task of estimating them. In a sense, this is a generalization of another misunderstanding, according to which one has to "estimate" uncertainties in order to approach as well as possible the unknown "true uncertainties". On this same line, it might be claimed (and has been in some quarters) that PDFs are more difficult to estimate than their standard deviations. These statements miss the genuine approach adopted in Supplement 1, as well as in the GUM. The PDFs (and the uncertainties in the GUM) are subjectively (not arbitrarily...) assigned by the experimenter, according to his views on the available information.

As the GUM provides guidance on the assignment of input standard uncertainties in most practical cases, so does Supplement 1 concerning the assignment of the input PDFs. In the latter case, some basic tools, namely, the Principle of Maximum Entropy and Bayes' theorem, give a powerful help, in the sense that the appropriate PDFs for the various situations arise naturally from their use. In this way, a reasonable harmonization can be expected, although a certain level of subjectivity is unavoidable [GUM 3.4.8]. What matters (and actually happens) is that subjectivity is confined to the formulation stage, whereas the calculation stage is, so to say, subjectivity-free.

4.2 Calculation Stage

Given the input PDFs (or the joint input PDF if the input quantities are not independent), the PDF for the measurand estimate can in principle be obtained analytically by using the theory of random variables. However, the calculations are difficult and in general cannot be carried out without resorting to numerical integration. Therefore, in the Supplement a different way is followed, that is, a numerical approximation to the distribution function for the output estimate is obtained by using the Monte Carlo method. From this numerical approximation the relevant quantities can be obtained, i.e., the expectation, the standard deviation and a coverage interval having a prescribed coverage probability.

The Monte Carlo method (MCM) is a tool which has proved its power in a wide range of applications. Since the method rests on the ability to automatically generate large samples of random numbers, it is ideally suited for computer implementation. There exists an immense literature on it and its applications.[7]

The implementation of the Monte Carlo Method for the propagation of PDFs is very intuitive and can be split in the following steps:

- For each of the N input quantities X_i, draw at random a value x_i among the possible values that the quantity can take, according to its PDF (or to the joint PDF of the input quantities). Such operation is known as *sampling from a PDF*, although the wording can be misleading.
- Use this first sample of N values $(x_1, \dots, x_N)$ (we will denote it as a $N \times 1$ column vector x_1) to evaluate model (1), thus obtaining a first possible measurand value $y_1 = f(x_1)$.
- Repeat M times the two steps above, thus obtaining M possible values y_r for the measurand $(r = 1, \dots, M)$.
- Use these M values to construct a numerical approximation $\hat{G}_Y(\eta)$ of the distribution function $G_Y(\eta)$ for the measurand.
- From $\hat{G}_Y(\eta)$ obtain the best estimate y and its standard uncertainty $u(y)$, as well as a suitable coverage interval at a stipulated coverage probability.

4.3 Comments

A number of issues in the preceding step-by-step procedure need to be addressed. First, the whole procedure rests, as already mentioned, on the ability to sample from a given PDF. The basis to do this is a generator of (pseudo) random numbers rectangularly distributed in the interval [0, 1]. Pseudo-random numbers distributed according to any probability distribution can be thus obtained by means of a suitable transformation algorithm based on a well-known theorem.[8] Supplement 1 gives detailed guidance on this issue, proposing some dedicated generators and tests for checking other generators the experimenter might choose.

The choice of the appropriate number M of Monte Carlo trials is another delicate point. A large number might involve an unacceptably long computing time, whereas a small one might provide unreliable results. Although the appropriate number is to some extent application-dependent, an adaptive procedure is suggested in Supplement 1 which automatically determines M according to the numerical tolerance the experimenter decides to accept.

The construction of the numerical approximation $\hat{G}_Y(\eta)$ of the distribution function $G_Y(\eta)$ requires a sorting of the M values y_r. In some circumstances (simple models and very large value for M, say 10^8 or more), this might result in heavier computation with respect to the construction of the numerical approximation $\hat{g}_Y(\eta)$ of the PDF $g_Y(\eta)$. However, the described procedure is normally preferred because the construction of a PDF requires a subjective decision concerning the cells widths.

The coverage interval at a stipulated probability level, the ultimate outcome of the whole procedure, requires detailed discussion. The output distribution in general is not symmetric. In these circumstances, the coverage interval is not unique, and the appropriate choice has to be made. Typically, a shortest or a probabilistically symmetric interval will be chosen, depending upon the specific application. These two intervals coincide for symmetric PDFs.

4.4 Comparing GUM and Supplement 1

The results provided by the procedure recommended in Supplement 1 (measurand estimate, standard uncertainty and coverage interval) are in general different from those obtained with the GUM procedure. This is due to several reasons.

First, the conditions under which the propagation of distributions is valid are by far more general than those in which the GUM framework is valid. Therefore, in case of discrepancy between the GUM results and the Supplement 1 results, the conclusion should be that the applicability conditions of the GUM framework do not hold and that the propagation of distributions should be preferred. Supplement 1 provides a careful analysis of the applicability conditions, as well as a comparison criterion for validation of the GUM results as concerns the endpoints of a coverage interval.

Second, even in those cases in which the conditions for the applicability of the GUM are met, the standard uncertainties provided by the two methods are unlikely to coincide. This happens because in Supplement 1 the PDF assigned to an input quantity X_i for which a sample of n indications is available is a scaled and shifted Student's t-distribution. This implies that the standard uncertainty $u(x_i)$ of the estimate x_i is $\sqrt{(n-1)/(n-3)}$ times larger than the corresponding standard uncertainty recommended by the GUM procedure.

This apparently minor feature has far-reaching implications. In Supplement 1 there is no longer a mixture of statistical and probabilistic concepts, as in the GUM, and the internal consistency of the procedure is improved with respect to the GUM procedure.
Since PDFs rather than standard uncertainties are propagated, the somewhat artificial classification in Type A and B evaluations does not apply any longer. Also the degrees of freedom for the input estimates, as well as the effective degrees of freedom for the output estimate, are no longer necessary. As a consequence, use of the Welch-Satterthwaite formula is avoided.

5 A CASE EXAMPLE

To demonstrate how the propagation of distributions compares to that of best estimates and standard uncertainties, we have applied both methods to a case example in the field of chemical analysis. This example concerns the evaluation of uncertainty in the preparation of a calibration standard for atomic absorption spectroscopy (AAS) from the corresponding high purity metal (in this example ≈ 1000 mg l^{-1} Cd in diluted HNO_3).[9] The analysis is carried out following both the GUM and the Supplement 1 procedures.

5.1 Model

The measurand is the concentration of the calibration standard solution, which depends upon the weighing of the high purity metal (Cd), its purity and the volume of the liquid in which it is dissolved. The concentration is given by

$$c_{Cd} = \frac{1000 \cdot P \cdot m}{V} \text{ mg l}^{-1} \tag{4}$$

where

c_{Cd} : concentration of the calibration standard $[\text{mg l}^{-1}]$
1000   : conversion factor from [ml] to [l]
P : purity of the metal given as mass fraction
m : mass of the high purity metal [mg]
V : volume of the liquid of the calibration standard [ml]

5.2 Assignment of the Input PDFs

We use here a specific Clause of Supplement 1 concerning the assignment of the appropriate PFD based on the available information about each input quantity in model (4). This information is taken from the relevant document.[8] Table 1 summarizes the assignments made.

Since the input variables are independent, a probability density function is assigned to each variable instead of a joint probability function, which would be appropriate for non-independent variables.

5.2.1 Purity P.   The purity P of the metal (Cd) is quoted in the supplier's certificate as (99.99 ± 0.01) %. P is therefore $0.999\,9 \pm 0.000\,1$.

Because there is no additional information about the uncertainty value, a rectangular distribution $R(a, b)$ is assumed for P with limits $a = 0.999\,8$ and $b = 1$. Hence, the best estimate for P is $x_P = 0.999\,9$ and the associated standard uncertainty $u(x_P)$ is given by

$$u(x_P) = \frac{0.000\,1}{\sqrt{3}} = 0.000\,058 \tag{5}$$

5.2.2 Mass m. The relevant mass m of cadmium is determined by a "tared weighing", giving the estimate $x_m = 0.100\,28$ g. The manufacturer identifies three uncertainty sources for the tared weighing: repeatability, readability (digital resolution) of the balance scale and uncertainty in the calibration function of the scale. The uncertainty associated with the mass of the cadmium is evaluated, using the data from the calibration certificate and the manufacturer's recommendations on uncertainty determination, as $u(x_m) = 0.05$ mg. This evaluation takes into account the three contributions earlier identified.

At this level of knowledge (neither the degrees of freedom for the stated standard uncertainty $u(x_m)$, nor a detailed model specifying how the three uncertainty contributions influence $u(x_m)$ are given), the appropriate probability distribution for m is a Gaussian $N(\mu, \sigma^2)$ with parameters $\mu = x_m = 100.28$ mg and $\sigma^2 = u(x_m)^2 = (0.05 \text{ mg})^2$.

Table 1 *Distributions for the input quantities of the considered model*

		Parameters			
Quantity	Distribution	μ	σ	a	b
P	$R(a, b)$			0.999 8	1
m	$N(\mu, \sigma^2)$	100.28 mg	0.05 mg		
V	$N(\mu, \sigma^2)$	100 ml	0.07 ml		

Probability distributions assigned to the input quantities P, m, V of model (4) on the basis of available information.

5.2.3 Volume V. The uncertainty of the volume V of the solution contained in the volumetric flask (having a nominal value $x_V = 100$ ml) has three major components: calibration, repeatability, and temperature effects. The three contributions are combined to give the standard uncertainty of the volume, leading to the value $u(x_V) = 0.07$ ml. Again, at this level of knowledge, a Gaussian probability distribution $N(\mu, \sigma^2)$ is assigned to V, having parameters $\mu = x_V = 100$ ml and $\sigma^2 = u(x_V)^2 = (0.07 \text{ ml})^2$.

5.3 Calculation applying the GUM Uncertainty Framework

The application of the GUM uncertainty framework (GUF) is based on a first-order Taylor series approximation to model (4). Model and derivative evaluations are made using exact derivatives, calculated at the best estimates x_P, x_m and x_V of the input quantities. Since the input quantities are independent and the purity P, which is the only non-normally distributed (i.e., having a rectangular distribution) has the smallest uncertainty, the conditions of validity of the Central Limit Theorem are fulfilled. Therefore, the output distribution for the estimate y of the measurand c_{Cd} can be considered as a Gaussian distribution with mean equal to y and standard deviation equal to the associated standard uncertainty $u(y)$. As a consequence, the expanded uncertainty U can be obtained by multiplying the standard uncertainty by a factor $k = 2$. The expanded uncertainty U represents the half-width of a 95% coverage interval symmetric about the output estimate.

The results are given in the first line of the first four columns of table 2. Three decimal digits are reported for comparison purposes only.

5.4 Calculation applying the Monte Carlo Method

We have treated the same example using an implementation of the Monte Carlo Method described in Supplement 1 and summarized in par. 4.2 of this paper. Using the information given in table 1 and running $M = 10^5$ Monte Carlo trials yields $y = 1002.704$ mg l^{-1}, where y is the average of the M outcomes y_r resulting from the evaluation of model (4) for the M input vectors $x_r = [x_P, x_m, x_V]_r^T$. The associated uncertainty $u(y)$, obtained as the standard deviation of the M output values y_r is $u(y) = 0.862$ mg l^{-1} and the coverage interval is $[1001.020, 1004.387]$ mg l^{-1}. This is obtained as the shortest 95 % coverage interval for the output distribution for y. Two further Monte Carlo runs, with $M = 10^6$, confirm the estimate and its standard deviation to the first two decimal digits. Table 2 lists the results in the first four columns. Three decimal digits are reported to highlight the variability in the results.

Table 2: *Results from applying GUF and MCM to the considered model*

Method	y /mg l^{-1}	$u(y)$ /mg l^{-1}	Shortest 95 % coverage interval /mg l^{-1}	$\delta = 0.005$		Is GUF validated?	$\delta = 0.05$		Is GUF validated?
				d_{low}	d_{high}		d_{low}	d_{high}	
GUF	1002.700	0.864	[1000.972, 1004.428]						
MCS $M = 10^5$	1002.704	0.862	[1001.020, 1004.387]	0.048	0.041	No	0.04	0.03	Yes
MCS $M = 10^6$	1002.701	0.863	[1001.011, 1004.391]	0.039	0.037	No	0.03	0.03	Yes
MCS $M = 10^6$	1002.700	0.864	[1001.008, 1004.392]	0.036	0.036	No	0.03	0.03	Yes

Application to model (4) of the GUM uncertainty framework (GUF) and the Monte Carlo Method (MCM), and corresponding comparison determining whether the coverage intervals obtained by the GUF and the MCM agree to within a stipulated numerical tolerance δ.

5.5 Comparing the two Methods

We have compared the GUM and Monte Carlo methods, according to the validation procedure suggested in Supplement 1. The comparison procedure determines whether the coverage intervals obtained by the GUM uncertainty framework and MCM agree to within a stipulated numerical tolerance δ. This numerical tolerance is related to what is regarded as a meaningful number of significant decimal digits in the expression of the standard uncertainty $u(y)$. The endpoints of the coverage interval obtained by the GUM uncertainty framework are compared with those provided by MCM. If both the absolute values d_{low} and d_{high} of the differences of the corresponding endpoints of the two coverage intervals are not larger than δ, then the comparison is favourable and the GUM uncertainty framework is considered validated in this instance.

In the specific example, two possible numerical tolerances, $\delta = 0.005$ and $\delta = 0.05$ were considered. These are obtained regarding as meaningful two or one significant digits in the uncertainty, respectively. Table 2 shows the absolute differences d_{low} and d_{high} between the endpoints of the three 95% coverage intervals obtained with MCM and the one obtained with GUM, for the two tolerances considered. For $\delta = 0.005$ the GUM uncertainty framework is not validated, since all the absolute differences d_{low} and d_{high} are larger than the tolerance. For $\delta = 0.05$ (that is, regarding as meaningful just one significant digit in the uncertainty) the coverage intervals obtained with the GUM procedure and MCM agree within the stipulated tolerance, and the GUM can be considered validated.

6 SUPPLEMENT 2: MODELS WITH ANY NUMBER OF OUTPUT QUANTITIES

This document is at an earlier stage of preparation.[10] Accordingly, only a brief summary of its content is given here.

In many measurement applications several measurands Y_i depend on a common set of input quantities X_i. The GUM is not very informative on this issue [GUM, 3.1.7], the only hint given there being that the scalar measurand and its variance are replaced by a vector measurand and its covariance matrix. However, the situation occurs frequently. Just as an example, this is the case of a set of mass standards, calibrated by subdivision from a single reference kilogram standard using a common set of balances.

In these cases, matrix notation is convenient. With this notation, the measurand is a vector $Y_{(P \times 1)}$, function of the input quantities $X_{(N \times 1)}$ according to

$$Y = f(X). \tag{6}$$

More complicated models are also encountered, especially in electrical metrology. These may involve complex input/output quantities, or may be implicit, that is, of the form

$$f(X,Y) = 0. \tag{7}$$

Different *multivariate* models can be classified according to their level of complexity. In Supplement 2, an exhaustive classification is given of the various possible models. This classification is useful for the appropriate solution. To give an example, from this broader viewpoint, the particular case covered in the GUM is the so-called univariate explicit model, which can be represented by

$$Y_{(1 \times 1)} = f(X_{(N \times 1)}). \tag{8}$$

A common feature to all multivariate models is that the uncertainty of the output estimate is no longer a number (a scalar), since the estimate itself is no longer a scalar.

Therefore, in the multivariate case a complete uncertainty description for P output estimates y_j implies evaluation of P standard uncertainties and $P(P-1)/2$ covariances from N input standard uncertainties and $N(N-1)/2$ input covariances. The appropriate mathematical tool for a concise description of the output uncertainty is the covariance matrix, or *uncertainty matrix* U_y, which is a straightforward generalization of the standard uncertainty $u(y)$.

To proceed in the generalization, a GUM-like solution for evaluating U_y is a first-order Taylor expansion of the measurement model which, under conditions similar to those holding for the scalar case, allows to propagate through the model the input uncertainty matrix U_x in order to obtain U_y.

Supplement 2 will give guidance on the appropriate solution for the various classes of general models, with examples taken from metrological practice.

A second issue is that also in the multivariate case it might be preferable, in some circumstances, to adopt a procedure able to obtain the joint PDF for the output vector by propagating through the model the joint PDF of the input vector. Supplement 2 will also describe an implementation of MCM for the multivariate case.

Finally, the problem of the construction of a multivariate region of confidence will be addressed.

7 FURTHER DOCUMENTS

The JCGM-WG1 has planned a number of further documents, addressing various topics of metrological interest.[2] These include the use of least-squares, the difficult art of suitably modelling an experiment, the use of uncertainties in conformity assessment and others.

References

1 BIPM, IEC, IFCC, ISO, IUPAC, IUPAP, and OIML. *Guide to the Expression of Uncertainty in Measurement*, 1995. ISBN 92-67-10188-9, 2nd Edn.

2 http://www1.bipm.org/en/committees/jc/jcgm

3 http://www1.bipm.org/en/cipm-mra/mra_online.html

4 B. D. Hall and R. Willink, *Metrologia*, 2001, **38**(1), 9.

5 M. Ballico, *Metrologia*, 2000, **37**(1), 61.

6 BIPM, IEC, IFCC, ILAC, ISO, IUPAC, IUPAP, OIML, *Guide to the Expression of Uncertainty in Measurement, Supplement 1- Propagation of distributions using a Monte Carlo method*, to be published.

7 C. P. Robert and G. Casella, *Monte Carlo Statistical Methods*, Springer-Verlag, New York 1999.

8 J. D. Gibbons and S. Chakraborti, 3rd Edn., *Nonparametric Statistical Inference*, Marcel Dekker, New York, 1992, ch. 2, p. 26.

9 EURACHEM / CITAC Guide CG 4. *Quantifying Uncertainty in Analytical Measurement*, 2000. ISBN 0-948926-15-5, 2nd Edn.

10 BIPM, IEC, IFCC, ILAC, ISO, IUPAC, IUPAP, OIML, *Guide to the Expression of Uncertainty in Measurement, Supplement 2- Models with more than one output quantity*, in preparation.

HOW TO COMBINE RESULTS HAVING STATED UNCERTAINTIES: TO MU OR NOT TO MU?

David L. Duewer

Analytical Chemistry Division, Chemical Science and Technology Laboratory, National Institute of Standards and Technology, Gaithersburg, Maryland 20899-8390 USA

1 INTRODUCTION

It is widely accepted that knowledge of the expected uncertainty of a measured value is necessary for judging the measurement's fitness-for-purpose.[1] Considerable effort continues to be expended on how to best estimate and report measurement uncertainty (MU) for chemical and biological measurands.[2-4] Several academic research groups are actively exploring ways of using MU estimates in multivariate data analysis.[5,6] It is therefore perhaps surprising that some experienced data analysts argue against using MU in even relatively simple univariate applications such as using interlaboratory data to assign the expected value of a measurand in particular materials.[7,8,9] But if MU is not expected to be useful for univariate tasks, can it be useful in multivariate applications? More generally, can estimates of MU provide quantitatively useful chemical information when there may be significant bias among the data analyzed, such as when measurements are from more than one source or are taken over a long period of time?

The primary concern of those who advise against using MU in establishing consensus values is that estimating MU is not yet sufficiently routine and standardized to ensure that different participants interpret and report the uncertainties in their values in a consistent manner. Any enquiry into the potential utility of MU for assigning consensus values to interlaboratory study materials therefore requires, at a minimum, that there be philosophically consistent MU estimates for all of the reported measurements and, to enable valid comparison, trustworthy reference values (RVs) for the studied measurands. Few, if any, "routine" interlaboratory studies currently meet this standard. Indeed, reliably estimating even short-term measurement precision from studies involving well-motivated but metrologically naïve participants is challenging.[10]

Beginning in the 1993, a number of very non-routine interlaboratory studies involving various chemical measurands have been performed under the auspices of the Comité Consultatif pour la Quantité de Matière (CCQM) of the Comité International des Poids et Mesures (CIPM). The number and diversity of these studies dramatically increased with the signing in late 1999 of the Mutual Recognition Arrangement (CIPM MRA) for national measurement standards and for calibration and measurement certificates issued by national metrology institutes (NMIs). Currently, data are publicly available for 150 measurands evaluated in 33 studies performed by the member organizations of the CCQM. These organizations have considerable expertise in chemical metrology and experience with MU evaluation. All of the samples used in the CCQM studies are thoroughly characterized for homogeneity and stability. Virtually all of the individual values are reported with a symmetrical uncertainty estimate of defined confidence. The source of every datum in these studies is fully attributed. Most importantly, reliable RVs – assigned either by gravimetric sample preparation or extensively debated expert review – are available for all measurands.

This report describes the use of these publicly available CCQM data to explore whether appropriately evaluated MU estimates can be useful in combining results from interlaboratory studies. Three different approaches to using MU are evaluated (weighting, bootstrap resampling, and mixture-models) relative to the performance of a number of commonly used or recently proposed evaluation metrics.

2 DATA

2.1 CCQM Key Comparisons and Pilot Studies

As of February 2006, data are publicly available from the 23 CCQM Key Comparisons (KCs) and 10 CCQM pilot studies listed in Table 1.[11,12] KCs are formal studies designed to establish the extent of measurement agreement among the participating CCQM member organizations and their official delegates for particular measurands. All of the measurement values submitted for a particular KC are made public with the formal acceptance of the study's Final Report. Depending on the type of sample and measurands investigated, some form of RV is established for all measurands. Many of these RVs are assigned through gravimetric preparation verified by measurement, others are assigned through a combination of consensus technical judgment and statistical analysis. No participant's KC result can be withdrawn from publication once it has been submitted; however, results that are technically suspect are generally not used in assigning RVs.

Pilot studies are relatively less formal investigations that may also involve expert but non-CCQM member participants. Pilot studies often are designed to investigate technical challenges in particular measurement processes or evaluate the relative metrological utility of alternate measurement technologies. They are sometimes conducted in parallel with KCs, using the same sample materials, timing, and shipping protocols, to allow interested parties to evaluate their measurement capabilities without the burden of public participation. While all pilot study results are available to all of the participants in the study, every participant must agree to publication for attributed study results to be made public. In addition to such separate publication, data for a few pilot studies have been published in the Final Report of a follow-on KC addressing similar measurands. While pilot studies are typically not as thoroughly catechized as are KCs, well-evaluated RVs are available for all of the measurands in the published pilot studies.

Table 1 lists the designation code ("Study"), a general description of the measurands ("Measurand(s)"), the number of participants in each study ("#Labs"), the total number of different measurands reported in the study ("#Sets"), and whether the RVs are assigned through gravimetric preparation of the sample materials. The number of participants is listed as a range for some studies that address multiple analytes, multiple samples, and/or different measurement conditions. In all, a total of 1294 individual measurement values are available.

Table 1 *CCQM Key Comparison and Pilot Studies Publicly Available as of 1-Feb-2006*

Study	Measurand(s)	#Labs	#Sets	Grav?
K01	Trace gases in N_2, components of natural gas	8 - 10	28	All
K02	Metals in natural water	9	2	
K03	Gases in auto emissions	13	3	All
K04	Ethanol in air	8	1	All
K05	pp'-DDE in fish oil	10	2	
K06	Cholesterol in serum	7	2	
K07	Volatile organic compounds in N_2	8	5	All
K08	Metal calibration solutions	12 - 13	4	All
K09	pH of phosphate buffer	4 - 10	20	
K10	Volatile organic compounds in N_2	8	3	All
K11	Total glucose in human serum	3	1	
K12	Creatinine in human serum	5	2	
K13	Metals in sediment	14	2	
K14	Calcium in serum	9	1	
K16	Components of natural gas	7 - 9	23	All
K17	pH of phthalate buffer	11	3	
K21	pp'-DDT in fish oil	8 - 9	2	
K24	Cadmium in rice	18	1	
K25	PCBs in sediment	7 - 9	5	
K27	Ethanol in aqueous matrix	2 - 9	7	6 of 7
K28	Tributyltin in sediment	10	2	
K31	Arsenic in shellfish	7	1	
K43	Metals in salmon	5 - 9	5	
P06	Cholesterol in serum	7	2	
P08	Glucose in serum	4	2	
P09	Creatinine in serum	5	2	
P13	Metals in artificial food digest	10 - 12	3	All
P17	PCBs in sediment	8 - 10	4	
P18	Tributyltin in sediment	10 - 12	2	
P29	Zinc in rice	14	1	
P32	Anion calibration solutions	9 - 11	2	All
P39	Metal in tuna fish	8 - 15	5	
P43	Dibutyltin in sediment	11	2	

2.2 Participants

Only signatories of the CIPM MRA and their official designees can participate in KC studies. Therefore, more than 95 % (1230 of the 1294 total) of the publicly available CCQM data are from the 31 national or international organizations listed in Table 2. The remaining 62 pilot study measurements are reported by 17 different academic, commercial, or governmental laboratories that have interest and expertise in the particular measurement systems studied.

2.3 Measurements

Measurement results for all KCs must be reported as values with an associated MU of defined confidence. For all of the data in the published CCQM studies, the confidence intervals are symmetric about the expected value and are expanded with the intent to include the true value with about 95 % confidence. The data for the ith participant in the jth data set can thus be denoted: $x_{ij} \pm U_{95}(x_{ij})$. These data are typically displayed using standard "dot-and-bar" graphs such as that shown in Figure 1.

Table 2 *NMI Participation in the Public CCQM Key Comparison and Pilot Studies*

Organization(s)	Representing	#Values
BAM, PTB, UBA	Germany	144
NIST	USA	122
VNIIM, VNIIFTRI	Russia	114
NMIJ	Japan	106
KRISS	South Korea	104
LGC, NPL	United Kingdom	96
NRCCRM	PR China	84
NMi-VSL	The Netherlands	76
LNE	France	70
NMIA	Australia	58
GUM	Poland	46
SMU	Slovak Republic	40
OMH	Hungary	35
IRMM	European Union	30
NRC	Canada	30
CENAM	Mexico	27
DPL, Radiometer	Denmark	12
EMPA, METAS	Switzerland	10
CSIR-NML	South Africa	8
IAEA	Global	5
IMGC	Italy	5
NPLI	India	4
CMI	Czech Republic	3
INMETRO	Brazil	2
SP	Sweden	1
		1232

Since all of the MUs in these CCQM studies are expressed as symmetric 95 % confidence intervals, the CCQM measurements can also be interpreted as normal kernel densities with expected value x_{ij} and standard deviation proportional to $U_{95}(x_{ij})$ but covering the true value with about 68 % confidence, $U_{68}(x_{ij})$.[9] For normal distributions

$$U_{68}(x_{ij}) \cong U_{95}(x_{ij})/2 . \qquad (1)$$

Just as each measurement kernel, $N(x_{ij}, U_{68}(x_{ij}))$, represents the expected probability density function (PDF) of the true location given the measurement, the sum of the n_j kernels defines a mixture-model PDF (MM-PDF) for the combined set of measurements[9]

$$\text{PDF}_j = \sum_{i}^{n_j} N\left(x_{ij}, U_{68}(x_{ij})\right) \Big/ n_j . \qquad (2)$$

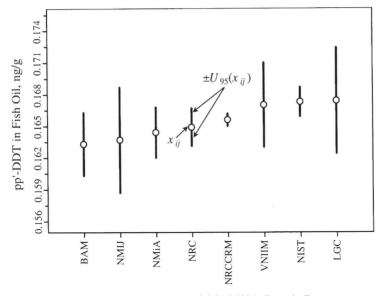

Figure 1 *Measurements with Stated Uncertainties as Dot-and-Bars, $x_{ij} \pm U_{95}(x_{ij})$*

This alternate kernel density and PDF representation of the CCQM measurements is displayed in Figure 2. The PDF_j at the right of the graph is a marginal distribution for the combined results.

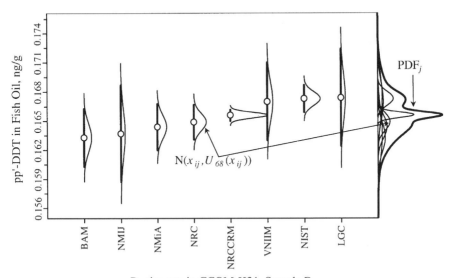

Figure 2 *Measurements with Stated Uncertainties as Kernel Densities, $N(x_{ij}, U_{68}(x_{ij}))$*

3 LOCATION METRICS

There are a very large number of summary statistics that are routinely used or have been proposed for estimating a measurand's "true value" from interlaboratory study results. Each of these location metrics makes somewhat different assumptions about the nature of the data being combined and summarized. The extent of agreement between the estimated location and "truth" depends upon how well the observed data match these underlying assumptions. Only a few, hopefully representative, metrics are examined in this report. Graphical representations of the philosophical basis for most of these metrics are presented elsewhere.[13]

Location metrics are often examined with respect to their *efficiency* (how rapidly the estimate converges upon the true value as the number of observations increases, assuming that all data accord with the metrics assumptions) and their *robustness* (how insensitive the metrics are to violations of the assumptions).[14] While there are a number of ways to quantitatively evaluate these properties for specific situations, in this report these two properties are only qualitatively discussed.

3.1 "Traditional" Metrics That Do Not Utilize Measurement Uncertainties

3.1.1 Mean. Used in conjunction with some form of "outlier" rejection, the mean or arithmetic average is the most commonly used location metric. The mean is readily calculated in the familiar closed form

$$\text{Mean}_j = \sum_i^{n_j} x_{ij}/n_j \ . \tag{3}$$

where n_j is the number of results to be combined and summarized. When all of the results are truly random draws from a population that is normally distributed with location μ and standard deviation σ, denoted $N(\mu,\sigma)$, then the mean is the most efficient of the location estimates. The expected difference between the mean of the sample and μ declines as the square-root of the number of observations increases. The mean is fairly insensitive to the detailed shape of distribution as long as the observed results are symmetrically distributed about μ. However, the mean is not robust to the presence of unrecognized "outlier" results or asymmetric sampling of the underlying population. Recognizing this sensitivity, the mean is generally used to summarize results after some form of "outlier" data rejection has been used to identify and remove suspect data.

3.1.2 Median. The median is the value that is in the center of a given set of results. It is the most common "robust" location metric and has been recommended for use with KC data.[15,16] The median is readily calculated from rank-ordered data

$$\text{Median}_j = x_{kj}, k = \text{int}(n_j/2)+1 \text{ when } n_j \text{ is odd}$$
$$= (x_{kj} + x_{k+1j})/2, k = \text{int}(n_j/2) \text{ when } n_j \text{ is even} \ . \tag{4}$$

Like the mean, the median assumes that all observed results (including outlier values) are symmetrically distributed about the μ of the majority population. The median is much less sensitive than the mean to the presence of a modest number of outliers. For truly normal data without outliers, the median is about 20 % less efficient than the mean.[14]

3.1.3 Shorth. The shorth ("shortest half") is the value that is in the center of the most compact half of the reported results. It is the univariate analogue of the multivariate minimum volume ellipsoid estimator and has been used in at least one non-chemical

KC.[17,18] The value of the shorth is generally similar to the median, except that it does not require that outliers are symmetrically distributed about the μ of the majority population and it is considerably less efficient than the median when outliers are either not present or are symmetrically distributed. Although it has no closed form, the shorth is readily calculated from rank-ordered data

$$Shorth_j = \frac{x_{k,j} + x_{k+m,j}}{2} \tag{5}$$

where m is $n_j/2$ when n_j is even and $int(n_j/2)+1$ when n_j is odd and k is the index in the range 1 to n_j-m that minimizes $x_{k+m,j} - x_{k,j}$. For some data, two or more values for k may yield the same minimum difference. In such cases, a "grand shorth" can be defined as the average of the individual estimates.

3.1.4 A15. The A15 is a location metric that achieves robustness by truncating "outlier" values to be no more than a set distance from the central location. It assumes that outliers are symmetrically distributed about the μ of the majority population. It has essentially the same efficiency as the mean in the absence of outliers. The A15 does not have a closed form but is readily evaluated via iteration using the rule

$$A15_j = \sum_i^{n_j} \frac{z_{ij}}{n_j} \tag{6}$$

$$z_{ij} = x_{ij} \; if \; |x_{ij} - A15_j| < c * MADe_j \; else \; z_{ij} = x_{ij} + sign(x_{ij} - A15_j) * c * MADe_j$$

where $MADe_j$ is a robust estimate of the standard deviation of the majority population calculated from the median absolute deviation of the x_{ij} from the median and c is an empirical constant (generally set to 1.5) that defines where "outlierness" begins in terms of number of $MADe_j$. Freeware spreadsheet software is available for the A15 estimate.[19]

3.1.5 H15. The H15 (also known as Huber's Estimate 2) is very similar to the A15 with the exception that the standard deviation of the majority population is iteratively estimated simultaneously with the H15. The H15 thus makes more complete use of the data at the expense of somewhat greater computational complexity. The H15 has been strongly recommended for use with interlaboratory results.[20,21] The iterative solution for H15 (and its associated standard deviation) is based on

$$H15_j = \sum_i^{n_j} z_{ij}/n_j$$

$$z_{ij} = x_{ij} \; if \; |x_{ij} - H15_j| < c * s_j \; else \; z_{ij} = x_{ij} + sign(x_{ij} - H15_j) * c * s_j \tag{7}$$

$$s_j = \beta \sqrt{\sum_i^{n_j} \frac{(z_{ij} - H15_j)^2}{n_j - 1}}$$

where β is a function of c that adjusts for the distortion introduced by the data truncation. Freeware spreadsheet software is available for the H15 estimate.[19]

3.1.6 L1½. The L1½ is a member of the class of "least power" location metrics recently proposed for use with KC and other interlaboratory data.[1] Given that the mean is the L2 metric and (when n_j is odd) the median is the L1 metric, L1½ has properties roughly "mid way" between the mean and the median. While not having a closed form, L1½ is readily solved by direct minimization

$$L_{1\frac{1}{2}j} = y \text{ where } \left(\sum_{i}^{n_j} |x_{ij} - y|^{\frac{3}{2}} \right)^{\frac{2}{3}} \text{ has minimum value .} \tag{8}$$

3.2 Uncertainty Weighting

The most common use of reported MU is in a weighting function that gives each of the x_{ij} more or less influence in a particular calculation based at least partly on the $U_{68}(x_{ij})$. Many different weighting functions have been proposed for use with interlaboratory studies. The following two variants have been used in the analysis of CCQM KCs.

3.2.1 WtU. The most commonly used weighted location metric is the simple "inverse-square" weighted mean. It is easily evaluated in closed form

$$\text{WtU}_j = \sum_{i}^{n_j} (w_{ij} x_{ij}) \Big/ \sum_{i}^{n_j} w_{ij}; \quad w_{ij} = \frac{1}{U_{68}^2(x_{ij})} . \tag{9}$$

As a given $U_{68}(x_{ij})$ becomes relatively small, regardless of the x_{ij} value, the x_{ij} becomes relatively influential; as the $U_{68}(x_{ij})$ becomes relatively large, the x_{ij} becomes relatively inconsequential. The robustness of the WtU thus depends upon there being a strong positive association between "outlier" x_{ij} and relatively large $U_{68}(x_{ij})$. When all of the $U_{68}(x_{ij})$ are equal, the WtU reduces to the simple mean and will have the same efficiency as the mean. When the $U_{68}(x_{ij})$ are not equal, the efficiency of the WtU is related in a complex fashion to the veracity of the $U_{68}(x_{ij})$.

3.2.2 WtMP. Mandel-Paule weighting is an early version of what is now a large group of weighted location metrics that enforce a limit to the influence of unrealistically small $U_{68}(x_{ij})$.[23] It has recently been shown to be an approximate maximum likelihood location estimate.[24] The WtMP does not have a closed form but is solved numerically by iteration

$$\text{WtMP}_j = \sum_{i}^{n_j} (w_{ij} x_{ij}) \Big/ \sum_{i}^{n_j} w_{ij}$$

$$w_{ij} = \frac{1}{U_{68}^2(x_{ij}) + s_j} \quad \text{where} \quad \sum_{i}^{n_j} \frac{(x_{ij} - \text{WtMP}_j)^2}{U_{68}^2(x_{ij}) + s_j} = n_j - 1 \tag{10}$$

where s_j can be interpreted as an estimate of the standard deviation of the combined data. The WtMP is about equally influenced by all x_{ij} with $U_{68}(x_{ij})$ smaller than s_j while being relatively little influenced by x_{ij} with $U_{68}(x_{ij})$ much bigger than s_j. Like the WtU, WtMP reduces to the mean when all of the $U_{68}(x_{ij})$ are equal and both the efficiency and robustness are related to the veracity of the $U_{68}(x_{ij})$ when the $U_{68}(x_{ij})$ are not equal.

3.3 Kernel Density Bootstrap Resampling

Bootstrap resampling has become a widely used method for estimating the value and variability of summary estimates when the observed data are not well approximated as belonging to one of the well-studied PDFs.[25] Bootstrap estimates are derived by analyzing a large number of "pseudo" datasets that are constructed by randomly and repeatedly drawing values from the observed data. What constitutes a "large number" is situation-specific, but is seldom less than a few hundred and may run into the millions. A "total median" location metric based upon the bootstrap concept has been proposed for use with KC data.[8]

A modified bootstrap method that uses MU has been used with multi-source chemical data; rather than resampling just from the x_{ij}, the pseudo-data are obtained by random sampling from the $N(x_{ij}, U_{68}(x_{ij}))$ kernel densities.[26] That is, each resampling is accomplished by generating a random but characteristic value from a randomly chosen kernel. This kernel-density bootstrap reduces to the traditional bootstrap when the $U_{68}(x_{ij})$ values are asserted to be zero.

Since pseudo-data sets are randomly generated, bootstrap calculations are never "exact". Two successive bootstrap estimates of the same value for the same data are likely to differ slightly, with the extent of agreement expected to improve as the number of pseudo-data sets evaluated is increased. There are also many different ways to summarize bootstrap results. While any or all of the traditional location metrics could be evaluated using the kernel-density bootstrap, only two such metrics are examined here.

3.3.1 BSmean. The BSmean is defined, for this study, as the mean of bootstrap means for pseudo-data sets of size n_j

$$\text{BSmean}_j = \frac{\sum_1^{NBS}\left(\sum_{i=1}^{n_j} R\left(x_{ij}, U_{68}\left(x_{ij}\right)\right)\Big/n_j\right)}{NBS} \tag{11}$$

where $R(x_{ij}, U_{68}(x_{ij}))$ is a random number drawn from a normal distribution having mean x_{ij} and standard deviation $U_{68}(x_{ij})$ and NBS is the number of pseudo-data sets generated. For this report, $NBS = 10,000$.

3.3.2 BSmedian. BSmedian is defined, for this study, as the median of bootstrap medians for pseudo-data sets of size n_j

$$\text{BSmedian}_j = \text{Median}\Big|_1^{NBS}\left(\text{Median}\Big|_{i=1}^{n_j} R\left(x_{ij}, U_{68}\left(x_{ij}\right)\right)\right). \tag{12}$$

The same 10000 pseudo-data sets generated for the BSmean were used to estimate BSmedian.

3.4 Mixture Model Probability Density Functions

The MM-PDF of a combined set of results represents "all" of the information provided by the reported expected values and their associated MUs. Visualizing the MM-PDF at the margin of an otherwise routine dot-and-bar graph helps to establish the shape and dispersion of the combined results (Figure 2). MM-PDFs have been proposed as a way to empirically establish 95 % confidence intervals on traditional location estimates for KC results.[9] Several different MM-PDF metrics have been proposed for use with CCQM KC data.[13] Unlike the kernel-density bootstrap estimates, the proposed MM-PDF metrics are in principle "exact" for a given set of combined results.

3.4.1 MMmode. The MMmode is a direct analogue of the mode, the most common value in a set of discrete values. It is the location at which the MM-PDF has greatest density; i.e., where the largest value is:

$$\text{MMmode}_j = y \text{ where } \sum_{i=1}^{n_j} \varphi\left(y, x_{ij}, U_{68}\left(x_{ij}\right)\right) \text{ has maximum value .} \tag{13}$$

In practice, the MMmode is interpolated from the MM-PDF as evaluated on a reasonably dense, equally-spaced grid.

3.4.2 MMmedian. The MMmedian is a direct analogue of the median. It is the location which divides the MM-PDF into two sections of equal area.

$$\text{MMmedian}_j = Y \text{ where } \frac{\sum\limits_{i=1}^{n_j} \int\limits_{-\infty}^{Y} \varphi\left(y, x_{ij}, U_{68}\left(x_{ij}\right)\right) dy}{n_j} = 0.5 \tag{14}$$

where $\varphi(y, x_{ij}, U_{68}(x_{ij}))$ is the probability density at y for a $N(x_{ij}, U_{68}(x_{ij}))$ kernel of unit area. In practice, the MMmedian is interpolated from the cumulative area of the MM-PDF as evaluated on a reasonably dense, equally-spaced grid.

3.4.3 MMsh/mid. The MMsh/mid is a direct analogue of the shorth (described above). It is the half-range location between the endpoints, YL and YH, of the shortest interval that contains one-half of the MM-PDF area

$$\text{MMsh/mid}_j = \frac{YL + YH}{2} \text{ for } \min(YH - YL) \text{ where } \frac{\sum\limits_{i=1}^{n_j} \int\limits_{YL}^{YH} \varphi\left(y, x_{ij}, U_{68}\left(x_{ij}\right)\right) dy}{n_j} = 0.5 . \tag{15}$$

Like the MMmedian, these endpoints are in practice interpolated from the cumulative area of the MM-PDF as evaluated on a reasonably dense, equally-spaced grid. As with the shorth, it is possible that two or more intervals may be equally compact and may, in principle, be similarly made unique by averaging the multiple solutions.

3.4.4 MMsh/med. The MMsh/med is very similar to the MMsh/mid but makes more complete use of the MM-PDF by finding the location of the half-*area* between the YL and YH endpoints

$$\text{MMsh/med}_j = Y \text{ where } \frac{\sum\limits_{i=1}^{n_j} \int\limits_{YL}^{Y} \varphi\left(y, x_{ij}, U_{68}\left(x_{ij}\right)\right) dy}{n_j} = 0.25 . \tag{16}$$

5 RESULTS AND DISCUSSION

5.1 Uncertainty Versus Bias

Weighting values (assigning more or less influence) on the basis of the magnitude of the associated MU implicitly asserts that the smaller the MU, the better the data; that is, that MU and measurement bias are positively correlated. Assuming that (1) the RVs stated in the Final Reports of the publicly available CCQM data sets are reasonable approximations to a "true" value for the measurands and (2) that all of the reported measurements can be validly interpreted as $N(x_{ij}, U_{68}(x_{ij}))$ kernels, it is possible to test this assertion.

The magnitude of measurement bias of a particular result is estimated as the absolute value of the difference between the expected result and the RV

$$\text{Bias}_{ij} = \left|x_{ij} - \text{RV}_j\right| \tag{17}$$

where RV_j is the assigned RV for the jth measurand. While these bias estimates can be directly compared to the $U_{68}(x_{ij})$ within each data set, none of the CCQM data sets are sufficiently large to make such comparisons very informative. However, it is possible to compare the bias and MU estimates for all data sets at the same time by normalizing the bias and MU estimates to have a common scale

$$\text{NormalizedBias}_{ij} = \frac{\left|x_{ij} - RV_j\right|}{s_j}; \quad \text{NormalizedMU}_{ij} = \frac{U_{68}\left(x_{ij}\right)}{s_j} \tag{18}$$

where s_j is a measure of the dispersion of the dataset. While there are many possible ways to estimate s_j, for this study it is defined from the central 50 % of the area of each PDF_j.[13] This and related PDF-based dispersion metrics estimate the total variance of the combined data.[9]

Figure 3 displays all 1294 of the normalized {Bias, MU} pairs for the currently available CCQM KC and pilot studies. The four symbols code the measurand type: • denotes pH, □ gases, + organic, and × inorganic. All values larger than 4.0 s_j are plotted at the right or top margins. There is little if any evidence for any strong relationship between bias and MU in these data.

The data do, however, speak to the veracity of the CCQM MU estimates. The vertical line at normalized bias of 2.0 s_j is a routine threshold for dividing values that are apparently unbiased (to the left of the line) from those that may be (to the right): by this metric, over 90 % of the CCQM measurements agree well with the assigned RV. The thick line of slope 0.5 is the threshold separating measurements that include the RV within their 95 % confidence (above the line) from those that do not (below): over 72 % of the CCQM measurements reported MU at least as large as appropriate. The thin line of slope 0.5 is an approximate division between apparently complete (below) and potentially over-estimated (above) MU for the relatively unbiased measurements: by this metric, MU is over-estimated for only about 7 % of the measurements. In contrast, there is evidence for significant but unrecognized bias in about 8 % of all measurements.

5.2 Reference Values Vs Location Metrics

The majority of the CCQM measurement values agree well with the assigned RVs and most MU estimates appear realistic. However, given the number and diversity of the measurands, the modest number of "outlier" values (the 10 % relatively biased), and both the apparently under- (28 %) and over-estimated (7 %) "outlier" MUs, these data should challenge any metric used to summarize the location of each set of combined results. Comparison of the RVs assigned by expert evaluation to the estimates produced by unguided application of the various metrics thus should help characterize some of the properties of the metrics.

5.2.1 Evaluation. The expected location of each of the 150 publicly available CCQM data sets was estimated with all 14 location metrics described above using spreadsheet software.[13] The redundancy in the resulting 15 row (14 location estimates plus the assigned RV) by 150 column (measurands) data matrix was removed using principal components analysis. The resulting 15 estimate by 14 abstract-factor matrix captures 100 % of the covariance information contained in all 2250 location estimates.[27]

Table 3 lists the scores for the first 10 factors. The variance explained by each nth abstract factor is indicated by its eigenvalue, λ_n; the proportion of the total variance, $\%Var_n$, explained by the factor is here $100*\lambda/2250$; and cumulative proportion of the variance explained by the n largest factors is given by $\%CumVar_n$.

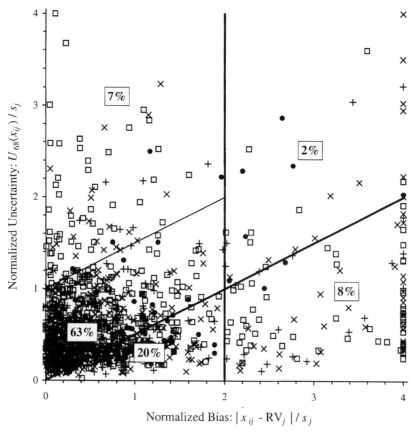

Figure 3 *Measurement Uncertainty as a Function of Measurement Bias*

5.2.2 Comparison of Metrics to Reference Values. Location metrics that respond similarly to the various challenges presented in the evaluation of the 150 measurands will have similar score coefficients, c_{mn}, where m is the metric (row) index and n is the factor (column) index. Close examination of Table 3 indicates that the pattern of c_{mn} for MMsh/mid and MMsh/med are indeed quite similar.

It is, however, much more convenient to quantitatively summarize the similarities of the different metrics with the Euclidean score distance, $D_{p,q}$:

$$D_{p,q} = \sqrt{\sum_{n=1}^{14} \left(c_{pn} - c_{qn} \right)^2} \ . \tag{19}$$

where p and q are the row designations for two metrics. Should the two metrics have exactly the same properties, the value of $D_{p,q}$ will be 0.0; the value of $D_{p,q}$ will increase as the similarity between the metrics' behavior decreases. For reference, the score distance between MMsh/mid and MMsh/med, $D_{14,15}$, is 4.7. Figure 3 displays the $D_{p,q}$ for the 14 location metrics examined (rows 2 to 15 of Table 3) relative to the RV (row 1). The metrics are sorted in order of decreasing similarity to the RV. These distances suggest two broad conclusions. First, none of the 14 metrics gives location estimates that are closely related to the RV: unguided statistics cannot replace expert evaluation. Second, of the metrics evaluated, the MMmedian is distinctly the most similar to the RV: chemical MU estimates can provide quantitatively useful information.

Table 3 *Scores (c_{mn}) for the 10 Most Significant Factors of the Location Estimates*

Metric					Abstract Factors					
	#1	#2	#3	#4	#5	#6	#7	#8	#9	#10
RV	-3.9	-6.9	14.0	-0.7	1.8	0.7	-1.0	0.1	-0.1	0.1
Mean	12.0	0.1	-1.1	-1.5	0.6	-2.1	-4.7	-0.9	-2.0	-1.7
Median	-1.9	2.8	0.2	-2.5	-2.5	-1.5	2.3	4.2	-2.1	1.2
Shorth	-11.0	7.1	-1.3	-8.0	7.6	-1.1	0.4	-0.8	0.7	-0.2
A15	3.0	1.9	0.7	-2.9	-4.5	-0.7	1.8	-0.8	0.5	0.8
H15	5.0	1.2	0.8	-2.9	-5.0	-0.5	0.7	-3.1	1.7	0.9
L1½	4.0	1.6	0.0	-2.2	-2.4	-1.4	-0.1	0.2	-1.1	-0.3
WtU	0.7	-12.8	-5.7	-1.2	2.9	1.8	4.2	-1.3	-1.4	0.2
WtMP	6.7	-5.8	-3.8	-1.9	1.3	0.8	-3.6	3.1	3.1	0.4
BSmean	14.9	5.9	1.3	6.7	6.1	0.5	1.1	-0.4	-0.2	1.4
BSmedian	0.3	3.5	1.1	4.3	-0.7	2.4	4.5	1.0	1.5	-2.2
MMmode	-11.1	-2.7	-1.9	7.3	-0.2	-7.4	-0.6	-0.5	0.7	0.2
MMmedian	-0.6	0.1	-0.5	-0.4	-2.4	0.4	0.3	0.3	-0.1	-1.5
MMsh/mid	-8.6	2.6	-1.9	2.6	-1.8	5.6	-3.3	-0.7	-0.7	0.9
MMsh/med	-9.4	1.3	-1.9	3.2	-0.8	2.6	-2.1	-0.3	-0.6	0.0
λ	884.8	372.5	262.5	233.5	176.4	113.7	100.7	42.7	28.4	16.2
%Var	39.3	16.6	11.7	10.4	7.8	5.1	4.5	1.9	1.3	0.7
%CumVar	39.3	55.9	67.5	77.9	85.8	90.8	95.3	97.2	98.5	99.2

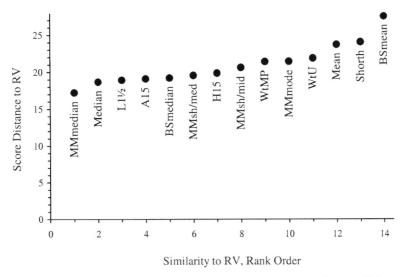

Figure 3 *Factor-Score Distance of Unguided Location Metrics to Reference Values*

5.2.2 Comparison of Metrics to the MMmedian. Figure 4 displays the $D_{p,q}$ of the RV and the various location metrics to the MMmedian. These values confirm that while the MMmedian estimates may be the most similar of the metrics evaluated to the RVs, they are much more closely related to those from most of the other metrics than to the RVs. The MMmedian appears to be most similar to the traditional metrics that are robust to symmetrically distributed "outlier" values (L1½, A15, Median, and H15) and least similar to the uncharacterized BSmean and the putatively robust but asymmetric Shorth. The behaviors of the other asymmetric metrics (MMsh/med, MMsh/mid, MMmode) are somewhat more similar than the Shorth to both the MMmedian and to the RVs, with the MMsh/med the most efficient of the asymmetric metrics for these basically symmetrically-distributed CCQM data.

The WtMP, WtU, and Mean metrics are about equally dissimilar to the robust estimates and to the RVs, confirming that measurement bias and MU are not sufficiently correlated in the CCQM data to identify relative bias from the relative MUs. The small difference between WtMP and WtU suggests that using MU to allocate influence is a basic fallacy rather than a deficiency in how such influence is allocated.

While the BSmedian estimates are roughly similar to those of the traditional robust metrics, they are not as close to the RV or the MMmedian as the median itself. In contrast, the BSmean estimates are the least similar to the MMmedian and the RVs of any metric. While chemical MU estimates provide information useful for combining results, as always the devil is in the details of how that information is used.

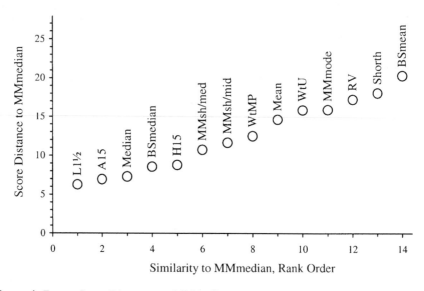

Figure 4 *Factor-Score Distances to MMmedian*

5 ACKNOWLEDGEMENT

I thank Reenie M. Parris, Kenneth W. Pratt, Michele M. Schantz, and Katherine E. Sharpless of NIST for their questions and gentle corrections; Aristides T. Hatjimihail of the Hellenic Complex Systems Laboratory for his encouragement; Richard G. Brereton of the

University of Bristol for his Multivariate Analysis spreadsheet freeware, and Steven L.R. Ellison of LGC for his RobStat spreadsheet freeware and for his deep insights into the use and abuse of chemical MU.

References

1   Eurachem. The fitness for purpose of analytical methods. A laboratory guide to method validation and related topics. 1998, www.eurachem.ul.pt/guides/valid.pdf
2   Eurachem. Quantifying uncertainty in analytical measurement, 2nd Ed. 2000, www.eurachem.ul.pt/guides/QUAM2000-1.pdf
3   G.E. O'Donnell, D.B. Hibbert. Treatment of bias in estimating measurement uncertainty. *Analyst*, 2005, **139**, 721.
4   CLSI. Expression of Uncertainty of Measurement in Clinical Laboratory Medicine (C51), *product in development.* www.clsi.org
5   M.N. Leger, L. Vega-Montoto, P.D. Wentzell. Methods for systematic investigation of measurement error covariance matrices. *Chemometrics Intell. Lab. Syst.*, 2005, **77**, 181.
6   M.S. Reis, P.M. Saraiva. Integration of Data Uncertainty in Linear Regression and Process Optimization. *AIChE J.*, 2005, **51**, 3007.
7   P.J. Lowthian, M. Thompson. Bump-hunting for the proficiency tester – searching for multimodality. *Analyst*, 2002, **127**, 1359.
8   M.G. Cox. A discussion of approaches for determining a reference value in the analysis of Key-Comparison data, NPL Report CISE 42/99. 1999. www.npl.co.uk/ssfm/download/nplreports.html
9   P. Ciarlini, M.G. Cox, F. Pavese. G. Regoliosi. The use of a mixture of probability distributions in temperature interlaboratory comparisons. *Metrologia*, 2004, **41**, 116.
10 D.L. Duewer, J. Brown Thomas, M.C. Kline, W.A. MacCrehan, R. Schaffer, K.E. Sharpless, W.E. May, J.A. Crowell. NIST/NCI Micronutrients Measurement Quality Assurance Program: Measurement repeatabilities and reproducibilities for fat soluble vitamin related compounds in human sera. *Anal. Chem.* 1997, **69**, 1406.
11 BIPM. kcdb.bipm.org/AppendixD/default.asp.
12 BIPM. www.bipm.fr/en/committees/cc/ccqm/pilot_cc.html
13 D.L. Duewer. A robust approach for the determination of CCQM key comparison reference values and uncertainties. Working document CCQM/04-15, BIPM, 2004. www.bipm.info/cc/CCQM/Allowed/10/CCQM04-15.pdf
14 C. Croux, G. Haesbroeck. Maxbias curves of robust location estimators based on subranges. *J. Nonparametr. Stat.,* 2002, **14**, 295.
15 J.W. Müller. Possible advantages of a robust evaluation of comparisons. *J. Res. Nat. Inst. Stds. Technol.* 2000, **105**, 551.
16 M.G. Cox. The evaluation of key comparison data. *Metrologia* 2002, **39**, 589.
17 P.J. Rousseeuw. Multivariate estimation with high breakdown point. In: W. Grossman, G. Pflug, I. Nincze, W. Wetrz (eds.), *Mathematical Statistics and Applications*, 1985, 283-297. Reidel, Dordrecht, The Netherlands.
18 A.H. Rose, C.-M.Wang, S.D. Byer, Round Robin for optical fiber Bragg grating metrology. *J. Res. Nat. Inst. Stds. Technol.* 2000, **105**, 839.
19 Analytical Methods Committee. RobStat.xla. 2002. www.rsc.org/Membership/Networking/InterestGroups/Analytical/AMC/Software/RobustStatistics.asp
20 Analytical Methods Committee. Robust statistics: A method of coping with outliers. AMC Technical Brief 6. 2002. www.rsc.org/Membership/Networking/InterestGroups/Analytical/AMC/TechnicalBriefs.asp

21 Analytical Methods Committee. Robust statistics - How not to reject outliers. Part 1. Basics. *Analyst* 1989, **114**, 1693.

22 F. Pennecchi, L. Callegaro. Between the mean and the median: the Lp estimator. *Metrologia*, in press

23 R.C. Paule, J.Mandel. Consensus values and weighting factors. *J. Res. Nat. Bur. Stds.* 1982, **87**, 377.

24 A.L. Rukhin, B.J. Biggerstaff, M.G. Vangel. Restricted maximum likelihood estimation of a common mean and the Mandel-Paule algorithm. *J. Stat. Plan. Infer.* 2000, **83**, 319.

25 P. Diaconis, B. Efron. Computer-intensive methods in statistics. *Sci. Am.* 1983, **248**, 116.

26 D.L. Duewer, B.R. Kowalski, J.L. Fasching. Improving the reliability of factor analysis of chemical data by utilizing the measured analytical uncertainty. *Anal. Chem.* 1976, **48**, 2002.

27 R.G. Brereton. Chemometrics : Data Analysis for the Laboratory and Chemical Plant, Wiley, Chichester, 2003.

UNCERTAINTY AND TRACEABILITY: THE VIEW OF THE ANALYTICAL CHEMIST

A. Sahuquillo and G. Rauret

Departament de Química Analítica, Facultat de Química, Universitat de Barcelona. Av. Diagonal, 647, E-08028 Barcelona (Spain)

1 INTRODUCTION

The responsibility of an analytical chemist is to provide useful and reliable information, based in analytical results, to help in decisions making in most cases with economic and social impact. One of the requirements for achieving this purpose is to obtain analytical results traceable to the basic units of measurement, and with a well established uncertainty.

When reporting an analytical result, the analyst must consider, on the one hand, the component corresponding to the experimentally obtained information about the magnitude of the measurand based on a measurement procedure, and on the other hand a second component related to the uncertainty. As the first component is concerned, it has to be assured that the analytical process used is providing results free from systematic errors which imply the assessment of bias. Further step is the verification of the traceability of the obtained result which is normally performed on the basis of a stated reference. According to the type of reference used, the level of traceability will be different, and the usual procedure applied in an analytical laboratory is the use of certified reference materials as primary methods are not always available.

An analytical laboratory claiming to work properly must nowadays follow different recommendations or good laboratory practices compilations. The most widely followed requirements, even for not accredited laboratories, are those stated in the guide ISO/IEC 17025 [1] where it is mentioned that laboratories must estimate and report uncertainty of measurements, that is the second component of the analytical result. The International vocabulary of basic and general terms in metrology (VIM) [2], under revision, defines the uncertainty as the parameter that characterises the dispersion of the quantity values that are being attributed to a measurand based on the information used. The assessment of uncertainty is thus considered an unavoidable practice and it is especially important for testing the potential equivalency between laboratories and with respect to reference values or legislated limits. This information can often prevent the unnecessary repetition of analyses. Uncertainty, moreover, is a parameter that is more powerful than reproducibility because it also helps to demonstrate traceability [3].

Different approaches for uncertainty evaluation are described in the literature. The *bottom up approach* based on the identification, quantification and combination of all sources of uncertainty of the measurement was adapted by EURACHEM [4] for chemical measurements. In the so-called *top-down approach*, the evaluation of uncertainty is based on a view of the laboratory from a higher level, that is, as a member of a population of laboratories which can be advantageous when information coming from collaborative trials is available [5]. There is another procedure for evaluating uncertainty based on the information generated in the laboratory from the validation process [6], where the laboratories get information about the bias and the intermediate precision of the analytical procedure. From a practical point of view it would be very useful for the analytical chemist to be able to evaluate uncertainty from the laboratory work performed during the validation step of a method of analysis.

In the present paper, the effect of different variables on the final estimation of uncertainty has been studied on the basis of data obtained during in-house validation process for the determination of two trace elements in environmental samples by using spectroscopic techniques: the analysis of mercury in sediments, and the analysis of chromium in plants, sediments, sewage sludges, fly ashes and soils. The considered variables were the total number of certified reference materials (CRMs) involved in validation and the concentration range. Moreover, the potential commutability of the uncertainty values obtained was ascertain for the analysis of chromium in matrices from different origin.

The conclusions of the study allow stating some recommendations from a practical point of view in order to be able to design a good validation strategy for providing suitable data for evaluating uncertainty.

2   EXPERIMENTAL

2.1.   Mercury determination in sediments

2.1.1. Certified reference materials. Five CRMs covering a range of mercury content from 0.81 to 132 mg·kg^{-1} were used in this work and results obtained for different spectroscopic techniques are reported elsewhere [7]: an estuarine sediment, IAEA-405 with a certified total Hg content of 0.81 mg kg^{-1}; BCR-277, an estuarine sediment with a content of 1.77 mg kg^{-1}; PACS-2, a marine sediment with a Hg content of 3.04 mg kg^{-1}; LGC 6156, a harbour sediment with a Hg content of 10.1 mg·kg^{-1}; and ERM®-CC580, estuarine sediment with a Hg content of 132 mg·kg^{-1}.

2.1.2. Analytical procedure. For sample digestion 1 g of sediment was extracted with the appropriate amount of *aqua regia* for 16 h at room temperature and further maintained under reflux conditions for 2 h at 130°C, including a special adsorption vessel for preventing Hg losses. The extracts were measured by inductively coupled plasma-mass spectroscopy (ICP-MS) using Rh as internal standard. This procedure was applied five-fold to each CRM in non-consecutive days and the obtained results are shown in Table 1. For each material, the certified value with the associated confidence range and the experimental mean values obtained with the relative standard deviation corresponding to the five analysis are included.

2.2.   Chromium determination in plant materials

2.2.1. Certified reference materials. Five CRMs covering a range of chromium content from 2.14 to 36.3 mg·kg^{-1} were used in this case: rye grass, BCR-281 with a certified total Cr content of 2.14 mg kg^{-1}; NIST SRM-1575, pine needles with a content of 2.6 mg kg^{-1}; BCR-100, beech leaves with a Cr content of 8.0 mg kg^{-1}; BCR-414, a plankton with a Cr content of 23.8 mg·kg⁻¹; and BCR-596, an aquatic plant (*trapa natans*) with a Cr content of 36.3 mg·kg⁻¹.

Table 1. *Obtained data [7] from method validation for Hg determination in sediments*

Type of matrix	CRM code	Certified value (mg Hg/kg)	Obtained result (n=5) (mg/kg) (% RSD)
Estuarine sediment	IAEA 405	0.81 ± 0.02	0.80 (4.4 %)
Estuarine sediment	BCR CRM 277	1.77 ± 0.07	1.8 (6.6 %)
Marine sediment	PACS-2	3.04 ± 0.10	2.90 (2.1 %)
Harbour sediment	LGC 6156	10.1 ± 0.82	10.6 (1.9 %)
Estuarine sediment	ERM-CC 580	132 ± 5.1	135 (1.3 %)

2.2.2. Analytical procedure. An opened-digestion was performed on 0.5 g of sample placed in a Teflon beaker in a sand-bath heated up to 120-150 °C. Subsequent aliquots of HNO_3 and a mixture of $HF:H_2SO_4$ (2:1) were added until the remaining residue did not have a siliceous aspect. The final residue was dissolved in diluted HNO_3 and made up volume with double deionised water. The solutions were measured by atomic absorption spectroscopy with electrothermal atomisation and Zeeman effect as background correction (ZETAAS). Calibration was run by standard addition method. The procedure was applied in triplicate to each CRM in non-consecutive days and the results obtained are shown in Table 2. For each material, the certified value with the associated confidence range and the experimental mean values obtained with the relative standard deviation corresponding to the three analysis are included.

Table 2. *Obtained data from method validation for Cr determination in plants*

Type of matrix	CRM code	Certified value (mg Cr/kg)	Obtained result (n=3) (mg/kg) (% RSD)
Rye grass	BCR CRM 281	2.14 ± 0.12	2.33 (13.5 %)
Pine needles	NIST SRM 1575	2.6 ± 0.2	2.4 (5.7 %)
Beech leaves	BCR 100	8.0 ± 0.6	7.3 (6.1 %)
Plankton	BCR 414	23.8 ± 1.2	23.0 (5.7 %)
Aquatic plant	BCR 596	36.3 ± 1.7	32.0 (14.6 %)

2.3. Chromium determination in sediments

2.3.1. Certified reference materials. Four CRMs covering a range of chromium content from 75 to 29600 mg·kg⁻¹ were used in this work: a pond sediment, NIES CRM n°2 with a certified total Cr content of 75 mg kg⁻¹; BCR-280, a lake sediment with a content of 114 mg kg⁻¹; BCR-320, a river sediment with a Cr content of 138 mg kg⁻¹; and NBS SRM-1645, a river sediment with a Cr content of 29600 mg·kg⁻¹.

2.3.2. Analytical procedure. An opened-digestion was performed on 0.5 g of sample placed in a Teflon beaker in a sand-bath heated up to 120-150 °C. Subsequent aliquots of HNO_3 and a mixture of $HF:HClO_4$ (2:1) were added until the sample was completely attacked. The final residue was dissolved in diluted HNO_3 and made up volume with double deionised water. The solutions were measured by atomic absorption spectroscopy with an air-acetylene flame (FAAS) and using 0.07 mol·l⁻¹ oxine as suppressor of interferences. Calibration graph was used for quantification. The procedure was applied in

triplicate to each CRM. The obtained results together with the information corresponding
to the CRMs used are shown in Table 3.

Table 3. *Obtained data from method validation for Cr determination in sediments*

Type of matrix	CRM code	Certified value (mg Cr/kg)	Obtained result (n=3) (mg/kg) (% RSD)
Pond sediment	NIES (CRM n°2)	75 ± 5	70 (1.3 %)
Lake sediment	BCR CRM 280	114 ± 4	112 (0.6 %)
River sediment	BCR CRM 320	138 ± 7	132 (2.2 %)
River sediment	NBS SRM-1645	29600 ± 2800	28705 (4.8 %)

2.4.   Chromium determination in sewage sludges, fly ashes and soil

2.4.1. Certified reference materials. Five CRMs with quite similar chromium content
were used in this case: two different sewage sludges, one of them from domestic origin
(high organic matter content), BCR-144R with a certified total Cr content of 104 mg kg^{-1},
and the other one from industrial origin, BCR-146R with a content of 196 mg kg^{-1}. The two
fly ashes materials were BCR-038 from pulverised coal and with a Cr content of 178 mg
kg^{-1}1, and NIST 1633b with a Cr content of 198.2 mg·kg^{-1}. Finally, a calcareous soil, BCR-
141R with a Cr content of 195 mg·kg^{-1} was also considered.

2.4.2. Analytical procedure. An opened-digestion was performed on 0.3 g of sample
placed in a Teflon beaker in a sand-bath heated up to 80 - 100 °C. After a gently treatment
with HNO_3, H_2O_2 was added and the mixture was allowed to react for 4-5 h with
occasional shaking. Further aliquots of HF:HClO$_4$ (3:1) were added until a clear residue
was obtained. The final residue was dissolved in diluted HNO_3 and made up volume with
double deionised water. The solutions were measured by atomic absorption spectroscopy
with a nitrous oxide-acetylene flame (FAAS) and calibration graph was used for
quantification. The procedure was applied five-fold to each CRM in non-consecutive days
and the obtained results are shown in Table 4 together with the information related to the
CRMs used.

Table 4. *Obtained data from the analysis of Cr in different matrices*

Type of matrix	CRM code	Certified value (mg Cr/kg)	Obtained result (n=5) (mg/kg) (% RSD)
Sewage sludge	BCR 144 R	104 ± 3	106 (27.1 %)
Fly ashes	BCR 38	178 ± 8	187 (1.5 %)
Calcareous soil	BCR 141R	195 ± 7	202 (4.5 %)
Sewage sludge	BCR 146R	196 ± 7	190 (9.9 %)
Fly ashes	NIST 1633b	198.2 ± 4.7	197 (1.86 %)

3   CALCULATION APPROACH FOR EVALUATING UNCERTAINTY

The combined uncertainty of trace elements determination in environmental samples can
be assumed to be due mainly to two contributions, as indicated in Equation 1: the
contribution arising from the accuracy, that is the bias assessment (u_{bias}), and the
contribution arising from the intermediate precision (u_{proc}) of the analytical procedure.

$$u = \sqrt{u_{bias}^2 + u_{proc}^2}$$ (Eq. 1)

With this assumption, there are other possible contributions to the combined uncertainty that are not considered. The sampling procedure can contribute significantly to the total uncertainty, even if this step is usually not under the responsibility of the analytical laboratory. On the other hand, the contributions due to sample pre-treatment for getting a homogeneous sample to be processed are not considered. Thus, the laboratory has to undertake different strategies for assuring the homogeneity such as drying, milling, or even increasing the final sample intake to be processed.

Three different approaches for evaluating the uncertainty budgets due to the bias and to the procedure were compared. A complete description of the theoretical basis involved is described elsewhere [7]. One of the approaches (approach A) is based on the individual calculation for each single CRM used in the validation step, providing an uncertainty value expressed in mg·kg[-1]. In the other two approaches (approaches B and C), calculations are based on the consideration of all the information obtained (*n* replicates analysed for *i* CRMs) pooled for obtaining a mean estimation in two different ways [7,8], providing a combined uncertainty values expressed in percentage.

In approach B, the contribution arising from bias (u_{bias}) is calculated considering two terms, the root mean square (RMS) including the pooled relative bias of all CRMs used in method validation, and the RMS of the pooled relative uncertainty of CRMs calculated in percentage. The contribution of the analytical procedure (u_{proc}) was estimated as the RMS of the pooled relative standard deviation obtained from the CRMs analysed. In approach C, the results from the individual replicates are normalised with respect to the certified value and the first contribution to u_{bias}, is calculated as the mean value of this set of normalised data. The second contribution corresponds to the uncertainty coming from the estimation of bias, and the third contribution to the bias uncertainty is calculated as the RMS of the pooled relative uncertainty of CRMs calculated in percentage. Secondly, the contribution of the analytical procedure (u_{proc}), is calculated from the RSD obtained from a set of normalised data where each individual measure is normalised with respect to the mean measured for each material. The expressions used for calculation in these two last approaches are summarised in Table 5.

A comparison of the results obtained for the three calculation approaches are shown in Figure 1 for Hg determination in sediments by ICP-MS, where results obtained for approach A has been expressed in percentage for better comparison.

No significant differences were observed for the final estimation of uncertainty provided by the three methods even if slightly low values were obtained from approach C. The main advantage of approaches B and C is that a direct uncertainty estimation expressed in percentage is obtained which would be very useful for an analytical laboratory. Thus, the effect of different variables on the uncertainty values obtained were undertaken for these two last approaches. The presented results correspond to approach C even if similar conclusions could be drawn when approach B was used for calculation.

Table 5. Expressions used for evaluating uncertainty from method validation data [7]

Approach	Equations for calculation
A	$$(u_{bias})_i = \sqrt{(bias)^2 + \left(\frac{s_{bias}}{\sqrt{n}}\right)^2 + (u_{CRM})_i^2} \qquad (u_{proc})_i = (s_{proc})_i$$
B	$$u_{bias} = \sqrt{RMS_{bias}^2 + RMS_{CRM}^2}$$ $$RMS_{bias} = \sqrt{\frac{\sum (relative\ bias)_i^2}{i}} \qquad RMS_{CRM} = \sqrt{\frac{\sum (u_{relative,CRM})_i^2}{i}}$$ $$u_{proc} = \sqrt{\frac{\sum (RSD)_i^2}{i}}$$
C	$$u_{bias} = \sqrt{(bias_{measured\ /\ certified})^2 + \left(\frac{s_{bias,\ measured\ /\ certified}}{\sqrt{n \cdot i}}\right)^2 + RMS_{CRM}^2}$$ $$bias_{measured\ /\ certified} = \frac{\sum \frac{x_{n,i} - x_{i,certified}}{x_{i,certified}}}{n \cdot i}$$ $$u_{proc} = RSD_{measured\ /\ mean\ measured}$$

i: number of CRMs used; n: number of replicates

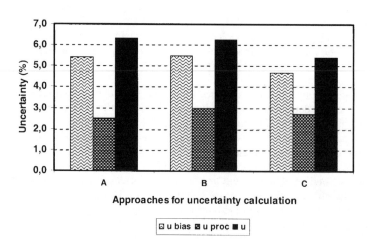

Figure 1. *Evaluation of uncertainty for Hg determination in sediments by ICP-MS.*

4 EFFECT OF THE NUMBER OF CRMs ON UNCERTAINTY EVALUATION

The effect of the number of CRMs involved in the uncertainty evaluation was studied for two different cases: Hg determination in sediments, considering a concentration range from 0.81 to 132 mg·kg^{-1}, and for Cr determination in plants in a concentration range from 2.14 to 36.3 mg·kg^{-1}. In both cases, after a sample digestion with a different acidic mixture, the trace element was determined in the extracts with an atomic spectroscopic technique, as described in the experimental section. The results obtained are given in Figure 2 A and B.

In each case, from five to two CRMs were used in the calculation of uncertainty budgets, maintaining the same concentration range. For each graph, the first group of results corresponds to the maximum number of CRMs considered and the last group to the results obtained when considering the two CRMs with the lowest and the highest trace element concentration for calculation. The remaining groups correspond to the different uncertainty budgets obtained for different combinations of four and three CRMs. The materials considered in each case are presented in Table 6.

Table 6. *CRMs used in uncertainty budgets evaluation*

Trace element	Presented results	CRMs considered in calculation
Hg (sediments)	5 CRMs	All
	4 CRMs (a)	IAEA 405, PACS-2, LGC 6156, ERM-CC 580
	4 CRMs (b)	IAEA 405, BCR CRM 277, PACS-2, ERM-CC 580
	3 CRMs (a)	IAEA 405, LGC 6156, ERM-CC 580
	3 CRMs (b)	IAEA 405, BCR CRM 277, ERM-CC 580
	2 CRMs	IAEA 405, ERM-CC 580
Cr (plants)	5 CRMs	All
	4 CRMs (a)	BCR CRM 281, BCR 100, BCR 414, BCR 596
	4 CRMs (b)	BCR CRM 281, NIST 1575, BCR 100, BCR 596
	3 CRMs (a)	BCR CRM 281, BCR 414, BCR 596
	3 CRMs (b)	NIST 1575, BCR 100, BCR 596
	2 CRMs	BCR CRM 281, BCR 596

As it can be seen in Figure 2, no important differences on the combined uncertainty values obtained were observed when considering from 5 to 3 materials in the validation step. Whereas for Hg determination in sediments the obtained values ranged from 4.5 to 5.2 %, for Cr determination in plants the uncertainty values were within 10.1 – 11.3 %. However, when considering only two CRMs for calculation, a combined uncertainty of 3.61 % and 12.9 % was obtained for Hg and Cr, respectively.

Thus, the effect of the number of CRMs used for calculation on the uncertainty values obtained, showed that a good estimation would be obtained when using at least three CRMs in the validation step, as the use of two CRMs can lead to an underestimation or an overestimation of uncertainty values.

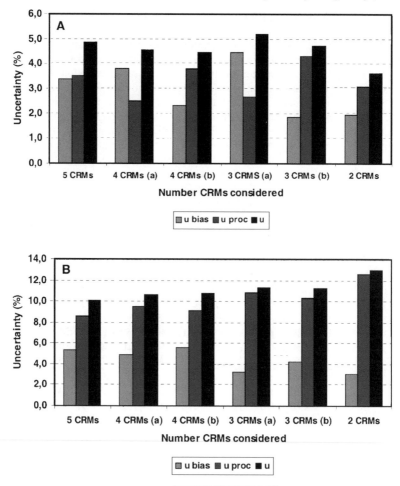

Figure 2. *Effect of the number of CRMs considered on evaluation of uncertainty. A) Hg determination in sediments; B) Cr determination in plants.*

5   EFFECT OF THE CONCENTRATION RANGE ON UNCERTAINTY EVALUATION

The second variable considered was the concentration range of the CRMs used for validation. For this study, data originated in the method validation of Cr determination in sediments and in plants was selected. In the first case, four different CRMs were available covering a concentration range of four orders of magnitude (75-29600 mg·kg^{-1}), and for Cr determination in plants, five CRMs were available with a concentration range from 2.14 to 36.3 mg·kg^{-1}. Uncertainty budgets were calculated for both matrices and elements, considering the three CRMs shown in Table 7 for two different concentration ranges in each case. The results obtained are shown in Figure 3A and B.

Table 7. *CRMs used in uncertainty budgets evaluation*

Trace element	Concentration range (mg/kg)	CRMs considered in calculation
Cr (sediments)	75 – 29600	NIES (CRM n°2), BCR 320, NBS 1645
	75 - 138	NIES (CRM n°2), BCR 320, BCR 280
Cr (plants)	2.6 –36	NIST 1575, BCR 100, BCR 596
	2.2 - 8	BCR 281, NIST 1575, BCR 100

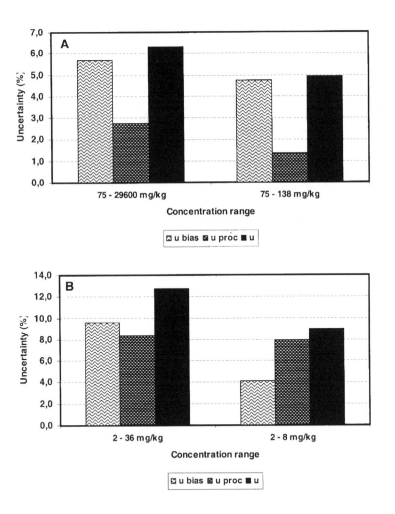

Figure 3.B. *Effect of the concentration range considered in the CRMs used in validation: A) Cr determination in sediments; B) Cr determination in plants.*

As it can be seen, in both cases, higher uncertainty budgets were obtained when the CRM with the highest trace element concentration was included in the calculation.

When considering the widest concentration range, the final uncertainty value obtained would be applicable to a wider number of samples which could be useful to an analytical laboratory and in any case, the analyst would always be in a safer position when reporting an analytical result. Moreover, the increase in the uncertainty budget associated to a lower concentration using this approach would only be from 1.4 to 3.8 % which can be perfectly acceptable in daily practice.

6   EFFECT OF THE TYPE OF MATRIX ON UNCERTAINTY EVALUATION

The last variable studied was the type of matrix involved in the uncertainty evaluation with the aim of ascertaining potential commutability of the uncertainty values obtained for one matrix to another type of sample arrived into the laboratory, when using the same sample pre-treatment and analytical technique for trace element determination.

The study was carried out for Cr determination in sewage sludges from domestic and industrial origins, in two fly ashes and in a calcareous soil. For these five CRMs, whereas Cr concentration was not so different (104 – 198 mg·kg^{-1}), the main difference regarding the chemical composition of the considered matrices was the organic matter content. The sewage sludge from domestic origin presented the highest organic matter content, whereas organic matter content in fly ashes was negligible. The sewage sludge from industrial origin and the calcareous soil presented low-intermediate organic matter content.

The first group of results presented in Figure 4 corresponds to the obtained values when considering the five materials altogether. In this case, the combined uncertainty value was 12.1 % with the most relevant uncertainty budget due to the analytical procedure. When considering for calculation only the two matrices with low organic matter content (soil and industrial sewage sludge) or the two matrices with negligible organic matter content (fly ashes), the final uncertainty values decreased to 7.4 % and 3.3 %, respectively. Thus, the type of matrix considered during the validation step may have an important role in the estimation of the uncertainty budgets associated.

Potential commutability of the uncertainty values estimated with one matrix to another one, would be only possible when a complete chemical characterisation of both matrices showed a high degree of equivalency at least for key parameters such as in this case, organic matter content. However, it has to be taken into account that this practice can only provide a rough estimation of uncertainty values associated to a sample, and a further study for definitive uncertainty calculation would be advisable focusing the matrix of interest.

7   CONCLUSIONS

From this study it can be concluded that evaluation of uncertainty from validation method data can be a suitable practice for analytical laboratories that no requires many additional work to the daily routine work. Even if the calculation method used for evaluation uncertainty values does not have an effect on the obtained results, the use of a calculation method providing uncertainty values expressed in percentages is the best recommended choice.

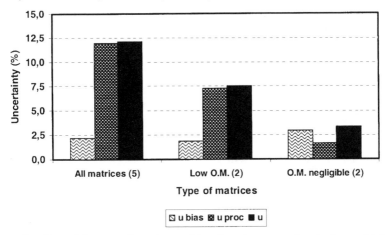

Figure 4. *Effect of the type of matrix considered on the uncertainty budgets obtained.*

For designing a useful validation step two main recommendations can be followed:
- Inclusion of three different CRMs
- Use of the widest available concentration range

Special attention must be paid to the selection of matrices to be considered in the validation, because commutability of the uncertainty values provided would be only possible in matrices with a very close chemical composition.

References

1   ISO/IEC 17025 General requirements for the competence of testing and calibration laboratories, 2005.
2   International vocabulary of basic and general terms in metrology (VIM), revision of the 1993 edition. International Organization for Standarisation, 2004 (draft).
3   EAL European co-operation for Accreditation of Laboratories. *The expression of Uncertainty in Quantitative Testing*. Pub. Reference EAL-G23, 1996.
4   EURACHEM/CITAC Guide CG 4. Quantifying Uncertainty in Analytical Measurement. 2nd edition, 2000.
5   Analytical Methods Committee, *Analyst,* 1995, **120**, 2303.
6   A. Maroto, R. Boqué, J.Riu and F.X. Rius, *Trac-Trends in Anal. Chem.,* 1999, **18**, 577.
7   A.Guevara-Riba, A. Sahuquillo, J.F.López-Sánchez and R. Rubio, accepted for publication in *Analytical and Bioanalytical Chemistry*, 2006.
8   B. Magnusson, T. Näykki, H. Hovind and M. Krysell, in *Handbook for calculation of measurement uncertainty in environmental laboratories*, 2nd edition, Finland, 2004.

PROPAGATING NON-NORMALLY DISTRIBUTED UNCERTAINTY – THE LJUNGSKILE CODE

A. Ödegaard-Jensen[1], G. Meinrath[2,3], Ch. Ekberg[1)]

[1] Chalmers University of Technology, Nuclear Chemistry Group, Göteborg/Sweden
[2] RER Consultants, Passau, Germany
[3] Technische Universität Bergakademie Freiberg, Freiberg/Germany

1 INTRODUCTION

The EURACHEM/CITAC Guide on quantifying uncertainty in analytical measurement [1] is widely used as a reference in the evaluation of measurement uncertainty budgets in chemical measurements. The EURACHEM guide is, however, limited to situations where the measurand can be expressed by a closed mathematical expression. Such situation occur mostly in quantifying amounts of single analytes. In more complex situations, especially those involving chemical reactions, the Normality assumption becomes highly questionable. Examples for such situations are measurements involving correlated quantities or influence factors and non-negligible covariance contributions [2].

Then, no a priori assumption concerning the distribution of a measurement uncertainty and its moments is feasible. The assessment of the measurement result must direct to an estimate of the empirical distribution of the measurement uncertainty budget. Because measurement results from one experiment may become auxiliary parameters in subsequent evaluations, these empirical distributions need to be documented and reported in a way that allows application in a subsequent evaluations. Some of these tasks are similar to those required in risk analysis and performance assessment, for instance of nuclear waste repositories in deep geological formations. The complexity of risk analysis codes most likely will even exceed those required for the evaluation of complete measurement uncertainty budgets. Especially stratified sampling techniques may become a helpful tool in chemical metrology.

Simulation of chemical speciation is an important step in a wide range of environmental and technical applications. Up to now, such simulations are based exclusively on mean value calculations without statement of the prediction uncertainty in the estimated amounts of the various chemical species. The simulation of speciation from published thermodynamic data requires the propagation of non-Normally distributed uncertainties. On the other hand, thermodynamic data are usually given without a meaningful estimate of measurement uncertainty [3-5]. Appropriate concepts in agreement with the GUM [6] have been proposed and demonstrated [7-9]. Some concepts available for reporting, applying and progressing empirical distribution functions (EDFs) will be presented in the following. As a basis of discussion the LJUNGSKILE speciation code [10] will be used, where some the key elements are implemented. The LJUNGSKILE code is at present the only available code to provide speciation calculations with associated output uncertainties.

2 PARAMETRIC AND NON-PARAMETRIC DISTRIBUTIONS

The Normal distribution, N, is fully characterised by its mean values, μ, and its variance, σ^2. It is often abbreviated as $N(\mu,\sigma)$

$$N(\mu,\sigma) = \frac{1}{\sigma\sqrt{2\pi}} e^{-\frac{(\mu-x)^2}{2\sigma^2}} \tag{1}$$

The mean value, μ, is calculated by the closed formula eq. (2):

$$\mu = \frac{1}{n}\sum_{i=1}^{n} x_i . \tag{2}$$

where n is the number of observations obtained by sampling from $N(\mu,\sigma)$. The standard deviation, s, is obtained via eq. (3):

$$\sigma^2 = \frac{1}{n-1}\sum_{i=1}^{n}(\mu - x_i)^2 . \tag{3}$$

In addition to the fact that the parameters μ and s can be conveniently calculated by closed formulas, eq. (2) and eq. (3), the Normal distribution does have a few other convenient properties. For instance, it is a symmetric monomodal distribution with the mean value corresponding to the maximum of the distribution and the median being equal to the mean. There are other distribution which are characterised by a few parameters, for instance the Poisson distribution, or the Weibull distribution. However, their forms are more complicated and the parameters cannot be calculated as simply as the mean μ and standard deviation σ of the Normal distribution.

In fact, the Normal distribution, eq. (1), does not give a probability but a probability density. The corresponding probability distribution is the so-called error function Φ, eq. (4):

$$\Phi(x) = \frac{1}{\sqrt{2\pi}} \int_{-\infty}^{x} e^{-\frac{t^2}{2}} dt . \tag{4}$$

Eq. (4) cannot be solved analytically but is given in a tabulated form or as an approximation, e.g. as Chebyshev polynomial. The error function eq. (4) is the cumulative Normal distribution. $\Phi(40,1)$ is shown in Fig. 1.

In a similar way, the empirical cumulative distribution function (EDF) can be obtained from any set of observations. The EDF F(E) is defined by eq. (5)

$$F(E) = \frac{1}{n}\sum_{j=1}^{n} H(E - E_j) \tag{5}$$

where H(u) is the unit step function jumping from 0 to 1 at u=0. The concept of the EDF is demonstrated using the measurement data given in Table 1 as an example. The resulting

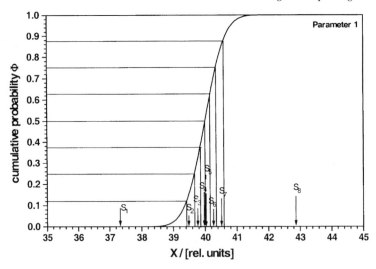

Figure 1: *Cumulative Normal distribution F(40,1) divided into m = 8 strata of equal probability. The S_i give the randomly selected representative values for each stratum i.*

Table 1 *empirical data (measured weight of water in a pipette with nominally 1 mL volume)*

No.	weight
	[g]
1	0.9903
2	0.9982
3	0.9993
4	1.0015
5	1.0023
6	1.0023
7	1.0028
8	1.0067
9	1.0079
10	1.0139

EDF is shown in Figure 2. The EDF answers the question: 'What is the probability to observe a value smaller than x'. To observe a value smaller 0.9903 g is zero (there is none), to observe a value smaller 1.0016 g is 0.4 (four out of ten observations are smaller) and so on.

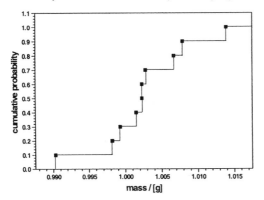

Figure 2: *EDF of the data given in Table 1.*

3 EFFICIENT SAMPLING FROM MULTIDIMENSIONAL DISTRIBUTIONS

Empirical distributions of complex measurement values can be obtained, e.g. from the methods outlined in refs. [7-9]. In order to propagate these EDFs in subsequent evaluations, it is necessary to sample from these distributions in a way that considers the full extend of the distributions, including the information about the position and extension of the tails. The following example will illustrate that merely random sampling, e.g. Monte Carlo strategy, will not work as soon a more than two or three parameters are involved.

Let's assume to have Normally distributed data. To observe with at least 99% probability at least one sample from the 95% percentile(s) of a single distribution, there must be

$$1-0.95^N > 0.99 \tag{6}$$

N = 90 samples drawn. With two distributions, there must be 1840 samples obtained, with three distributions there must be 36840 samples and so on. Monte Carlo is, obviously, not effective for higher-dimensional problems. Here, Latin Hypercube sampling (LHS) is an alternative. LHS is a method of stratified sampling [11,12]. LHS requires the EDF to be divided into at least m strata of equal probability, where m is the total number of distributed parameters in the simulation. From each stratum i, one value, S_i, is selected randomly to represent the stratum in the subsequent simulation(s). Such a stratification is shown in Figure 1. Figure 3 shows a Latin Hypercube lattice with m = 8, where the numbers in each field represent a stratum of a parameter's EDF. In each input vector, no stratum number appears twice. The same holds true for each row. Stratification warrants that each section of the EDFs is properly considered in the simulation. The LHS strategy also avoids that some strata are considered more often than others. Hence, LHS is a balancing strategy. Further details can be found elsewhere [12].

4 AN IMPLEMENTATION OF LHS: THE LJUNGSKILE CODE

The LJUNGSKILE code has been made publicly available [13]. LJUNGSKILE code allows to simulate species concentrations as a function of various parameters in aqueous solutions. It provides an LHS shell for the speciation code PHREEQC [14], developed by USGS. It also allows to compare Monte Carlo and LHS sampling strategies. Due to the lack of EDFs for almost all thermodynamic formation constants reported up to now in literature, Normal and rectangular distributions can be specified. A typical output from the LJUNGSKILE code is shown in Figure 4, where the simulated speciation in an Fe(III) solution with $2 \cdot 10^{-9}$ mol dm^{-3} Fe is given.

	P_1	P_2	P_3	P_4	P_5	P_6	P_7	P_8
V_1	2	8	3	6	1	5	7	4
V_2	6	3	1	5	8	2	4	7
V_3	7	5	8	3	4	1	2	6
V_4	5	1	4	8	7	6	3	2
V_5	1	7	5	4	2	8	6	3
V_6	3	2	7	1	6	4	8	5
V_7	8	4	6	2	3	7	5	1
V_8	4	6	2	7	5	3	1	8

Figure 3 *A Latin Hypercube lattice. Note that no number appears twice in each column and in each row. P_i are the distributed parameters, while V_i represent the input vectors to the simulation. Hence, in order to obtain an estimate of the mean value of the output distribution, LHS requires a minimum of m repetitions of the simulation, given m parameter distributions. In practice, a higher number of repetitions (m = 50 to m = 500) will be performed, requiring the EDFs to be stratified into m strata.*

The output distributions are given as modified Box plots. The skewed output distributions (note the log ordinate scale) are readily visible. Such speciation calculations play an important role, e.g. in the derivation of sorption parameters [15]. Even moderate measurement uncertainty in the thermodynamic parameters easily render such sorption data void.

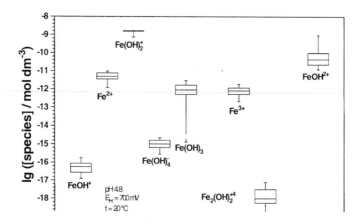

Figure 4 *A LJUNGSKILE simulation of the likely composition of an solution with a total Fe concentration of $2 \cdot 10^{-9}$ mol dm^{-3}. The output distributions are represented as modified Box plots where the central bar gives the median, the box encloses the upper and lower 68% percentiles and the whiskers give the 95% percentiles. The non-Normality of the output distributions is readily visible.*

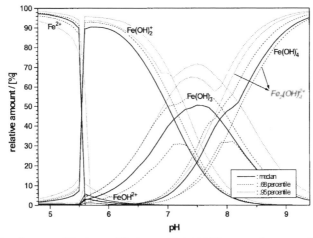

Figure 5 *Distribution of Fe in equilibrium with Fe(OH)$_{3a}$ as a function of pH. The input uncertainties have been expressed by uniform distributions. Median values, .68 and .95 percentiles are given. For Fe$_2$(OH)$_4^{2+}$, the median is always zero, but the upper percentiles are significantly larger than zero. Note that some species may have relative amounts between zero per cent and 80 per cent (e.g. Fe$_2$(OH)$_4^{2+}$; notwithstanding the rather poor evidence for the very existence of such a solution species).*

5 CONCLUSIONS

The tasks in deriving a complete measurement uncertainty budget for measurement values from complex situations have some similarity to risk analysis. The influence quantities need to be expressed by meaningful distributions. These distributions form the input para-meters of a detailed numerical model of the situation of interest. While in risk assessment the distributions often result from subjective judgement, they can be derived from other measurement in chemical metrology. The measurement of pH [16] and thermodynamic data [17] are practical examples.

A practical implementation demonstrating the application of LHS is the LJUNGSKILE speciation code [10,13]. At present, Normal distributions and uniform distributions are built-in parametric distributions. There is, however, no problem to implement empirical distributions, e.g. expressed by Chebyshev polynomials. Another point of consideration is the much more complicated information that must be communicated. The lack of statistical education beyond Normal distributions and Pearson correlation coefficients in most chemistry curricula may hamper the educated application of advanced metrological concepts. The 'metrological chemist' may supplement the other chemical faculties in future.

References
1 EURACHEM/CITAC Quantifying Uncertainty in Chemical Measurement. EURACHEM/CITAC 2000
2 G. Meinrath, Chemometrics Intell. Lab. Syst. 2000, **51**, 175
3 J. Bjerrum, G. Schwarzenbach, L.G. Sillèn, Stability Constants of Metal Ion Complexes, Special Publication 6, Royal Chemical Society, London, UK (1975)

4 M. Bond, G. T. Hefter, Critical Survey of Stability Constants and Related Thermodynamic Data for Fluoride Complexes in Aqueous Solution, IUPAC Chemical Data Series 27, Pergamon Press, Oxford, UK (1980)

5 R. M. Smith, A. E. Martell, Critical Stability Constants Vol. 6, Plenum Press, New York, USA (1989)

6 ISO, Guide to the Expression of Uncertainty in Measurement. 2nd ed. 1995. ISO Geneva/CH

7 G. Meinrath, S. Lis, Fresenius J. Anal. Chem. 2001, **369**, 124

8 G. Meinrath, Fresenius J. Anal. Chem., 2001, **369**, 690

9 G. Meinrath, A. Kufelnicki, M. Świątek, Accred. Qual. Assur. 2005, **10**, 494

10 Ödegaard-Jensen, C. Ekberg, G. Meinrath, Talanta, 2004, **63**, 907

11 M. D. McKay, R. J. Beckman, W. J. Conover, Technometrics 1979, **2**, 239

12 G. Meinrath, Ch. Ekberg, A. Landgren, J.O. Liljenzin, Talanta, 2000, **51**, 231

13 http://www.geo.tu-freiberg.de/software/Ljungskile/index.htm

14 D. L. Parkhurst, Water Resources Investigation Report 95-4227, US Geological Survey, Lakewood/CO, USA (1995)

15 G. Meinrath, B. Merkel, A. Ödegaard-Jensen, Ch. Ekberg, Acta hydrochim. hydrobiol. 2004, **32**, 154

16 G. Meinrath, M.F. Camoes, P. Spitzer, H. Bühler, M. Mariassy, K. Pratt, C. Rivier, this proceedings

17 G. Meinrath, S. Lis, A. Kufelnicki, this proceedings

A SYSTEMATIC APPROACH IN THE EVALUATION OF UNCERTAINTY IN ANALYTICAL CHEMISTRY. APPLICATION TO ICP-AES ANALYSIS

P. Carconi, R. Gatti, G. Zappa, C. Zoani

ENEA - Casaccia Research Centre (Rome) - Dpt. Biotechnologies, Protection of Health and Ecosystems

1. INTRODUCTION

Despite the effort of the EURACHEM/CITAC Guide (1) to adapt ISO (2) principles to chemical measurements, a simple and practical procedure for estimating measurement uncertainty is not available for analytical chemistry laboratories yet.

The EURACHEM *bottom-up* approach for measurement uncertainty estimation involves the identification of uncertainty sources starting from the basic expression used to calculate the measurand. Several applications of the *bottom-up* approach in analytical chemistry (3,4,5) show some difficulties mainly for the sampling and the sample preparation phases, where the ISO approach (1,2) is generally inapplicable. In fact it is often impossible to recognize all uncertainty sources and to establish how they contribute to the uncertainty budget.

Therefore we propose a systematic approach that consists in a preliminary examination in order to distinguish, within the analytical procedure, the proper measurement process from the sample physical-chemical transformations that accompany it. By carrying out this break down, it is possible to apply different uncertainty evaluation approaches optimized for the two occurrences. It is widely recognized that is convenient to group "type A" uncertainties (2) in a single term in order to replace the run-to-run variability of each influence factor with the overall variance of the method. However, until we don't have a homogeneous and stable sample, it is impossible to group all the random effects for measuring "type A" uncertainty by repetitions of the experimental procedure and to quantify it as a standard deviation of the measured values. From that comes the necessity to perform the break down of the analytical procedure at the level of test portion aquirement. Also concerning the measurement of "type B" uncertainty, both recovery experiments and bias studies based on certified reference materials (CRMs), are performed on the test portion. Indeed, recovery experiments require a homogeneous sample and CRMs are usually available in "ready to use" form and not in "natural" form.

Finally a practical application of the proposed approach to Inductively Coupled Plasma Atomic Emission Spectrometry Analysis (ICP-AES) is introduced.

2. EVALUATION OF MEASUREMENT UNCERTAINTY AFTER THE BREAK DOWN OF ANALYTICAL PROCEDURE

In order to apply the proposed systematic approach for the distinction between proper measurement process and the sample physical-chemical transformation, at the beginning we have to break down the analytical procedure in two steps. In the first step we consider all the phases from the sampling up to the acquirement of a homogenous sample, that the test portion is picked out from. Then in the second step we consider the subsequent analytical phases up to the final result.

These two steps don't always match exactly with the above-mentioned distinction, but the first step always includes a sample physical-chemical transformation and the second step always includes the measurement process. The first step could also include some intermediate measurements that directly affect the result (ex: mass or volume measurements during the sample fractioning or the phase separation). On the other hand, second step could include a physical-chemical transformation as sample pre-treatment (ex.: sample dissolution) and often includes a sample phase transformation during the instrumental analysis (ex: atomization in atomic spectroscopic analysis).

After the break down of the analytical procedure we have to analyze each phase of the two steps, in order to identify the uncertainty sources. It is necessary to distinguish the uncertainties arising from the measurements of the parameters that explicitly appear in the basic expression used to calculate the measurand (named afterward as *base parameters*) from the other uncertainty sources. In fact random effects have to be considered only in the first case and the uncertainty deriving from calibration of the equipment employed for measuring the aforesaid parameters has to be considered always.

Moreover the *bottom-up* approach is directly applicable only to uncertainties arising from the measurements of *base parameters* and from the measurements of other parameters that directly affect the values of the *base parameters* (ex: measurement of the ambient temperature that affects the volume value by the coefficient of volume expansion).

Indeed the application of the *bottom-up* approach to uncertainties deriving from different sources, usually involves a complex pathway starting from the identification of all influence parameters that affect the sample physical-chemical transformations. Then we have to establish the relationships that occur among the influence parameters, any induced change of sample physical-chemical state and the result. Considering that it is often very difficult to fix a mathematical function for these relationships, the related quantification of the uncertainties is seldom feasible by this way.

An alternative pathway is to evaluate the occurrence of systematic effects and use the uncertainty related to this evaluation for quantifying "type B" uncertainties (2).

3. "TYPE B" UNCERTAINTY EVALUATION

Systematic effects are usually estimated by recovery studies (4). However, a general approach for this estimation doesn't exist yet. In order to develop a pathway for the evaluation of "type B" uncertainties in analytical chemistry, it is necessary to examine the whole analytical procedure considering possible systematic effects and afterward estimate the uncertainties related to bias evaluation or correction.

The systematic effects could be classified in three groups:
-   additive effects (plus or minus);
-   multiplicative effects;
-   worsening of representativeness.

P00FM428V

UNIVERSITY OF STRATHCLYDE

Ship To: 28995002 C

UNIVERSITY OF STRATHCLYDE
CURRAN BUILDING
101 ST JAMES ROAD
GLASGOW
STRATHCLYDE
G4 0NS

Bill To: 28995002

UNIVERSITY OF STRATHCLYDE
CURRAN BUILDING
101 ST JAMES ROAD
GLASGOW
STRATHCLYDE
G4 0NS

ISBN
0−85404−848−0

Qty
1

Sales Order
C 4655487 10

Customer P/O No
0L273 : UNSTRAS0 −102

Cust P/O List
99.95 GBP

Fund:

Title: Royal society of chemistry: special

Format: Cloth/HB
Author:
Publisher: Royal Society of Chemistry,
Volume: 307
Edition:

Year: 2006.

Order Specific Instructions

COUTTS LIBRARY SERVICES: UK 005304161 LGUK001R001 UKRWLG12

Additive effects, as contaminations or analyte losses, could arise from interactions (I) between the sample and the environment, as well as between the sample and the contact materials, other than reagent contaminations. These effects may occur in many phases of both *step 1* and *step 2* of the analytical procedure. In the phase of instrumental analysis, additive effects could be also produced by the increase or the decrease of the zero instrumental line due to experimental condition variations (C) or matrix effects (M).

Multiplicative effects are mainly caused by matrix (M) or by experimental condition variations (C) occurring during the sample preparation, the test portion pre-treatment or the analyte measurement. Worsening of representativeness may occur as a consequence of an inappropriate sampling or sub-sampling procedure (P). Effects due to un-calibration (U) of the equipment employed for measurement of *base parameters* or caused by an inappropriate procedure (P) of data processing could also occur. Table 1 shows the possible systematic effects that occur during the two steps of the analytical procedure.

Table1: Possible systematic effects in the analytical procedure

Causes: I = Interactions with environment, contact materials and reagent contamination U = Un-calibration of measuring equipment C = experimental Condition variations P = inappropriate Procedure		Additive effects		Multiplicative effects	Worsening of representativeness
		+	-	x	P
STEP 1 (from sampling to test portion)	Sampling	I			P
	Sample transport and storage	I	I		
	Sample preparation without fractioning or phase separation	I	I		
	Sample fractioning or phase separation	I	I	U,C,M	
	Sub-sampling (test portion picking out)	I			P
STEP 2 (from test portion to result)	Test portion size measurement (mass or volume)			U	
	Test portion pre-treatment and dilution	I	I	U,C,M	
	Analyte measurement	I,U,C,M	I,U,C,M	U,C,M	
	Data processing	P	P	P	

The occurrence of the bias effects could be evaluated by:
- employment of reference materials (primary or matrix-CRMs);
- analysis of procedure blanks;
- comparison of results with those of reference methods;
- recovery studies on spiked samples, spiked blanks or previously studied material.

In any case the evaluation is a comparison between two results or one result and a reference value, by which comes the decision on the occurrence of a significant effect and if a consequent correction should be applied. According to systematic effects are absolute (additive) or relative (multiplicative), we have to consider a difference (*b*) or a ratio (*R*) between the observed (C_{obs}) and the reference (C_{ref}) values:

$$b = \overline{C}_{obs} - C_{ref} \qquad [1]$$

$$R = \frac{\overline{C}_{obs}}{C_{ref}} \qquad [2]$$

The bias uncertainty [$u(b)$ or $u(R)$] is then estimated by combining the standard uncertainties of the two values:

$$u(b) = \sqrt{\frac{s^2(C_{obs})}{n} + u^2(C_{ref})} \qquad\qquad\qquad [3]$$

$$u(R) = R \cdot \sqrt{\frac{s^2(C_{obs})}{n \cdot \overline{C}_{obs}^2} + \left(\frac{u(C_{ref})}{C_{ref}}\right)^2} \qquad\qquad [4]$$

where $s(C_{obs})/\sqrt{n}$ is the standard deviation of the mean of the n replicate measurements of C_{obs}.

In agreement whit the ISO Guide (2), bias corrections should be applied for all significant systematic effects. The significance is tested as follows:

$$|b| \leq k \cdot u(b) \qquad\qquad\qquad\qquad\qquad\qquad [5]$$

$$|R - 1| \leq k \cdot u(R) \qquad\qquad\qquad\qquad\qquad [6]$$

As k value, is frequently adopted the two-tailed value from Student's distribution.

When the bias is statistically significant, all results should be corrected (C_{cor}) by b or R values:

$$C_{cor} = C_{obs} - b \qquad\qquad\qquad\qquad\qquad [7]$$

$$C_{cor} = \frac{C_{obs}}{R} \qquad\qquad\qquad\qquad\qquad\qquad [8]$$

In both cases (bias significant or not) the uncertainty of b [$u(b)$] or R [$u(R)$] should be included in the evaluation of the combined standard uncertainty of the result.

Most of the chemical measurements are relative measurements based on comparison between the sample signal and a reference signal arising from a known amount of analyte. As a consequence, the phases that must undergo to bias evaluation are strictly related to the point where the reference material is introduced within the analytical procedure, other than how much the reference material closely matches to the sample. In all relative measurements, we employ one or more reference materials (calibrant RMs) during instrumental analysis. Therefore the need of bias evaluation for this phase only depends on the type of reference material(s) employed for the calibration (pure substances or closely matched matrix materials). The choice among the different possibilities for evaluating systematic effects strictly depends on the availability of reference materials and reference methods and their uncertainties. For instance, matrix-CRMs don't cover all analytical requirements and their uncertainties are very large as a consequence of the certification procedure that usually involves inter-laboratory studies. On the other hand, the employment of a pure substance as reference material, although it permits to narrow the uncertainty of bias evaluation, doesn't allow to cover the overall measurement process.

4. APPLICATION TO ICP-AES ANALYSIS

In order to test the proposed systematic approach in practice, we consider the analysis of nickel (Ni) in a vegetable sample by ICP-AES. In this case the break down is placed immediately before the dissolution step, after the acquirement of a homogeneous and stable sample by which the test portions are picked up. The basic expression used to calculate the measurand (C_{Ni}) is:

$$C_{Ni} = \left[f_d \cdot \frac{C_{std}}{I_{std}} \cdot I_{sample} \cdot \frac{V_1}{m} \right] \qquad [9]$$

In the *step 1* we have only one measurement, that is the determination of the dry mass fraction (f_d). In the *step 2* the measurement process starts from the test portion and involves the measurement of the test portion mass (m) and the test solution volume (V_1), besides the measurements of the mean emission intensity signals of the n_1 replicates of the test solution (I_{sample}) and the standard solution (I_{std}).

It is convenient to group "type A" uncertainties into three clusters:

- "type A" uncertainty related to the determination of the dry mass fraction (f_d) by measurement of the dry mass (m_d) and the wet mass (m_w):

$$u_{typeA}(f_d) = f_d \cdot \sqrt{\left(\frac{s(m_d)}{m_d} \right)^2 + \left(\frac{s(m_w)}{m_w} \right)^2} \qquad [10]$$

- "type A" uncertainty related to the determination of sensitivity (C_{std}/I_{std}) by measurement of the emission intensity signal (I_{std}) of a standard solution (C_{std}), prepared from a *mother* solution (C_m), diluted V_2 to V_3:

$$u_{typeA}\left(\frac{C_{std}}{I_{std}} \right) = \left(\frac{C_{std}}{I_{std}} \right) \cdot \sqrt{\left(\frac{u_{typeA}(C_m)}{C_m} \right)^2 + \left(\frac{1}{n_1} \right)\left(\frac{s(I_{std})}{I_{std}} \right)^2 + \left(\frac{s(V_2)}{V_2} \right)^2 + \left(\frac{s(V_3)}{V_3} \right)^2} \qquad [11]$$

- "type A" uncertainty related to replicate test solution preparation and measurement of its emission intensity signal ($\overline{I}_{sample}$), could be expressed as the standard deviation of the replicate "preparation-measurement" (s_{rep}) of the test solution:

$$u_{typeA}\left(\overline{I}_{sample} \cdot \frac{V_1}{m} \right) = s_{rep} \qquad [12]$$

where V_1 is the test solution volume and m is the test portion mass.

For "type B" uncertainty evaluation, we have considered the following effects:

➤ **uncertainty of the certified reference solution concentration**
the uncertainty related to the concentration value of the *mother* reference solution has been deduced from a previous work (5);

➤ **un-calibration of measurement equipments (balance and volumetric equipments)**
the calibration of measurement equipments has been checked by IS traceable reference masses. Uncertainty has been estimated by combining the uncertainties of the reference value and the observed value, according to [4]. For checking the calibration of volumetric equipments (volumetric flasks, micropipettes), the reference value has been obtained gravimetrically by water density data (6);

➤ **contaminations from sample contact materials**
contaminations from sample contact materials during *step 1* phases (sample storage, milling, homogenization, drying and sub-sampling) has been evaluated by a standardized loop-procedure of washing-leaching. This procedure consists on replicate washing of the apparatus until the concentration value in the leaching solution reaches a constant value or becomes less than the detection limit (D.L.). The test was stopped because D.L. was reached; from the D.L. in the leaching solution, by an appropriate conversion factor the uncertainty related to this evaluation procedure has been calculated. Contamination from sample contact materials during *step 2* has been evaluated by preparing and measuring a procedure blank. The uncertainty related to this

bias evaluation has been quantified by combining the standard deviations of the mean
(*n* replicates) concentration values for sample (C_{sample}) and for procedure blank (C_{blank}):

$$u\left(Contamination_{step2}\right) = \sqrt{\frac{s^2\left(C_{sample}\right)}{n} + \frac{s^2\left(C_{blank}\right)}{n}}$$ [13]

> **spectral interferences**

the occurrence of line spectral interferences has been evaluated by comparison between
the analysis performed at two different wavelengths (λ_1 and λ_2). The difference between
the two concentration mean values $\left(\overline{C}_{\lambda1} - \overline{C}_{\lambda2}\right)$ has been tested for it's significance
by *test t*, employing, as standard deviation, that associated with the difference itself. In
any case, also when the test shows no significant difference ($t < t_{crit}$), uncertainty related
to this bias evaluation has been calculated as a combination of the standard uncertainties
of the two mean concentration values (*n* replicates):

$$u\left(\overline{C}_{\lambda1} - \overline{C}_{\lambda2}\right) = \sqrt{\frac{s^2\left(C_{\lambda1}\right)}{n} + \frac{s^2\left(C_{\lambda2}\right)}{n}}$$ [14]

Background interferences have not been specifically evaluated because the method
works in background correction mode, therefore the "type A" uncertainty related to this
correction is already included in the terms $s(C_{sample})$ or s_{rep} considered in [12] and [13];

> **matrix interferences on sensitivity**

the occurrence of matrix interferences on sensitivity has been evaluated by standard
addition and the uncertainties have been estimated by combining standard uncertainties
of the two mean values (*n* replicates):

$$u\left(\frac{\overline{C}_{obs}}{\overline{C}_{ref}}\right) = \sqrt{\frac{s^2\left(C_{obs}\right)}{n \cdot \overline{C}_{obs}^2} + \frac{s^2\left(C_{ref}\right)}{n \cdot \overline{C}_{ref}^2}}$$ [15]

As an alternative, all "type B" uncertainties could be grouped by employing an appropriate
CRM like for example LGC Strawberry leaves 7162 (Ni certified value: 2.6 mg/kg;
uncertainty: 0.7 mg/kg). Considering the physical-chemical state of the CRM (dehydrated,
milled), it is possible to use this procedure only for the *step 2*. However, in this case the
associate uncertainty should be estimated applying [4] and that would involve a very high
uncertainty value.

5.   CONCLUSIONS

The "break down" of analytical procedure permits to group all the random effects for
measuring "type A" uncertainty by repetitions of experimental procedure and to quantify it
as a standard deviation of the measured values. Bias evaluation is the most convenient way
to estimate "type B" uncertainty caused by possible occurrence of systematic effects. For
evaluating bias effects caused by sample contamination from environment or contact
materials, it could be convenient to have high pure matrix reference materials (HP-RMs).
These HP-RMs could be also conveniently used for recovery tests by spiking them.

Moreover the application of the systematic approach to the ICP-AES analytical
techniques points out the need to develop a specific software to calculate the measurement
uncertainty. This software, that would have to consider beyond to the run-to-run variations
also the evaluation of the bias occurrences, should supply a defined calculation procedure,
operating interactively with the instrumental software and accepting also external inputs as
calibration data, reference values and so on.

References

1   S.L.R. Ellison, M. Rosslein, A. Williams (eds.), *Eurachem/CITAC Guide 2nd ed.*, 2000.
2   *ISO, Guide to Expression of Uncertainty in Measurement*, Geneva, 1993.
3   E. Desimoni, B. Brunetti, *Annali di Chimica*, 2005, **95**.
4   V. J. Barwick, S.L.R. Ellison, *Analyst*, 1999, **124**,981.
5   P. Carconi, R. Gatti, N. Portaro, P. Sangiorgio, G. Zappa. *ENEA RT/2003/35/BIOTEC.*
6   E.W. Lemmon, M.O. McLinden, D.G. Friend, *NIST Standard Reference Database Number 69*, 2001.

SELF-REFERRING PROCEDURE FOR THE FULL CALIBRATION OF A DYNAMIC GAS DEVIDER.

R. Beltramini Boveri

Be.T.A. Strumentazione Srl – Borgolavezzaro (No) - Italy

1   INTRODUCTION

Gaseous reference materials may be prepared "just in time" by diluting pure gases or certified gas mixtures. The concentration uncertainty of the mixture prepared by dilution is a combination of the input gases concentration uncertainty and the dilution process uncertainty : we want to talk about this last point introducing a new way to consider the dilution uncertainty.

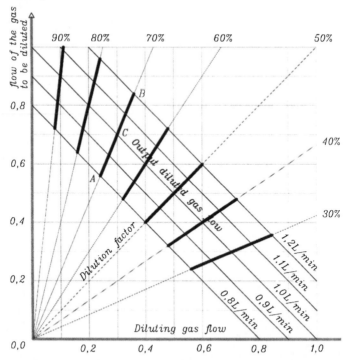

Observing the img. 1, it's evident that only one couple of flows (diluting gas and gas to be diluted), identified by point "C" for dilution rate 70%, allows the production of a defined diluted gas flow at a defined dilution factor (and then defined output concentration). People working on gas analysers calibration know that while dilution factor uncertainty reduction is very important, the amount of produced gas flow may have a big tolerance without affecting the calibration result. The location of the points in the segment A-B has the propriety that the ratio between the coordinates is constant (e.g. 70%) and the total diluted flow is in the range 1,2-0,8 L/min. That means not one only couple of flows defines a dilution rate, or better : a low uncertainty of absolute flow values is sufficient but not necessary condition to have a low dilution factor uncertainty.

In this paper we move the attention from absolute flows to flows ratio and then from flow regulators traceable calibration to flow regulators matching . To keep running our strategy, it was necessary to realise a new pneumatic design and a new calibration procedure which may be defined "self-referring" : in fact it uses an internal reference (not traced) and internal flows comparisons.

In our calibration procedure, zero and non linearity errors of all the internal Mass Flow Controllers (MFCs) are detected and compensated but the sensitivities are matched to that one of an internal MFC selected as reference : the sensitivity (gain) of this MFC stays unknown, but the ratio with the sensitivities of the other MFCs is perfectly controlled in all the points of the range.

This paper will introduce the hardware and the calibration procedure of such a gas dilution device. Short term errors of the MFCs (noise) and flow coefficients of the MFCs versus the various chemical components are not considered in this presentation.

This device, now prototyped, not only allows full calibration without a traceable reference, but is capable to produce diluted gas in the ppm range, diluting pure gases. As known, pure gases are generally available with better accuracy and longer stability than pre-mixed.

2   THE DILUTING DEVICE

As drafted in img. 2, it works with 6 MFCs (3 with lower range and 3 with higher range) configured as three simple blenders : each one is composed by one MFC of higher range

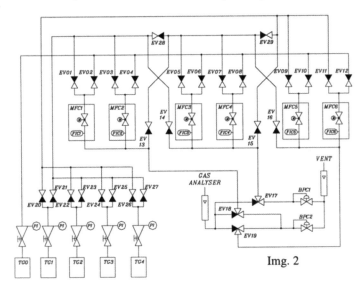

Img. 2

and one MFC of lower range. In our prototype (but it's not strictly required) higher range is 1 litre/min and lower range is 0,125 litres/min.: avoiding to use each MFC under 10% of his range (to get good accuracy results), the dilution rate of each elementary blender can reach the value 1: 80.

Elementary blenders, and even individual MFCs of the diluting device are joint one each other through a solenoid valve network which allows series, parallel and mix type connections.

The particular construction of the device gives interesting benefits both in accuracy and flexibility :

a)   the accuracy takes big advantages from the self-referring calibration procedure (later described) that may be repeated at low cost in short time and everywhere when some repetitive error is suspected.

b)   The flexibility of the device allows a very wide range of diluted flows and dilution factors.   Parallel or series combination of elementary dilution units gives the capability to apply dilution rates values from 1:1up to 1:512.000.

In this case, a concentration of few ppm may be produced by diluting pure gases.

3   THE SELF-CALIBRATION PROCEDURE

Self-calibration is a sequence of activities, controlled by the PC software, in which one gas only (generally pressurised air or nitrogen) is used as source.

The steps sequence includes :

•   realising the pneumatic interconnections between MFCs
•   setting the working point of the MFCs used as flow regulators
•   reading the flow measured by MFCs used as meters (with open valve)
•   calculating the unknown flow errors.

Doing each step (after the first), the measuring errors already calculated in the previous steps are accounted to correct the set points: In each step, an equation comes from comparing the flow or flows sum regulated by one or two MFCs and resultant flow measured by another MFC (used with open valve). Known and unknown measuring errors (zero error, sensitivity mismatching and non linearity errors) are included in those equations : unknown errors must be calculated solving the equation or the equations system.

Symbols used :
$s.p.(MFC1)$ = set point given to MFC1
$R1(MFC1)$ = flow reading by MFC1 in the step 1 of the cycle.
$Z1$ = zero error of MFC1
$S1$ = sensitivity error (at end of scale) of MFC1
$A1$ = non linearity error at 12.5% of the range for MFC1
$B1$ = non linearity error at 25.0% of the range for MFC1
$C1$ = non linearity error at 37.5% of the range for MFC1
$D1$ = non linearity error at 50.0% of the range for MFC1
$E1$ = non linearity error at 62.5% of the range for MFC1
$F1$ = non linearity error at 75.0% of the range for MFC1
$G1$ = non linearity error at 87.5% of the range for MFC1

3.1 Zero errors calculation

closing all the solenoid valves, the flow through each MFC must be zero. If the flow measurement is not zero, the read value is the zero error of the involved MFC.
$R1(MFC1) = Z1$; $R1(MFC3) = Z3$; $R1(MFC5) = Z5$

3.2 Sensitivity mismatching error calculation

The three higher range MFCs are considered now and one of them (MFC1)is assumed to be the internal reference ($S1 = 0$). The three considered MFCs are then series connected (img. 3) and MFC1 set point is set to end of scale + Z1 while the other two (MFC3 and MFC5) are set with fully open valve. Measurements R2(MFC3) and R2(MFC5) and relevant errors are compared with the flow regulated by MFC1

1 litre/min. $= R2(MFC3) -Z3 - S3$; *1 litre/min.* $= R2(MFC5) - Z5 - S5$

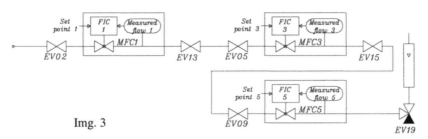

Img. 3

3.3 Non linearity errors calculations

The procedure checks linearity errors at 12.5, 25.0, 37.5, 50.0, 62.5, 75.0, 87.5 % of the range for each MFC. The pneumatic circuit is configured by solenoid valves in a way that the flows regulated by two MFCs (e.g. MFC1 and MFC3) are conveyed into a third MFC (e.g. MFC5).

Just as an example, we will see (img. 4) the first steps of the procedure :

Two MFCs receive a set point 50% corrected by known errors (zero and sensitivity) and the third measures the flows *sum*

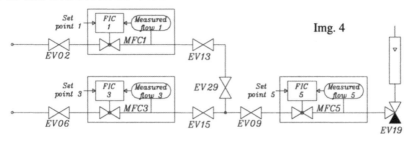

Img. 4

$s.p.(MFC1) = 0.5+Z1$; $s.p. (MFC3) = 0.5+Z2+0.5xS2$
A) $(0.5-D1) + (0.5-D3) = R3(MFC5)-Z5-S5$
$s.p.(MFC3) = 0.5+Z3+0.5xS3$; $s.p. (MFC5) = 0.5+Z5+0.5xS5$
B) $(0.5-D3) + 0.5-D5) = R3(MFC1)-Z1$
$s.p.(MFC1) = 0.5+Z1$; $s.p. (MFC5) = 0.5+Z5+0.5xS5$
C) $(0.5-D1) + (0.5-D5) = R3(MFC3)-Z3-S3$

From the equations system A-B-C, the values D1, D3, D5 (non linearity errors at 50% of the range) may be calculated.

The next steps of the calibration procedure are very similar to the above step :

3b) Two MFCs receive a set point 25.0% (to calculate B1, B3, B5)

3c) Two MFCs receive a set point 12.5 % (to calculate A1, A3, A5)

3d) one MFC get a s.p.12.5% and the second 25% (to calculate C1, C3, C5)

3e) one MFC get a s.p.12.5% and the second 50% (to calculate E1, E3, E5)

3f) one MFC get a s.p.25.0% and the second 50% (to calculate F1, F3, F5)

3g) one MFC get a s.p.37.5% and the second 50% (to calculate G1, G3, G5)

3.4 considering the lower range MFCs

Up to now, all the long term errors has been calculated for higher range MFCs.

Zero error calculation of lower range MFCs is identical to higher range MFCs.

To match the sensitivity error of lower range MFCs to that of higher range, MFC1can be used with s.p. = 12.5% of the range (corrected by all the known errors) as flow reference to find and correct sensitivity mismatching errors of MFC4 and MFC6. MFC3 or MFC5 may be used for the same correction on MFC2.

Non linearity errors calculation procedure for MFC2, MFC4 and MFC6 is very similar to that described for higher range MFCs.

4 CONCLUSION

Apparently, a conventional traceable calibration may have, compared to this procedure, the disadvantage to add at the uncertainty chain the uncertainty of the primary or traceable reference, which may be different for higher and lower range MFCs .

In a few weeks the first release of the controlling software will be available and data collection for statistical analysis will start.   The analysis of the variance in the calibration coefficients will help understanding the effect of the noise, that will be reduced by increasing the number of flow measures readings during the calibration procedure.

We expect that this device will be considered as a "primary tool" for high accuracy and high flexibility just in time reference gas mixtures preparation. (see prototype at img. 5).

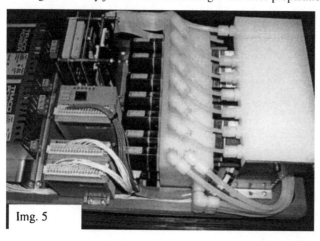

Img. 5

INVESTIGATION OF UNCERTAINTY RELATED TO MEASUREMENTS OF PARTICULATE ORGANIC POLLUTANTS

Angelo Cecinato[1], Catia Balducci[1], Alessandro Di Menno Di Bucchianico[2]

[1]Istituto Inquinamento Atmosferico CNR, Monterotondo Scalo RM, Italy
[2] APAT, Rome, Italy

1 INTRODUCTION

It is consolidated that the organic fraction of airborne particulates exploits a negative influence on ambient quality, due to its burden of long-term toxic compounds (e.g., polynuclear aromatic hydrocarbons PAH, pesticides and chlorinated dioxins). This is the reason why a number of investigation are conducted over the world, aimed to identify and quantify these chemicals and, finally, to control ambient pollution and preserve the health of populations. Nevertheless, several sources of biases and errors are known to be able to affect chemical determinations. In many cases, they give rise to wide uncertainty which impoverishes scientific meaning of investigations. Measurement uncertainty is a crucial topic whenever regulations or decisions about territory, human mobility and activity have to be adopted. For instance, PAH are regulated in Italy by concentration of benzo(a)pyrene (BaP). The threshold of 1.0 ng m^{-3} must not be exceeded in urban and "polluted" areas for yearly average. Also passing over the intrinsic limit of this approach, which neglects contribution to toxicity given by other PAC, the uncertainty arising from collection and chemical evaluation of BaP can reach 100%. Although a severe control for the entire procedure adopted would be necessary, it is often restricted to true analysis, whilst the reproducibility of measurement is not further investigated, e.g. by co-locating more collection devices or through parallel collection in sites formally similar, but experiencing different border situations.

2 EXPERIMENTAL

In one experiment conducted at Montelibretti (a semi-rural area located 30 km NE of Rome, Italy), two medium-flow lines (operating flow rate: 38 L min^{-1}) were set for collecting respirable and pulmonary fractions of airborne particles (PM$_{2.5}$ and PM$_{10}$, respectively). Particles were retained onto Teflon filters, held at finely controlled micro-ambient conditions (air temperature and pressure, pressure drop, aspiration flow, gas leak absence). Organic burdens of the two aerosol size fractions (including *n*-alkanes, PAH and fatty acids) were investigated to assess the *whole* uncertainty related to collection and analysis.

For this purpose, airborne particulates were collected over 24-h intervals from January to February and July up to September 2005. After sampling, loaded filters were pooled in groups according to the $PM_{2.5}$ to PM_{10} mass ratio, since we verified that its value varied concurrently with meteo-climatic situations. Chemical determinations of the target groups were based upon solvent extraction (in soxhlet with dichloromethane and acetone 4:1), fractionation (by column chromatography on neutral alumina and elution with 100% of iso-octane followed by iso-octane and dichloromethane 3:2) and CGC-MSD analysis. To determine acids, a sample aliquot was processed for esterification with BF_3 and propanol.

The former period of investigation was dedicated to test the representativeness of sampling location with respect to typology of site and to verify the repeatability and reproducibility of the method adopted. For this purpose, two identical sampling apparatuses were co-located, the former in a *clean* environment, the latter close to a small car parking zone inside the Research Area of our Institute.

The latter testing set was a true sampling of organic particulates under varying environmental conditions, aimed to highlight differences existing *intra-* and *inter* the PM_{10} and $PM_{2.5}$ fractions. In this case, the same sampling apparatus, equipped with two independent aspirating lines and both the PM_{10} and $PM_{2.5}$ inertial impactor heads, allowed to collect the airborne particulates exactly at the same site.

3   RESULTS AND DISCUSSION

The results of the winter tests are reported in Figure 1, where the average aerial concentrations of PAH and *n*-alkanes are shown. They indicate that:
-   an accurate choice of particulate collector location must be made (e.g., to prevent the influence of spotty emission sources); otherwise samples result to not be representative of ambient environment;

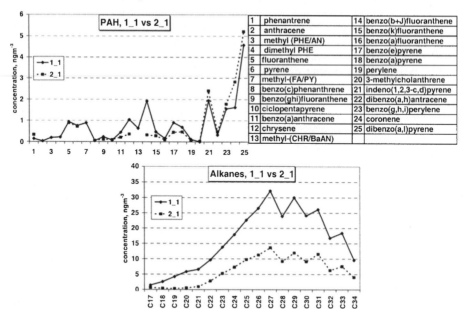

Figure 1, *average aerial concentrations of PAH and n-alkanes recorded in January and February 2005, by using two PM_{10} sampling lines*

- the repeatability of measurements can be retained within 10%, when an appropriate set of internal reference compounds is adopted for chemical analysis.

Table 1 shows the average burdens of n-alkanes (A), fatty acids (B) and PAH (C) in both respirable and pulmonary particulates over two daily date sets, respectively of September and August (*Sp* samples were collected during an event of saharian dust transport over our region, while *Ag* group refers to days when the $PM_{2.5}/PM_{10}$ ratio was close to 0.39). It is worth noting that not only the total contents of target pollutants, but also the respective distribution patterns of congeners within the respective groups were different. By looking to the whole data sets (not reported here), it was clear that any lack of control of operating conditions leaded to unexpected results, including apparent incongruities. For instance, in some cases the $PM_{2.5}$-associated concentrations of compounds seemed to be larger than the corresponding ones present in PM_{10}; that occurred when the filter pre-cleaning and analytical internal standard spiking were not properly conducted.

Table 1, *average concentrations of n-alkanes (A), fatty acids (B) and PAH (C) in PM_{10} and $PM_{2.5}$, observed in two airborne particulate sets collected in September (Sp) and August (Ag) 2005 at Montelibretti (Rome, Italy). Symbols: C_n indicates the n-alkanes with n C atoms. A_n indicates the n-alkanoic and alkenoic acids with n C atoms. PAH: Fa=Fluoranthene, Py=Pyrene, Bz(c)Phe=Benzo(c)phenantrene, Bz(a)An=Benzo(a)anth-racene, Chr=Chrysene, Bz(b)Fa=Benzo(b)fluoranthene, Bz(K+J)Fa=Benzo(k+j) fluoranthene, Bz(e)Py=Benzo(e)pyrene, Bz(a)Py=Benzo(a)pyrene, Pe=Perylene, In(1,2,3-cd)Py= Indeno(1,2,3-cd)pyrene, Bz(ghi)Pe=Benzo(ghi)perylene, Dibz(a,i)Py=Dibenzo-(a,i)pyrene.*

ALKANES A					ACIDS B					POLICYCLIC AROMATIC HIDROCARBON C			
	Sp (ng m⁻³)		Ag (ng m⁻³)			Sp (ng m⁻³)		Ag (ng m⁻³)			Sp (ng m⁻³)		Ag (ng m⁻³)	
	PM_{10}	$PM_{2.5}$	PM_{10}	$PM_{2.5}$		PM_{10}	$PM_{2.5}$	PM_{10}	$PM_{2.5}$		PM_{10}	$PM_{2.5}$	PM_{10}	$PM_{2.5}$
C_{18}	1.11	1.77	2.32	1.03	A_{15}	3.98	4.77	13.47	7.16	Fa	0.21	0.19	0.07	0.09
C_{19}	2.30	3.08	3.73	1.77	$A_{16:1}$	5.28	7.42	19.68	8.04	Py	0.18	0.17	0.06	0.07
C_{20}	2.71	3.32	3.75	1.88	A_{16}	40.60	21.16	79.40	59.76	Bz(c)Phe	0.01	0.01	0.01	0.00
C_{21}	2.99	3.43	3.69	1.84	A_{17}	1.61	1.57	4.44	2.60	Bz(a)An	0.05	0.06	0.03	0.03
C_{22}	3.19	3.15	3.88	1.69	$A_{18:1}$	22.02	17.73	54.14	23.71	Chr	0.17	0.17	0.09	0.07
C_{23}	2.94	2.67	3.21	1.32	A_{18}	19.51	15.30	34.44	31.91	Bz(b)Fa	0.11	0.11	0.09	0.08
C_{24}	2.51	2.00	2.79	1.07	A_{19}	0.38	0.31	0.40	0.20	Bz(k+J)Fa	0.11	0.12	0.1	0.1
C_{25}	2.16	1.34	1.89	0.80	A_{20}	2.68	1.65	4.22	2.02	Bz(e)Py	0.26	0.26	0.22	0.17
C_{26}	2.21	1.65	1.35	0.74	A_{21}	0.49	0.23	0.05	0.02	Bz(a)Py	0.08	0.08	0.05	0.07
C_{27}	3.54	1.72	2.03	1.27	A_{22}	3.27	2.25	3.89	2.22	Pe	0.04	0.03	0.02	0.01
C_{28}	3.56	1.66	2.19	1.33	A_{23}	0.77	0.57	0.05	0.00	In(1,2,3,-cd)Fa	0.02	0.02	0.02	0.01
C_{29}	7.08	3.23	3.93	2.74	A_{24}	4.18	3.06	3.58	3.48	Dibz(a,h)An	0.05	0.06	0.05	0.06
C_{30}	4.49	2.24	2.54	1.73	A_{25}	0.78	0.47	0.00	0.00	In(1,2,3,-cd)Py	0.14	0.13	0.1	0.09
C_{31}	4.65	2.6	3.03	1.78	A_{26}	3.32	1.82	5.31	0.00	Bz(ghi)Pe	0.19	0.11	0.11	0.08
C_{32}	2.28	0.82	1.37	0.81	A_{27}	0.00	0.00	0.00	0.00	Dibz(a,l)Py	0.08	0.06	0.05	0.00
C_{33}	1.32	0.71	0.89	0.45	A_{28}	2.58	0.71	1.92	0.00	Cor	0.10	0.08	0.05	0.04

MICRONUCLEUS TEST IN PERIPHERAL ERYTHROCYTES OF FISH: VARIABILITY IN MICROSCOPE SCORING

Daniela Conti, Sabrina Barbizzi, Vanessa Bellaria, Alessandra Pati, Stefania Balzamo, Maria Belli

Agenzia per la Protezione dell'Ambiente e per i Servizi Tecnici (APAT)- Servizio Laboratorio, Misure e Attività in campo, Via di Castel Romano 100, 00128 Rome, Italy

1. INTRODUCTION

The micronucleus (MN) test is considered to be a reliable and useful tool for assessing chromosome damage induced by pollutants in different aquatic organisms[1-8]. Fish are frequently used as test organisms, as they can be easily handled in the laboratory and easily exposed to genotoxicants[1,9,10]. In addition, fish react to chemicals like higher vertebrates[1,11,12] and generally, the frequency of the micronucleus in fish is assessed in peripheral blood erythrocytes[1-3, 7-9,13,14-17].

Several variables have to be considered to compare micronucleus test results performed in different laboratories. The most important is the "between-scorers" experience on microscope scoring of the slides, as the use of personalized criteria for selecting the different cell types represent the most relevant source of variability. The scorer's experience and his capability play a relevant role in discriminating the true and false micronucleus and in identifying micronuclear and nuclear boundaries[18-20]. To this end, in the frame of a slide-scoring exercise, performed involving 34 laboratories from 21 countries (HUMN project)[18], a considerable "within" and "between" laboratories variation was observed in the MN frequency and micronucleated cells.

The present study is aimed at the evaluation of the following variables on microscope scoring for micronucleus tests in fish with circulating erythrocytes:
1. Minimum number of peripheral erythrocytes that should be scored for micronucleus frequency in each specimen;
2. Individual cell counting uncertainty (single skilled scorer)[21];
3. "Within-laboratory" cell counting reproducibility.

A micronucleus test in fish, with circulating erythrocytes, was organized using the marine fish *Dicentrarchus labrax* L. (Sea bass). This organism has already been used for acute toxicity[22,23], and genotoxicity testing[24,25] of environmental agents. Sea bass specimens were exposed in the laboratory to water polluted with benzene. Benzene, used as a positive control in genotoxicity experiments, has demonstrated to be able to induce a micronucleus in fish erythrocytes[3, 26, 27]. During the test, a set of scoring criteria for normal and micronucleated Giemsa stained erythrocytes, and several microphotos was provided to the scores, to assist them during the cell counting on the slides.

2 MATERIALS AND METHOD

2.1 Fish exposure and micronucleus test

Juvenile *Dicentrarchus labrax* L. specimens (average weight 3.2±0.5 g and average length 7.4±0.5 cm) were supplied from the saltwater fish farm "Valle Ca' Zuliani" located in Rovigo (Italy). Fish were acclimated under laboratory conditions at 20±1°C, 12/12 h dark/light mode in well-aerated and filtered saltwater (salinity 20±1‰; pH 7.9) for 2 weeks and fed up to 24 hours before the beginning of the experiment.

Ten fish specimens were then exposed for 6 days to salt water polluted with benzene [Sigma, CAS 71-43-2] at a concentration of 10 ppm. Additional 10 fish were kept in normal saltwater, as control group. Due to the volatility of benzene, the salt water polluted with benzene was renewed each day. At the end of the exposure, blood samples were taken through the caudal vein and smeared on to clean microscope slides. After fixation with pure ethanol (20 minutes) and staining with 10% Giemsa solution (15 minutes), the slides were examined by light microscopy (Nikon Eclipse E 600) under an oil immersion lens (1000X magnification).

The frequency of micronucleated erythrocytes (FEMN‰) was obtained from the ratio between the total number of micronucleated erythrocytes (EMN) and the total number of erythrocytes analysed (ETOT) per 1000.

2.2 Scoring criteria

The scoring criteria for the peripheral erythrocyte micronucleus assay are shown in table 1.

Cell distribution on the slide	
	Micronuclei (fig. 3)
Fish erythrocyte examination should be performed in the areas of the slides where the cells are well and uniformly distributed.	• Should be smaller than the main nucleus (from 1/10 to ad 1/30 of the main nucleus) (Al-Sabti, 1995)
A correct cell distribution is obtained when at least 50% of the erythrocytes observable in one field, has the characteristics listed below (fig.1).	• Should be round or oval shaped
	• Usually they have the same staining intensity as the main nucleus, but occasionally staining may be more or less intense
Erythrocytes (fig. 2)	• They are non-refractile and can be readily distinguished from artefacts such as staining particles
• The cells should be oval shaped	
• The nucleus should be oval or round shaped and centrally located	• They are not linked or connected to the main nucleus (fig. 4)
• The nucleus should be dense and uniformly stained	• They may touch the main nucleus , but the micronuclear boundary should be distinguishable from the nuclear boundary
• The nucleus should have the nuclear membrane intact	
• The cells should have a cytoplasmic boundary or membrane intact and be clearly distinguishable from the cytoplasmic membrane of surrounding cells	• *They may be present in cells with nuclear abnormality. In this case, if the analysis also involves the count of NA, the cell that has both an MN and an NA should be record as a cell with MN and separately as a cell with NA.*
• The cells may touch, but the cytoplasmic boundary of each cell should be distinguishable	
• Erythrocytes containing nuclear abnormalities (NA) should not be considered or should be separately recorded.	

2.3 Counting criteria

2.3.1 Minimum number of peripheral erythrocytes. The minimum number of peripheral erythrocytes that should be scored in each specimen for micronucleus frequency, was determined following this procedure: one slide of the control group and one of the benzene treatment (randomly chosen) were scored by repeated counts (from 1000 erythrocytes up to 10000) three times by a single experienced scorer.

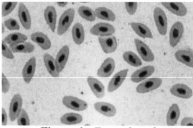 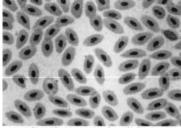

Figure 1 *Examples of correct erythrocyte distribution on the slide*

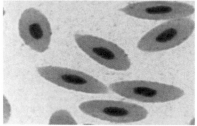 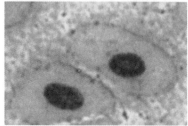

Figure 2 *Normal peripheral erythrocytes of Sea bass*

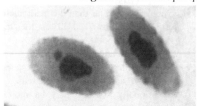 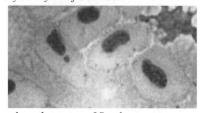

Figure 3 *Micronuclei in peripheral erythrocytes of Sea bass*

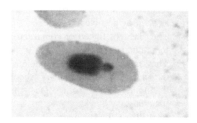 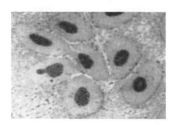

Figure 4 *False micronuclei in peripheral erythrocytes of Sea bass*

2.3.2 Individual counting uncertainty. The individual counting uncertainty was assessed by repeated counts of erythocytes performed by a single experienced scorer (named Alfa, α) on the same slide. This slide, from benzene exposure (benzene slide), was divided into five areas, as shown in figure 5. The X and Y coordinates of each area were well established.

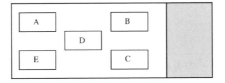

Figure 5 *Scoring areas*

The count of all five areas was repeated 5 times (X_1, X_2, X_3, X_4, X_5) over 9 days according to the following scheme:

1st day	(X_1)	A	B	C	D	E
3rd day	(X_2)	E	D	C	B	A
5th day	(X_3)	C	B	A	E	D
7th day	(X_4)	E	A	D	B	C
9th day	(X_5)	D	C	B	A	E

The range time between counts of different areas in the same day was 1 hour.

2.3.3 Within-laboratory counting reproducibility. The within-laboratory counting reproducibility was assessed by repeated counts performed on the same benzene slide by three different scorers, with different level of experiences, using the same microscope. The second and the third scorer were named Beta (β) and Gamma (γ). Three counts of the five areas (X_1, X_2, X_3 as the previous scheme) were carried out.

2.4 Statistical analysis

In order to evaluate the minimum number of erythrocytes that should be scored, the FEMN values obtained counting from 1000 to 10000 cells, were compared by using non parametric ANOVA (Kruskal-Wallis Rank test).

The individual counting uncertainty and the within-laboratory counting reproducibility were expressed in terms of Relative Standard Deviation (RSD: standard deviation of population from a sample of *n* results divided by the mean of that sample)[21]. The statistical evaluation is reported in detail in tables 4, 5, and 6.

3 RESULTS AND DISCUSSION

3.1 The minimum number of erythrocytes

Table 2 reports the FEMN obtained from repeated counting (from 1000 to 10000 cells) performed on the control slide and on the benzene treated slide.

In both treatments, the non parametric ANOVA test revealed that the FEMN values obtained counting from 2000 to 10000 cells are not significantly different. This suggests that the minimum number of erythrocytes that should be scored for micronucleus frequency is 2000.

Statistically significant differences were found comparing FEMN values obtained from counting 1000 cells with FEMN values obtained from counting from 2000 to 10000 cells.

Table 2 *Frequencies of micronucleated erythrocytes obtained from set of repeated countiung*

ETOT	Control slide			Benzene slide		
	FEMN			FEMN		
	I° count	II° count	III° count	I° count	II° count	III° count
1000	3.0	2.0	3.0	10.0	8.0	9.0
2000	1.5	1.4	1.5	13.0	12.5	12.5
3000	1.7	1.3	1.7	11.7	12.3	13.0
4000	1.3	1.5	1.5	12.8	12.0	12.3
5000	1.2	1.4	1.4	13.6	12.6	11.8
6000	1.3	1.5	1.3	13.3	12.5	13.2
7000	1.1	1.3	1.3	13.6	12.3	11.4
8000	1.3	1.4	1.5	13.0	13.0	12.0
9000	1.2	1.4	1.4	13.0	12.3	12.3
10000	1.5	1.4	1.4	13.3	12.9	12.5

To confirm the adequacy of 2000 cells, another set of repeated counts (1000, 2000, 4000 and 6000 cells) was carried out by the experienced scorer only on the same benzene treated slide following the statistical scheme reported in table 3. The coefficients of variation, shown in table 4, show a greater variability in the counts of 1000 erythrocytes.

Table 3 *Frequency of micronucleated erythrocytes obtained from repeated counting of the benzene treated slide*

	1000 cells			2000 cells			4000 cells			6000 cells		
	I°	II°	III°	I°	II°	III°	I°	II°	III°	I°	II°	III°
	10.0	7.0	17.0	11.5	11.5	12.5	12.0	12.8	12.0	11.8	11.7	12.2
	14.0	15.0	14.0	12.5	13.0	12.5	11.5	13.0	12.5	12.3	11.8	12.2
	8.0	12.0	9.0	11.5	12.5	11.5	12.0	12.5	12.0	11.8	12.2	12.0
	11.0	16.0	17.0	11.5	12.5	11.5	12.3	12.3	11.5	12.2	12.2	12.3
	18.0	10.0	17.0	12.5	11.0	13.0	12.3	12.0	12.0	12.0	12.3	11.7
M	12.2	12.0	14.8	11.9	12.1	12.2	12.0	12.5	12.0	12.0	12.0	12.1

Note: M: mean

Table 4 *Summary statistics*

1000 cells				2000 cells				4000 cells				6000 cells			
M_G	SD	SE	CV	M_G	SD	SE	CV	M_G	SD	SE	CV	M_G	SD	SE	CV
13.0	3.66	0.95	28.19	12.1	0.65	0.17	5.40	12.2	0.41	0.11	3.33	12.0	0.23	0.06	1.92

Note: M_G: grand mean; SD: standard deviation of M_G; SE: standard error of M_G; CV: coefficient of variation % of M_G

These results indicate counting 2000 cells as minimum number of erythrocytes in each specimen. Al Sabti and Metcalfe[1] reported scoring at least 1000 circulating erythrocytes from each specimen to evaluate clastogenicity. On the other hand, Metcalfe[28] proposed to increase this number to 4000 cells for detecting genotoxicity with weak clastogens. Other authors prefer to analyze 1000 peripheral erythrocytes for each animal[25, 29-31]. Al-Sabti[2] Grisolia and Starling[32], Bahari et al.[14] and Nepomuceno[15], suggest analysing respectively 2000, 3000, 4000 and 6000 erythrocytes for each fish.

In a previous test performed in our laboratory it was shown that the time necessary to an experienced scorer to analyze 2000 cells is about 30 minutes. Since micronucleus assay and in situ genotoxicity studies require the examination of a large number of specimens, it can be concluded that the counting of 2000 cells could be considered a good compromise between the counting feasibility and the representativeness/significativity of the result.

3.2 Individual counting uncertainty

Table 5 shows the FEMN values (from FEMN1 to FEMN 5) obtained from repeated counting of 2000 cells performed at different times on a benzene treated slide by the Alfa scorer. The table also reports the statistical parameters for estimating the mean of variances ($\sum RSD^2/n$) and the individual counting repeatability (RSD_α). The mean value of individual repeatability of counting for scorer Alfa is 2.8%.

3.3 Within-laboratory counting reproducibility

Table 6 shows the FEMN values (from FEMN1 to FEMN 3) obtained from repeated counting of 2000 cells performed at different times on a benzene treated slide by Alfa, Beta and Gamma scorers.

The within-laboratory counting reproducibility, expressed as Relative Standard Deviation $RSD_{\alpha+\beta+\gamma}$ was obtained as follows:

$$RSD_{\alpha+\beta+\gamma} = \sqrt{RSD^2\alpha + RSD^2\beta + RSD^2\gamma} \Big/ 3 = 0.25$$

The within-laboratory reproducibility, obtained by the individual counting uncertainty of the three scorers with different level of experience, is 25%.

Table 5 *Estimate of individual counting repeatability for the Alfa scorer*

SLIDE	FEMN 1	FEMN 2	FEMN 3	FEMN 4	FEMN 5	M	SD	RSD^2	$\sum$RSD^2/n	RSD$_\alpha$
AREAS								(SD/M)^2	(n= 5)	$\sqrt{\sum RSD^2/n}$
A	11.5	12.0				11.8	0.4	0.0012	0.0008996	**0.030**
B	12.5	12.5				12.5	0.0	0.0000		
C	11.5	12.0				11.8	0.4	0.0012		
D	11.5	11.0				11.3	0.4	0.0012		
E	12.5	13.0				12.8	0.4	0.0010		
A	11.5	12.0	12.5			12.0	0.5	0.0018	0.0008536	**0.029**
B	12.5	12.5	13.0			12.7	0.3	0.0006		
C	11.5	12.0	11.5			11.7	0.3	0.0007		
D	11.5	11.0	11.5			11.3	0.3	0.0007		
E	12.5	13.0	12.5			12.7	0.3	0.0006		
A	11.5	12.0	12.5	11.5		11.9	0.5	0.0018	0.0008018	**0.028**
B	12.5	12.5	13.0	12.0		12.5	0.4	0.0010		
C	11.5	12.0	11.5	11.5		11.6	0.2	0.0003		
D	11.5	11.0	11.5	11.0		11.3	0.3	0.0007		
E	12.5	13.0	12.5	12.5		12.6 .	0.2	0.0003		
A	11.5	12.0	12.5	11.5	12.0	11.9	0.4	0.0012	0.0006908	**0.026**
B	12.5	12.5	13.0	12.0	12.5	12.5	0.4	0.0010		
C	11.5	12.0	11.5	11.5	11.5	11.6	0.2	0.0003		
D	11.5	11.0	11.5	11.0	11.0	11.2	0.3	0.0007		
E	12.5	13.0	12.5	12.5	12.5	12.6	0.2	0.0003		

Note: FEMN1, FEMN2, FEMN3, FEMN4, FEMN1, FEMN5: frequencies of micronucleated erythrocytes obtained from x1, x2, x3, x4 e x5 counts respectively; RSD: relative standard deviation; n: number of areas.

Tab. 6 – Estimate of the within-laboratory counting reproducibility

Scorer	Slide Areas	FEMN1	FEMN2	FEMN3	M	SD	RSD^2	$\sum RSD^2_\alpha/n$	$RSD\alpha$	$RSD_{\alpha+\beta+\gamma}$
ALFA	A	11.5	12.0	12.5	12.0	0.5	0.0018	0.0009	0.0293	**0.25**
α	B	12.5	12.5	13.0	12.7	0.3	0.0006			
	C	11.5	12.0	11.5	11.7	0.3	0.0007			
	D	11.5	11.0	11.5	11.3	0.3	0.0007			
	E	12.5	13.0	12.5	12.7	0.3	0.0006			
								$\sum RSD^2_\beta/n$	$RSD\beta$	
BETA	A	10.5	10.0	10.5	10.3	0.3	0.0008	0.0040	0.0632	
β	B	11.0	12.0	10.0	11.0	1.0	0.0083			
	C	12.0	11.0	11.0	11.3	0.6	0.0026			
	D	10.5	10.5	10.5	10.5	0.0	0.0000			
	E	8.5	10.0	10.0	9.5	0.9	0.0083			
								$\sum RSD^2_\gamma/n$	$RSD\gamma$	
GAMMA	A	7.0	8.0	7.0	7.3	0.6	0.0062	0.1829	0.4276	
γ	B	10.0	3.5	13.0	8.8	4.9	0.3023			
	C	12.0	7.0	7.0	8.7	2.9	0.1110			
	D	1.5	2.5	5.5	3.2	2.1	0.4323			
	E	5.0	3.0	4.0	4.0	1.0	0.0626			

Note: FEMN1, FEMN2, FEMN3: frequencies of micronucleated erythrocytes obtained from x1, x2 and x3 counts respectively

This value is most probably attributable to the different level of experience of the three scorers, Alfa being a highly skilled scorer, Beta a scorer with six months of experience and Gamma a scorer in training. As shown in table 6, scorer Gamma produced different repeated counts on the same slide area and also the FEMN values are often lower than those of the scorers Alfa and Beta. In fact, the estimate of the within-laboratory reproducibility obtained only considering the counts of the experienced scorers (Alfa and Beta), decreases to 4.9%.

It is possible to conclude that, although the counting of micronucleated erythrocytes in fish is a simple procedure, it requires skilled scorers. In our laboratory, a training module is used to train scorers for about one month. In addition, to maintain the individual scoring capacity in the long-term, periodical (every three months) slide-scoring exercises are performed.

Acknowledgements

The Authors would like to thank the Laboratorio Ittiologico, ARPA Ferrara (Italy), in particular Mr. Fernando Gelli and Mr. Luciano Pregnolato who kindly supplied the fish necessary to perform this experiment.

References

1. K. Al-Sabti, C. D. Metcalfe. Fish micronuclei for assessing genotoxicity in water. Mutation Res. 343 (1995) 121-135.
2. C. Bolognesi, E. Landini, P. Roggieri, R. Fabbri, A. Viarengo. Geneotoxicity biomarkers in the assessment of heavy metal effects in mussels: experimental studies. Environ. Mol. Mutagen. 33 (1999) 287-292.
3. K. Al-Sabti. Chlorotriazine reactive Azo Red 120 textile dye induces micronuclei in fish, Ecotoxicol. Environ. Safety 47 (2000) 149-155.
4. F. Ayllon, E. Garcia-Vazquez. Induction of micronuclei and other nuclear abnormalities in European minnow Phoxinus phoxinus and mollie Poecilia latipinna: an assessment of the fish micronucleus test, Mutation Res. 467 (2000) 177-186.
5. D. R. Dixon, A. M: Pruski, L. R. J. Dixon and A. N. Jha. Marine invertebrate eco-genotoxicology: a methodological overview. Mutagenesis 17 (2002) 495-507.
6. T. Cavas and S.E Gozuraka. Micronuclei, nuclear lesions and interphase silver-stained nucleolar organizer regions (AgNORs) as cyto-genotoxicity indicators in Oreochromis niloticus exposed to textile mill effluent, Mutation Res. 538 (2003) 81-91.
7. E. de la Sienra, M. A. Armienta, M. E. Gonsebatt. Potassium dichromate increases the micronucleus frequency in the crayfish Procambarus clarkii, Environ. Pollution 126 (2003) 367-370.
8. T. Cavas, N. N. Garanko, V. V. Arkhipchuk. Induction of micronuclei and binuclei in blood, gill and liver cells of fish sub chronically exposed to cadmium chloride and copper sulphate, Food and Chemical Toxicol. 43 (2005) 569-574.
9. B. Gustavino, K. A. Scornajenghi, S. Minissi, E. Ciccotti. Micronuclei induced in erythrocytes of Cyprinus carpio (teleostei, pisces) by X rays and colchicine, Mutation Res. 494 (2001) 151-159.
10. D.A. Powers. Fish as model system, Science 246 (1989) 352-358
11. J.H. Yang, P.T. Kostecki, E.J. Calabroco and L.A. Baldwin. Induction of peroxisome proliferation in rainbow trout exposed to ciprofibrate, Toxicol. Appl.Pharmacol. 104 (1990) 476-482.
12. P.C. Washburn and R.T. Di Giulio. Stimulation of uperoxide production by nitrofurantoin, p-nitrobenzoic acid and m-dinitrobenzene in hepatic microsomes of three species of freshwater fish, Environ. Toxicol. Chem. 8 (1989) 171-180.
13. A. Buschini, A. Martino, B. Gustavano, M. Monfrinotti, P. Poli, C. Rossi, M. Santoro, A. J. M. Dorr, M. Rizzoni. Comet assay and micronucleus test in circulating erythrocytes of Cyprinus carpio specimens exposed in situ to lake waters treated with disinfectants for potabilization, Mutation Res. 557 (2004) 119-129.
14. B. Bahari, F. M. Noor, N. M. Daud. Micronucleated erythrocytes as an assay to assess actions by physical and chemical genotoxic agents in Clarias Gariepinus, Mutation Res. 313 (1994) 1-5.
15. J. C. Nepomuceno, I. Ferrari, M. A. Sanò and A. J. Centino. Detection of micronuclei in peripheral erythrocytes of Cyprinus carpio exposed to metallic mercury, Environ. Mol. Mutagen. 30 (1997) 293-297.
16. V. L. Maria, A. C. Correia and M. A. Santos. Anguilla anguilla L. biochemical and genotoxic responses to benzo(a)pirene, Ecotoxicol. Environ. Safety 53 (2002) 86-92.

17. C. Russo, L. Rocco, M. A. Morescalchi and V. Stingo. Assessment of environmental stress by the micronucleus test and the comet assay on the genome of teleost populations from two natural environments, Ecotoxicol. Environ. Safety 57 (2004) 168-174.

18. M. Fenech, W.P. Chang, M. Kirsch-Volders, N. Holland, S. Bonassi, E. Zeiger. HUMN project: detailed description of the scoring criteria for the cytokinesis-block micronucleus assay using isolated human lymphocyte cultures, Mutation Res. 534 (2003) 65-75.

19. K. L. Radack, S. M. Pinney, G. K. Livingston. Sources of variability in the human lymphocyte micronucleus assay: a population-based study, Environ. Mol. Mutagen. 26 (1995) 26-36.

20. K. Brown, A. Williams, H. R. Withers, K. T. Ow, R. Grey, C. Amies. Sources of variability in the determination of micronuclei in irradiated peripheral blood lymphocytes, Mutation Res. 389 (1997) 123-128.

21. DD ENV ISO/TR. Water quality – Guidance on validation of microbiological methods, 13843:2001.

22. F. Gelli, A. M. Cicero, P. Melotti, A. Roncarati, L. Pregnolato, F. Savorelli, D. Palazzi, L. Mariani, G. Casazza. Impiego di stadi larvali e giovanili di Dicentrarchus labrax (L) in saggi biologici: valutazione della qualità di acque marine, salmastre e di sedimenti attraverso test acuti, Biol. Mar. Mediterranea 10 (2003) 137-200.

23. L. Mariani, L. Manfra, C. Maggi, F. Savorelli, R. Di Mento, A. M. Cicero. Produced formation waters: a preliminary study on chemical characterization and acute toxicity by using fish larvae (Dicentrarchus labrax L., 1758), Fresenius Envir. Bull. 13 (12a) (2004) 1427-1432.

24. C. Gravato and M.A. Santos. Dicentrarchus labrax biotransformation and genotoxicity responses after exposure to a secondary treated industrial/urban effluent, Ecotoxicol. Environ. Safety 55 (2003) 300-306.

25. C. Gravato and M.A.Santos. Genotoxicity biomarkers' association with B(a)P biotransformation in Dicentrarchus labrax L., Ecotoxicol. Environ. Safety 55 (2003) 352-358.

26. T. Cavas, S. E.- Gozukara. Evaluation of the genotoxic potential of lambda-cyhalothrin using nuclear and nucleolar biomarkers on fish cells, Mutation Res. 534 (2003)93-99.

27. D. Conti, V. Bellaria, M. Belli, F. Gelli, L. Pregnolato, F. Savorelli, L. Pantaleoni. Determinazione della concentrazione di non-effetto a 9 giorni e test dei micronuclei per valutazioni di genotossicità dei disperdenti degli idrocarburi petroliferi su Dicentrarchus labrax L.. Atti del 35° Congresso Società Italiana Biologia Marina (SIBM) – Genova 19-20 luglio 2004, pag 193.

28. C. D. Metcalfe. Induction of micronuclei and nuclear abnormalities in the erythrocyte of mudminnows (Umbra limii) and brown bullheads (Ictalurus nebulosus), Bullettin Environ. Contam. Toxicol. 40 (1988) 489-495.

29. S. Masuda, Y. Deguchi, Y. Masuda, T. Watanabe, H. Nukaya, Y. Terao, T. Takamura, K. Wakabayashi, N. Kinae. Genotoxicity of 2 [2-(acetylamino)-4-[bis(2-hydroxyethyl)amino] –5-methoxyphenyl]-5-amino-7-bromo-4-chloro-2H-benzotriazole (PBTA-6) and 4-amino-3,3'-dichloro-5,4'-dinitro-biphenyl (ADDB) in goldfish (Carassius auratus) using the micronucleus test and the comet assay, Mutation Res. 560 (2004) 33-40.

30. M. A. Campana, A. M. Panzeri, V. J. Moreno, F. N. Dulout. Genotoxic evaluation of the pyrethroid lambda-cyhalothrin using the micronucleus test in erythrocytes of the fish Cheirodon interruptus interruptus, Mutation Res. 438 (1999) 155-161.

31. A.M.Farah, B. Ateeq, N.M. Ali and W. Ahmad. Evaluation of genotoxicity of PCP and 2,4-D by micronucleus test in freshwater fish Channa punctatus, Ecotoxicol. Environ. Safety 54 (2003) 25-29.

32. C. K. Grisolia, F.L.M.R.F. Starling. Micronuclei monitoring of fish from Lake Paranoà, under influence of sewage treatment plant discharges, Mutation Res. 491 (2001) 39-44.

EVALUATION OF INTRINSIC UNCERTAINTY IN THE k_0 – NAA

Tinkara Bučar and Borut Smodiš

Jožef Stefan Institute, Jamova 39, SI-1000 Ljubljana, Slovenia

1 INTRODUCTION

In quantifying measurement uncertainty using the relative-standardised neutron activation analysis (NAA), where element standards are co-irradiated with a sample, the parameters considered usually include weighing of samples and standards, counting statistics, interferences, neutron flux gradients, counting geometries, etc. In quantifying measurement uncertainty using the k_0-based neutron activation analysis (k_0-NAA), where tabulated experimentally determined values for certain nuclear data are being applied, some additional parameters should be considered in addition to the above-mentioned ones:

- First, the compound nuclear constant k_0, resonance integral to 2200 m/s cross section ratio Q_0 and the effective resonance energy $\overline{E}_r$. The values for the first two ones have been experimentally determined, and the values for the third one compiled from the literature. The parameters are occasionally re-evaluated, critically reviewed and published in a tabulated form[1]. The values have associated uncertainties that should be taken into account when calculating the overall uncertainty. In this respect, they contribute to an "intrinsic uncertainty" of the k_0-NAA, since they are inseparably included in all calculations following analytical measurements.
- Second, the parameters f and α, describing the shape of neutron fluence rate. They are characteristic for a given reactor/irradiation site and therefore influence the analytical result in a pre-determined way for the experimental facility used. Furthermore, they also influence the value $Q_0(\alpha)$, which is used in actual calculations.

This paper addresses the uncertainty introduced into the measurement due to the k_0 and Q_0 factors, $\overline{E}_r$, and the parameters f and α given by the specific irradiation conditions where the analysis is performed.

2 UNCERTAINTY CALCULATION

In estimating the k_0-NAA – based measurement uncertainty the model for uncertainty calculation was developed taking into account appropriate uncertainty propagation. In the model based on original approach developed by the founders of the k_0-NAA[2], the mass fraction of a measurand (c_a) in a sample irradiated by whole spectrum reactor neutrons, using the k_0 –NAA is calculated by[3]:

$$c_a = \frac{A_{sp,a}}{A_{sp,Au}} \frac{1}{k_{0,Au}(a)} \frac{f + Q_{0,Au}(\alpha)}{f + Q_{0,a}(\alpha)} \frac{\varepsilon_{p,Au}}{\varepsilon_{p,a}},$$ (1)

The subscript "a" refers to the measurand and "Au" refers to the co-irradiated gold monitor. A_{sp} is the specific count rate, $k_{0,Au}(a)$ is the k_0 factor of a versus Au, f is thermal-to-epithermal neutron fluence rate ratio and ε_p is the full-energy peak detection efficiency.

$$Q_0(\alpha) = (Q_0 - 0.429)\overline{E}_r^{-\alpha} + \frac{0.429}{(2\alpha+1)0.55^{\alpha}},$$ (2)

where α is the parameter describing the shape of the epithermal neutron fluence distribution.

In our calculations the experimental data for f and α measured at the TRIGA Mark II reactor of the Jozef Stefan Institute[4] was taken into account with the values: $\alpha = -0.0113$ (1 ± 40.7 %) and $f = 28.4$ (1 ± 2.82 %). For the parameters k_0, Q_0 and $\overline{E}_r$, values and uncertainties were taken from an IUPAC electronic database[5,6]. In calculating standard uncertainties the rectangular distribution[7] for these parameters was applied.

This paper addresses only the intrinsic uncertainty involved in element analysis by measuring particular activated radionuclides/reactions and the uncertainty due to the above-mentioned neutron fluence parameters. By the term "intrinsic" we understand the unavoidable uncertainty introduced into the measurement due to the k_0, Q_0 and $\overline{E}_r$ factors. Uncertainty propagation factors were calculated for all the above-mentioned uncertainty sources and uncertainty distributions assigned to them according to the GUM[7] recommendations. The uncertainties are expressed in relative way to make them comparable.

The relative combined standard uncertainty of $c_a = c_a(x_i)$ is

$$u_{c,rel} = \sqrt{\sum_i \left(Z(x_i)\, u_{rel}(x_i)\right)^2},$$ (3)

where $Z(x_i) = \dfrac{x_i}{c_a}\dfrac{\partial c_a}{\partial x_i}$ are uncertainty propagation factors and $u_{rel}(x_i)$ is the relative standard uncertainty of the i-th parameter x_i. Since we do not evaluate the expression (3) over all possible parameters i, the partial combined uncertainty dealt with in this paper is denoted by u_p and is not equal to the total combined uncertainty. It consists of intrinsic uncertainty u_{int} and the uncertainty due to the fluence rate parameters u_{flux}, so that $u_p^2 = u_{int}^2 + u_{flux}^2$, where

$$u_{int}^2 = \left(Z(Q_0)\, u_{rel}(Q_0)\right)^2 + \left(Z(k_0)\, u_{rel}(k_0)\right)^2 + \left(Z(E_r)\, u_{rel}(E_r)\right)^2,$$ (4)

$$u_{flux}^2 = \left(Z(\alpha)\, u_{rel}(\alpha)\right)^2 + \left((Z(f)\, u_{rel}(f)\right)^2.$$ (5)

The uncertainties were calculated using a software package specially developed for this purpose. The calculated u_p, u_{int} and u_{flux} for different target nuclides are presented in Figure 1. The nuclides are sorted by increasing uncertainty u_p. The nuclide numbers are explained in Table 1. Please note that the nuclide orders are different for Figures 1-4 due to better visualisation of the graphs. The orders for all four Figures are shown in Table 1.

The intrinsic uncertainty u_{int} was calculated for the nuclides with known parameters and the results are presented in Figure 2. Two extreme cases of partial combined uncertainty u_p were found for our neutron irradiation conditions: determination of ^{153}Sm and ^{99}Mo. The values found were 0.48 % and 2.83 %, respectively, and the contributions of the individual parameters are presented in Figures 5 and 6.

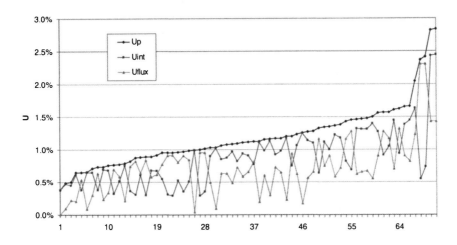

Figure 1 *Partial combined uncertainies u_p and relative contributions due to u_{int} and u_{flux}. Nuclides are sorted by increasing values of u_p; see Table 1.*

3 CONTRIBUTIONS TO INTRINSIC UNCERTAINTY

Contributions of the k_0 and Q_0 uncertainties to the intrinsic uncertainty are presented in Figure 2. The contribution of $\overline{E}_r$ is small and is not shown in the figure. Nuclides with lowest intrinsic uncertainties (^{59}Fe, ^{51}Cr, ^{65}Zn, ^{60}Co and ^{66}Cu) also have low uncertainy contribution due to Q_0, whereas the two nuclides with highest intrinsic uncertainties (^{99m}Tc and ^{99}Mo) have very large contribution of the Q_0 uncertainties. It is therefore instructive to analyse the contribution of the Q_0, which is: $Z(Q_0)u_{rel}(Q_0)$.

The uncertainty propagation factor $Z(Q_0)$ can be calculated from Equations (1)-(3):

$$Z(Q_0) = \frac{Q_0}{A + Q_0}, \text{ where } A = \left(f + \frac{0.429}{(2\alpha + 1)0.55^\alpha} \right) \cdot E_r^\alpha - 0.429. \qquad (6)$$

The function A has very narrow range that varies slowly with $\overline{E}_r$ ($\overline{E}_r$ from 1.5 to 13000 yields the values of A between 26 and 28). Consequently, A can be regarded as a constant even for different nuclides and may be fitted to the calculated data. For $\alpha = -0.0113$ and $f = 28.4$, the fitted value of A is 27.43. The value of $Z(Q_0)$ for different nuclides (Q_0 ranging from 0.53 to 251.6) is between 0 and 1, and is increasing with the value of Q_0.

The contribution of the k_0 factor, as shown in Figure 2, is equal to the relative uncertainty of k_0, $u_{rel}(k_0)$, since the $Z(k_0) = 1$.

The analysis of Q_0 contributions is shown in Figure 3. Some nuclides have very large $Z(Q_0)$ values and small uncertainties of Q_0, or vice versa, so the contributions are unpredictable and should be calculated for each nuclide. Generally, the lowest contributions have nuclides with small values of $Z(Q_0)$. If the tabulated Q_0 data would have lower uncertainties, the partial relative uncertainty would improve significantly for several nuclides, e.g., ^{188}Re, ^{188m}Re, ^{110m}Ag ^{75}Se, ^{111m}Pd, ^{99}Mo, ^{99m}Tc (numbers 40, 49, 59, 65, 69-

71). It is interesting to note that nuclides ^{188}Re (No. 40) and ^{188m}Re (No. 49) have the largest Q_0 uncertainties but their contributions are low due to the small propagation factors $Z(Q_0)$. On the other hand, the nuclide ^{97m}Nb (No. 48) has the lowest Q_0 uncertainty, but the largest $Z(Q_0)$ value.

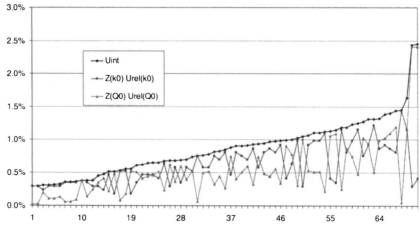

Figure 2 *Intrinsic uncertainies u_{int} and relative contributions due to k_0 and Q_0 factors. Nuclides are sorted by increasing value of u_{int}; see Table 1. Note that $Z(k_0) = 1$.*

4 CONTRIBUTIONS TO THE FLUENCE RATE UNCERTAINTY

Contributions to uncertainty due to the fluence rate parameters f and α are presented in Figure 4. The nuclides with the highest u_{flux} values are ^{97}Nb and ^{97m}Nb, with relatively high contributions from both f and α. The values for $u_{rel}(f)$ and $u_{rel}(\alpha)$ are the same for all the nuclides, so the main difference in u_{flux} values for various nuclides is due to the $Z(\alpha)$ and $Z(f)$ values. $Z(f)$ can be treated similarly as $Z(Q_0)$ in the Equation 6 and approximated by a simpler function of Q_0. However, this issue is out of scope of the present paper.

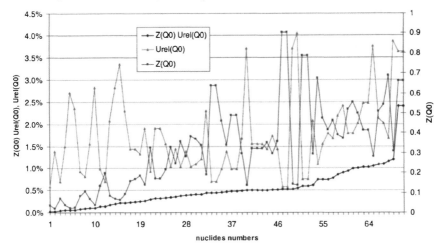

Figure 3 *Influence of $Z(Q_0)$ and $u_{rel}(Q_0)$ on the Q_0 - component of the partial relative uncertainty. Nuclides are sorted by increasing values of $[Z(Q_0) \, u_{rel}(Q_0)]$; see Table 1.*

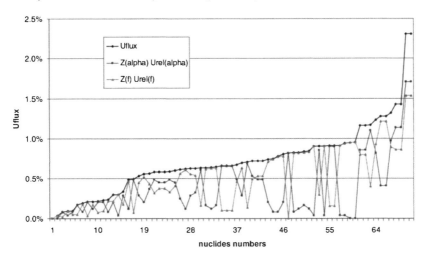

Figure 4 *Relative contributions to flux uncertainty. Nuclides are sorted by increasing u_{flux}; see Table 1.*

5 EXAMPLES

In Figures 5 and 6 all contributions to the u_p are presented graphically for the two extreme cases under the conditions studied: ^{153}Sm and ^{99}Mo. The former nuclide has the lowest and the latter one the largest value. In both cases the contribution due to the Q_0 – related values is the largest, as shown numerically in Table 2.

Table 1 *Nuclides orders as numbered in Figures 1-4. N – nuclide number in the listed figure.*

N	Fig. 1	Fig. 2	Fig. 3	Fig. 4	N	Fig. 1	Fig. 2	Fig. 3	Fig. 4
1	Au-198	Fe-59	Fe-59	Au-198	37	Rb-86	Br-82	Pd-109	Gd-159
2	Sm-153	Cr-51	Cr-51	Ag-110m	38	Cs-134m	Au-199	Pd-109m	Sr-87m
3	W-187	Zn-65	Nd-147	Re-186	39	Nb-94m	Rh-105m	Ir-194	Pd-111m
4	In-116m	Co-60	Mn-56	Tm-170	40	Se-75	U-239	Re-188	Y-90m
5	Ga-72	Cu-66	Na-24	Sm-153	41	Nb-95	Nb-95	Rh-105	Zr-95
6	Re-186	Sr-87m	Cl-38	Hf-180m	42	Sm-155	Sn-117m	Rh-105M	Nb-95
7	In-114M	Mn-56	Ge-75m	Cs-134m	43	Tb-160	Pd-109	Ru-105	Zn-69m
8	S-123m	Na-24	Zn-69m	In-116m	44	Cl-38	Rh-104	Re-186	Ag-108
9	Ho-166	Zn-69m	Co-60	Ir-194	45	Re-188m	Sm-155	Cs-134m	Hf-181
10	Pa-233	Au-198	Cu-66	W-187	46	Hf-180m	Nb-94m	Rb-86	Ge-75m
11	Sr-87m	Ga-72	Sr-87m	Tb-160	47	Sb-124	Ag-110m	Nb-97	Rb-86
12	Pd-109m	Sn-123m	Ga-72	Ho-166	48	Ru-105	Rb-88	Nb-97m	Co-60
13	Ir-194	W-187	Hf-181	Se-75	49	In-113m	Tm-170	Re-188m	Nd-147
14	Zn-69m	Sm-153	Zn-65	In-114m	50	Ag-108	Ru-97	Sm-155	Zn-65
15	Co-60	I-131	I-131	Pa-233	51	Re-88	Re-188m	Np-239	I-131
16	Re-188	Pd-109m	Ag-108	Rh-104	52	I-128	Ru-105	U-239	Cu-66
17	Zn-65	Ge-75m	Pm-149	Au-199	53	Y-90m	Cs-134m	Pa-233	Rb-88
18	G-77m	Hf-181	Sn-123m	Ga-72	54	Sn-113	Ag-108	Sn117m	Mn-56
19	Pm-149	Nb-97m	Y-90m	Sb-122	55	Np-239	Se-75	I-128	Tc-101
20	Hf-181	Re-188	Er-171	Sb-124	56	Lu-176m	Sb-124	Br-82	Mo-101
21	Cu-66	In-116m	Tm-170	Ge-77m	57	Rh-105	Y-90m	Rb-88	Fe-59
22	Fe-59	In-113m	Nb-95	Pd-109m	58	Gd-159	Tb-160	Tb-160	Cl-38
23	Ge-75m	In-114m	Zr-95	Pd-109	59	Sb-122	I-128	Ag-110m	Na-24
24	Mn-56	Re-186	Nb-94m	I-128	60	Tc-101	Hf-180m	Gd-159	Cr-51
25	I-131	Pm-149	W-187	Nb-94m	61	U-239	Tc-101	Lu-176m	In-113m
26	Ag-110m	Pa-233	Ge-77m	Pm-149	62	Ru-97	Gd-159	Ru-97	Sn-113
27	Cr-51	Ge-77m	Au-198	Re-188m	63	Pd-111m	Rh-105	Mo-101	Ru-97
28	Na-24	Np-239	Ho-166	Lu-176m	64	Sn-117m	Lu-176m	Tc-101	Sn-125m
29	Au-199	Ho-166	Au-199	Sn-123m	65	Mo-101	Mo-101	Se-75	Np-239
30	Tm-170	Nb-97	In-116m	Br-82	66	Nd-147	Sb-122	Sb-124	U-239
31	Er-171	Cl-38	Sm-153	Er-171	67	Sn-125m	Pd-111m	Sb-122	Sn-117m
32	Br-82	Ir-194	Rh-104	Re-188	68	Nb-97m	Nd-147	Sn-125m	Tc-99m
33	Rh-104	Rb-86	In-113m	Sm-155	69	Nb-97	Sn-125m	Pd-111m	Mo-99
34	Zr-95	Zr-95	Sn-113	Rh-105m	70	Tc-99m	Tc-99m	Mo-99	Nb-97m
35	Pd-109	Sn-113	In-114m	Ru-105	71	Mo-99	Mo-99	Tc-99m	Nb-97

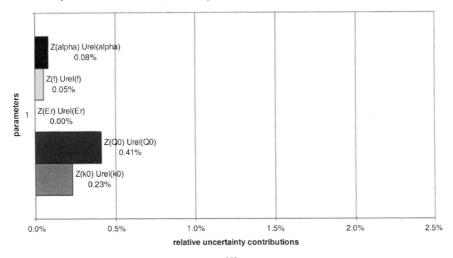

Figure 5 *Partial combined uncertainty u_p in ^{153}Sm determination using the k_0-NAA.*

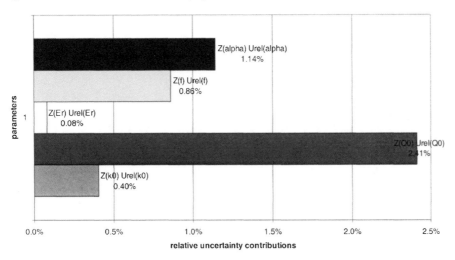

Figure 6 *Partial combined uncertainty u_p in ^{99}Mo determination using the k_0-NAA.*

6 CONCLUSIONS

In this work the various contributions to the partial combined uncertainty, u_p, of the k_0-NAA have been assessed. The u_p was defined as a combination of the: (1) intrinsic uncertainty, consisting of the uncertainties contained within the nuclear constants used[6] and (2) uncertainty due to the specific neutron fluence rate conditions defined by the irradiation facility used. From the model equation uncertainty propagation factors were calculated and various contributions combined according to the ISO[7] recommendations. Altogether 71 target nuclides were considered and individual parameter contributions evaluated both in terms of their uncertainties as well as uncertainty propagation factors. The calculated relative u_p values were found to be in the range of 0.48 % for [153]Sm to 2.83 % for [99]Mo. In both cases the main component was uncertainty due to the Q_0 – related values.

Table 2 *Numerical values for the contributions to partial combined uncertainty for the extreme cases of* [99]*Mo and* [153]*Sm.*

	[99]Mo			[153]Sm		
	Z	u_{rel}	$|Z*u_{rel}|$	Z	u_{rel}	$|Z*u_{rel}|$
E_r	241 eV	-0.007	11.61%	0.08%	8.53 eV	-0.004	0.64%	0.00%
k_0	8.46E-05	1	0.40%	0.40%	2.31E-01	1	0.23%	0.23%
Q_0	53.1	0.663	3.64%	2.41%	14.4	0.339	1.21%	0.41%
f	28.74	-0.305	2.82%	0.86%	28.74	0.019	2.82%	0.05%
α	-0.0113	0.028	40.71%	1.14%	-0.0113	0.002	40.71%	0.08%
u_{int}				2.45%				0.47%
u_{flux}				1.43%				0.10%
u_p				2.83%				0.48%

References

1   F. De Corte, A. Simonits, *At. Data Nucl. Data Tables*, 2003, **85**, 47.
2   A. Simonits, F. de Corte, J. Hoste, *J. Radioanal. Chem.*, 1975, **24**, 31.
3   F. De Corte, *Habilitation Thesis*, University of Gent, 1987.
4   R. Jaćimović, *PhD Thesis*, University of Ljubljana, 2003.
5   V. P. Kolotov, F. De Corte, *Compilation of k_0 and related data for NAA in the form of electronic database*, version v.4, 2002, IUPAC project number: 2001-075-1-500
6   V. P. Kolotov, F. De Corte, *Pure Appl. Chem.*, 2004, **76**, 1921.
7   *Guide to the Expression of Uncertainty in Measurement*, International Organization for Standardization (ISO), Geneva, 1995, ISBN 92-67-10188-9.

REPORTING OF UNCERTAINTY IN ENVIRONMENTAL MONITORING OF RADIONUCLIDES

B. Varga and S. Tarján

Department of Radiochemistry, National Food Investigation Institute, Mester u. 81. Budapest 1095 Hungary

1 INTRODUCTION

The main aim of this paper to summarise the problems may occur during the determination of radionuclides at the current environmental level through the uncertainties of different stages of the work. The required goal during the uncertainty analysis to find the components might be decreased or at least keep them at a low, well defined level.

Department of Radiochemistry lays big emphasis on quality assurance work in the frame of the Monitoring Network of Ministry of Agriculture and Regional Development (furthermore network). After the changes applied in last year from 19 laboratories remained only 13 and the central laboratory. The main task of the network is controlling of the food and agricultural production and their circumstances from radioactivity point of view. The goal is to form a consistent well-maintained countrywide database of anthropogen and natural radioisotopes in foodstuffs, feeding materials and different environmental matrixes with well defined spatial data. For the sake of the cause the following actions are regularly taken:
- regular trainings are organised for the whole network at least once a year, where each problem is discussed, some exercises are held and solutions are harmonised;
- laboratories are accredited according to MSZ EN ISO/IEC 17025 - give guidance for the accreditation procedure and for the regular revisions,
- organization of intercomparison and proficiency runs; mainly „natural samples" were chosen, from which large amounts could be collected and their activities were about or slightly above the typical environmental levels.

Fortunately the specific activity level of anthropogen radionuclides in foodstuffs is so low, that one of the main point to agree when are the results different enough from the background value and where is the decision threshold in the case of different types of methods determining the radioisotopes. The principle is never lose information, never write "non detected" or even better give the value and its uncertainty even if the result is lower than the decision threshold and has high uncertainty.

The network use the harmonised methods during the routine work, the uncertainty analysis were detailed together to make use of the occasion to think together. The important agreement during the building up the common database, never use other than k=1 for the reporting of combined uncertainty value.

The philosophy of handling uncertainties in running the countrywide network to decrease the components as small as possible or at least keep them in the same well controlled level.

In this paper the considerations are shown through only one chosen sample-type, namely sorrel and spinach samples which are treated together from the point of view of their radionuclide content.

2 UNCERTAINTIES OF DIFFERNT METHODS APPLIED IN THE NETWORK

2.1 Sampling

Uncertainty of sampling is not taken into account, however it is the field sampling, rather than analytical procedure and measurement of radioactivity of different isotopes that might appear as the largest source of the uncertainty of the results. For environmental monitoring it is often quoted that an analysis can never be of better quality than the sample upon which it is made. Robust statistical analysis of variance could be one possible and useful tool for determination of uncertainty of field sampling, because this method rely on the accommodation of outlying values rather than their rejection, weighting of the observation can give a realistic approach. Some efforts were made to determine the uncertainty of sampling of the soils in the case of classification of a given territory in Wirksworth, Derbyshire, UK[1], where the area of sampling target was 30600 m². There is another reference site for trace elements in Pozzulo del Friuli, Udine, Italy[2], where an international intercomparison was held with participation of some members of ALMERA (Analytical Laboratories for Measuring of Environmental Radioactivity) network in the frame of IAEA/SIE/01, this area measures 10000 m². Participants applied their own sampling protocols, there is a possibility to compare them and give an estimation for the uncertainty of the sampling. These evaluations are very useful, when the area of the contaminated land is similar to them and the borders are well determined, like in the case of an accident, which has only local effect.

Only a little part of the country is outside of the emergency-planning zones of 30 km, 80 km and 300 km taken into account the nuclear power plants in neighbouring countries also. The network needs another approach for various reasons, the main sampled media of the network are different foodstuffs, feedstuffs and soil. Of course there are some of them that are collected from the whole country, but there are some which has well defined production area depending on the soil properties, configuration of the terrain and the meteorological conditions. On the other hand, when data come from the countrywide network sometimes it is difficult to distinguish between the uncertainties of the sampling and the natural variation of the radioisotopes depending of spatial variation of soil properties and meteorology also. For this reason every laboratory has a different monitoring program, which contains common sampled media and they have some own samples depending on the food-production of controlled area. To follow the time-dependencies this monitoring program contains sampling points for example for milk, fodder, sorrel and spinach, which are the same from decades. Last year this idea was in several events disturbed in the case of fodder because of the imported additives. During the continuous monitoring the main governing effects of the spatial variations are discovered and if somebody has a feeling about getting the strange result, there is the possibility to check data by sorting the database either by sample-type, location or any other aspects immediately.

In the case of emergency situation the most important thing to find the borders of the contaminated area. If it is defined, there is a possibility to plan the sampling taken into account the capacity of the network and the required uncertainty of measurements.

2.2 Sample preparation

All kinds of food samples are dried at 105°C with the programmed increase of the temperature. The following investigation is valid for plant samples. The uncertainty of this step depends on the uncertainty of the weighing in the case of drying oven works properly, its effect is negligible. Determination of dry content of samples means measurement of fresh samples two times, because the required amount for the analysis usually 5 kg and maximum weight possible to put on the PB3002 balance is 3100 g, and it is put to at least two parts in the drying oven, weighing back also two times. The relative standard uncertainty of this step is 0,02 %. Sometimes this preparation is enough for moss, mushroom, fodder samples and of course for soil, but only air drying is applied in the last case.

Fortunately the specific activity of antrophogen isotopes in different elements of food chain is so low in Hungary, that generally all kinds of samples are ashed except the above mentioned ones. The collected sample amount has to be rather large to get reasonable limit of detection. After a programmed, moderate heating up, the optimal end-temperature is 450°C. Because of the well-controlled furnace the uncertainty of weighing is taken into account again. The relative standard uncertainty of ashing takes 0,23 %, where drying was also taken into account.

The usual way is to combine this small uncertainty of the sample preparation of individual analysis. As an example a short investigation carried out for sorrel and spinach samples collected in last ten years. This large database proved the expectations regarding the natural variability of the dry and ash content of the samples (Table 1).

Table 1 *Statistical characterisation of dry-content and ash-content of sorrel and spinach data from the period of 1994-2004*

Parameter	Determination of dry content	Determination of ash content
Number of data	1427	1432
Minimum	4,30 %	0,29 %
Maximum	45,50 %	7,87 %
Average	10,82 %	1,80 %
Empirical standard deviation, s	5,06 %	0,75 %
Number of data in case of Average±1s	1360	1368
Number of data in case of Average±2s	1285	1148

The relative empirical standard deviation is 47 % in the case of dry content of the given dataset. This difference of three orders of magnitude between the relative standard uncertainty of the determination of dry-content and the natural variability of the dry-content of the same sample-type is remarkable. The investigation of the population shows that into the interval determined by average value and the empirical standard deviation: 5,75-15,88 % fell 90 % of the whole dataset and into the interval determined by average and 2-times of empirical standard deviation: 0,69-20,94 % fell 95 % of the data (Figure 2). In the case of ashing the relative empirical standard deviation is in same range as the previous one, but the difference is only two order of magnitude between the relative standard uncertainty of the determination of ash-content and the natural variability of the ash-content. In this case into the range of 1,05-2,56 % fell 80 % of data and into the range of 0,30-3,31 % fell 96 % of the investigated dataset.

2.3 Determination of specific activity of ^{137}Cs by gamma-spectrometry

In the case of gamma-spectrometry using HPGe detectors the sources of standard uncertainties can be grouped according to their origin[3].

Uncertainties may occur during the preparation are the following: drying and ashing of the sample, weighing of the sample to a sample holder, sample inhomogeneity. The last contributor usually do not have any role, because the whole amount of ash are measured, and the representativeness to the whole amount of sample is also could leave out because of the same reason.

The following contributor is the uncertainty of energy and efficiency calibration. Energy calibration can be executed very precisely, its contribution usually not taken into account, linearity of modern ADC is considered extremely good. The efficiency calibration for a given geometry has big importance. Standards and reference materials used for calibration are in the same geometry as the sample. These are available in every laboratory of the network where gamma-spectrometry is performed. Of course the origin of these materials the same, it has an advantage to have calibration harmonised. As the intercomparison runs with invited external laboratories organised by Department of Radiochemistry showed the network have only the benefit of the harmonised calibration.

The most complex part of the uncertainty budget comes from the measurement of sample. For the routine monitoring measurements the network has different standards in required geometries. The uncertainty has the following contributions: counting statistics, reproducibility of the geometry, self-attenuation, random and true coincidences, net peak area determination, dead time effects and decay time effects. Generally the uncertainty due to counting statistics is one of the most important ones. For the reason of decreasing the effect of the positioning, all detectors have own sample holder frame tightly fitted onto the detectorcap. They have some changeable parts depending on the diameter of the used sample holders and some other positioning rings to keep the exact distance between the sample and the top of the detector. The substraction of instrumental background and uncertainty of corrected net peak area is obtained from the fitting. Environmental samples produce generally low count rates, therefore the random coincidence correction factor could not be taken, and good pile-up rejection might be useful. Dead time effects are also negligible. Correction factor for decay between sampling and measurement is done automatically by the software, but generally never more than one month elapsed after the sample taking to measurement. For the investigated isotope decay time effect during the measurement is negligible, if not the software can handle also. Nuclides decaying through the cascades can cause coincidence-summing especially in high efficiency semiconductor detectors. Self-attenuation correction is not taken into account by calculations, rather than to have standards of same density as the routinely measured samples.

At last nuclear data also have uncertainties. In a few cases gamma emission probabilities could be the major contributor to the final value. [137]Cs is not this kind of isotope, because the parameters are well known and widely investigated and used. The uncertainty of half-life is still small compared with other sources.

Altogether 16 HPGe detectors are working continuously in the network. All of them are cylindrical, their parameters are different, depending on the year of installation, n- or p-type detectors with 15-30 % relative efficiency, some of them are in ultra low background cryostat, some in low background and some only in normal. While they are installed in different laboratories in different part of the country, the measuring circumstances are also different (materials of the house, ventilation, stability of the electric supply), which may have effect on the background of the detector. The shieldings of the detectors are also different. From the mentioned facts come a range for the uncertainty despite a concrete value, when the goal is to know the uncertainty of ^{137}Cs-determination at the network level, especially in case of so low specific activity in the Hungarian environment. For determining this range, data of the last five years was chosen[4], because the transport process governing the plant contamination is very slow nowadays and they are regarded stationary, furthermore there was any contamination of ^{137}Cs during the investigated period, data from all the five years could treat as one homogeneous database.

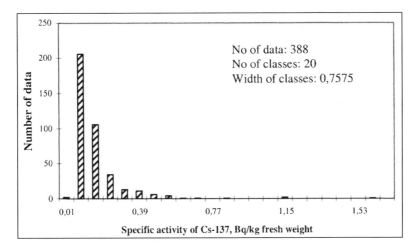

Figure 1 *Distribution of the specific activity of ^{137}Cs in sorrel and spinach samples from the years 2000-2004*

The same statistical analyses can not be performed as it was done in the case of dry-content and ash-content, because the distribution of data is completely different. The distribution of dataset does not follow the normal distribution (Figure 1), skewness is 4,70 and kurtosis is very high 34,4 despite of 3 and 0 as in case of normal distribution. The values of arithmetic mean: 0,12 Bq/kg fresh weight, median: 0,08 Bq/kg fresh weight and mode: 0,06 Bq/kg fresh weight show also, that in this concentration range of this antrophogen isotope does not give normal distribution. In this case robust statistic can give more reliable description about the dataset, because it is designed to cope with wide dataset, without resorting to removal of data from the dataset; robust mean: 0,092 Bq/kg fresh weight and the Median of all Absolute Distances (MAD): 0,057 Bq/kg fresh weight. Robust analysis was performed by HISTO designed by Z. Radecki and A. Trinkl (provided by IAEA), where the algorithm was reproduced from ISO 5725-5.

After the characterisation of the dataset may perform the description of the uncertainty taking into account the whole data coming from the countrywide network. Figure 2 shows well that in such a low specific activity for every result might be given an uncertainty range. After the curve fitting, namely the power-functions, the highest and the lowest uncertainty values were calculated and ranges are given in the Table 2. for the critical range from the point of view of dependence of the installation and parameters of the detectors.

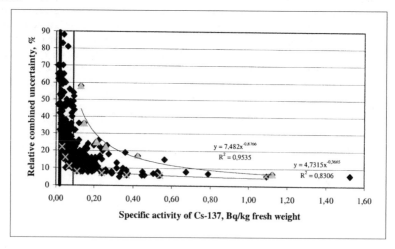

Figure 2 *Relative combined uncertainty of the specific activity of ^{137}Cs in sorrel and spinach samples from the years 2000-2004 in the network*

Below the 0,6 Bq/kg fresh weight the dominant component is the uncertainty of counting statistic. Above this value two main contributors are the uncertainty of calibration source and the uncertainty of efficiency curve fitting, the experimental values shows that the uncertainty is not worse than 7%. Calculation of combined uncertainty from the components taking into account the parameters of the different detectors also gives the value not worse than 7 %, above 0,6 Bq/kg fresh weight. Figure 2 gives another important information determined empirically, the range of minimal detectable specific activity 0,02 Bq/kg fresh weight for the better detectors and 0,07 Bq/kg fresh weight for the older.

From these results the conclusion could be drawn that at the reporting level, determined by 2000/473/EURATOM the network is able to give the specific activity of ^{137}Cs generally in food samples with uncertainty better than 10 %, which is acceptable. It is important to keep in mind that these values valid from the measurements of ashed samples.

Table 2 *The range of relative combined uncertainty of specific activity of ^{137}Cs in sorrel and spinach samples determined by the network*

Specific activity, Bq/kg fresh weight	Uncertainty range, %
0,1	11-56
0,2	9-31
0,3	7-21
0,4	7-17
0,5	6-14
0,6	6-12

2.3 Determination of specific activity of [90]Sr

Summary of the method: after usual sample preparation (drying and ashing) using $Sr(NO_3)_2$ carrier, leaching with HNO_3, some cleaning steps are included followed by precipitation of Sr as $SrSO_4$. The specific activity is determined after ^{90}Y in-growth by low background alpha-beta proportional counter.

The following uncertainty contributors may have effect the final value associated with the result: weighing, uncertainty of determination of chemical yield, uncertainty of calibration source, uncertainty of the determination of the efficiency of the detector, uncertainty of the evaluation of the count-rate.

The approach is similar as in case of ^{137}Cs, data from the years 2000-2004 was investigated.

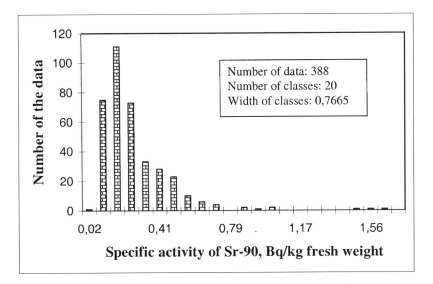

Figure 3 *Distribution of the specific activity of ^{90}Sr in sorrel and spinach samples from the years 2000-2004*

The values of arithmetic mean: 0,24 Bq/kg fresh weight, median: 0,18 Bq/kg fresh weight and mode: 0,14 Bq/kg fresh weight show also, that in this concentration range of this antrophogen isotope does not give normal distribution. The robust statistic gives the following: robust mean: 0,20 and MAD: 0,12 Bq/kg fresh weight. Robust analysis was performed by HISTO also.

The distribution of dataset does not follow the normal distribution (Figure 3), however the skewness: 2,95 is close to the normal distribution, but the kurtosis is rather high 16,28.

In the case of ^{90}Sr the experimentally defined minimal detectable specific activity is 0,06-0,12 Bq/kg fresh weight, depending on the equipment and of course the ash content of the sample (Figure 4). In this case the leading role of counting statistics in the uncertainty budget is till 0,4 Bq/kg fresh weight, above this value the uncertainty of the calibration source and the determination of efficiency have main importance. In the Table 3 are listed the uncertainty ranges for different given specific activities. There is one remarkable fact, that the practice in the network is always measuring 1 g from the final precipitate for a preparatum to be measured by low background alfa-beta counter. This way the effect of the uncertainty of the chemical yield is eliminated.

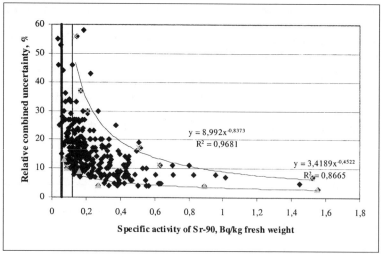

Figure 4 *Relative combined uncertainty of the specific activity of ^{90}Sr in sorrel and spinach samples from the years 2000-2004 in the network*

Table 3 *The range of relative combined uncertainty of specific activity of ^{90}Sr in sorrel and spinach samples determined by the network*

Specific activity, Bq/kg fresh weight	Uncertainty range, %
0,2	7-35
0,4	5-19
0,6	4-14
0,8	4-11
1,0	3-9
1,2	3-8

3 CONCLUSION

In the case of the investigated radionuclides the uncertainty ranges coming from the data of the whole network decreased with the increasing specific activity, as it was expected. In both cases an maximal uncertainty value could be given, for example we can say, that the network is able to measure these radionuclides above 1 Bq/kg fresh weight with not higher than 10 % relative combined uncertainty, if the sample is ashed food. The specific activity for both radionuclides is below this in Hungarian environment, therefore the agreed way to report countrywide data to the decision-makers in normal circumstances is to give the minimum, maximum and median values as a summary. From the position of the median value in a given range the conclusion may be drawn about the shape of distribution also by first sight. As an example the Figure 5 shows the current situation of ^{137}Cs in food-chain in Hungary.

The goal is to have uncertainty of a measurement at least so small, that the spatial variation of the radionuclide could be followed. The network made and makes effort to reach this goal, to have reliable dataset describing the situation of the country well.

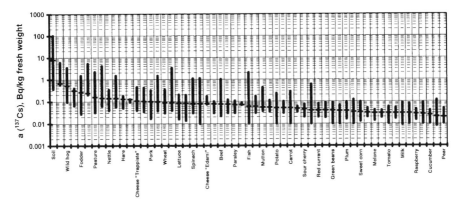

Figure 5 *Specific activity of ^{137}Cs in food-chain in Hungary, decreasing order of sample type is the following:*

Moss	Lettuce	Red currant
Soil	Poultry	Grape
Boletus edulis*	Spinach	**Green beans**
Wild hog	Sorrel	Cherry
Red pepper	Cheese "Edami"	Plum
Fodder	Cheese "Pannonia"	Green peas
Milk powder	Beef	**Sweet corn**
Pasture	**Ewe-milk**	Green pepper
Macrolepiota*	Parsley	Melon
Nettle	Goats' milk	Cauliflower
Red deer	Fish	Tomato
Hare	Cheese "Gomolya"	Asparagus
Cheese of goat	Mutton	Milk
Cheese "Trappista"	Turkey	Onion
Alfalfa	Potato	Raspberry
Pork	Duck	Peach
Deer	Carrot	Cucumber
Wheat	Goose	Apple
Rabbit	Sour cherry	Pear
	Cabbage	

In normal case, when kept an eye on the guideline of 2000/473 EURATOM and the mentioned uncertainty evaluation, the capacity of the network is the following:
- determination of radiostrontium: 2000 sample pro year,
- gamma-spectrometry by HPGe detectors: 3500 sample pro year,
- gross alpha measurement: 2000 sample pro year,
- alpha-spectrometry (U-isotopes, Pu-isotopes, ^{241}Am): 100 sample pro year,
- tritium: plants - 50 sample pro years, water – 300 sample pro year

- the number of the analysis of ^{241}Pu is determined by the capacity of the source preparation for alpha-spectrometry, there is no limit from the side of liquid scintillation spectrometry.

In nuclear emergency the normal requirements for sensitivity could not be fulfilled, usually keeping two orders of magnitude worse is enough for the decisions and suggestions for countermeasures. During the trainings and the intercomparisons this responses of the laboratories are also tested and reporting the results with associated uncertainty is also required. Without this checking the reliability of the results is questionable during an emergency situation. The following capacity should be achieved with extra staff for the administration and sample preparation:

- determination of radiostrontium: 100 sample pro week,
- gamma-spectrometry by HPGe detectors: 500 sample pro day,
- scintillation gamma-spectrometry: 2000 sample pro day,
- gross alpha measurement: 80 sample pro day,
- tritium in water 50 sample pro day.

References

1 M. H. Ramsey and A. Argyraki, *The Science of the Total Environment* 1997, **198**, 243
2 P. de Zorzi, M. Belli, S. Barbizzi, S. Menegon and A. Deluisa, *Accreditation and Quality Assurance*, 2002, **7**, 182
3 IAEA-TECDOC-1401: *Quantifying uncertainty in nuclear analytical measurements*, International Atomic Energy Agency, Vienna, 2004
4 *Yearly reports of the Radiological Monitoring Network of the Ministry of Agriculture and Regional Development from 1995 to 2004*, Budapest

THE USE OF REFERENCE MATERIALS IN INTERNATIONAL REFERENCE MEASUREMENT SYSTEMS AND FOR COMPARISON OF ANALYTICAL DATA

Hendrik Emons

Institute for Reference Materials and Measurements (IRMM), Joint Research Centre, European Commission, Retieseweg 111, 2440 Geel, Belgium

1 INTRODUCTION

Demands for international comparability of measurement results are stipulated by many interests. They include the requirements for implementing and monitoring legislation and regulations, for facilitating and promoting international trade, for ensuring safety and well-being of citizens in areas such as food and health care, and for monitoring the environment or industrial processes. Consequently the range of measurement targets has been significantly increased during the last decades and is covering now a huge number of chemical, biochemical and biological parameters.

One can categorize the general activities to obtain comparable measurement results into two principal approaches: standardization and metrology. The first one aims to achieve data comparability by the development and application of harmonized measurement methods (or more precisely: measurement procedures). This route is coordinated on a global scale by ISO (International Organization for Standardization) and is supported by the various regional and national standardization bodies. Legislation makes use of corresponding ISO standards and guides by prescribing specific measurement methods as the only accepted ones for particular measurements in regulatory frameworks. The second approach which is called here "metrology" is directed towards obtaining measurement results which are traceable in the metrological sense to commonly agreed reference points and measurement units and which are accompanied by properly estimated measurement uncertainties. One could characterize this approach also as the vision to achieve "the ultimate" analytical data for a given measurement problem. In recent years legislation in the EU and other countries has taken up this general approach in various areas such as environmental monitoring and food control by introducing the prescription of method performance characteristics, and not of specific methods anymore, in corresponding regulation for the implementation and monitoring of directives. It is acknowledged that the two principal approaches characterised briefly here are representing actually "borderline cases" for many real-world measurement problems to be solved in chemical, biochemical and biological analysis.

Moreover, the question has to be answered how to design and to implement such approaches for achieving measurement results which are comparable in time and between laboratories worldwide. Basically reliable measurement data are required for a large variety of parameters of interest, independent on the specific measurement method and instrument used and independent on the "human factors". It became obvious for almost all analytical tasks in chemistry, biochemistry and biology, that common materials are needed to achieve the desired international harmonization. Such reference materials are in the focus of this contribution and will be further explained in the following.

2 WHAT ARE REFERENCE MATERIALS?

A general scheme of the so-called total analytical process is presented in Figure 1. From that, one can deduce the types of reference materials (RM) which are required to performing properly the analysis of the sample of interest.

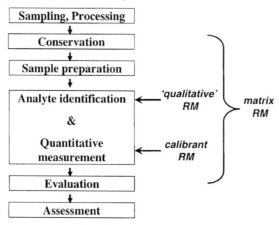

Figure 1 *Schematic overview of the total analytical process and the required reference materials (RM)*

An analysis which is directed only to qualitative properties of a sample (such as the chemical identity of a sample component) needs a reference material which allows to comparing this qualitative property in the sample and in the reference material during the application of the same measurement procedure. A typical example would be an RM consisting of a pure chemical compound with a well defined structure used in a specific chromatographic procedure for comparing its retention time with that of the unknown sample component.

Most of the quantitative measurements steps have to be calibrated because of the lack of completely known mathematical equations to calculate the relation between the targeted quantity in the sample and the measurement signal. Such calibration materials are almost indispensable for chemical and biochemical measurements. Moreover reliable measurement results on the majority of the real-world samples can only be obtained if appropriate quality control measures are applied for most steps of the analytical process. Therefore, so-called matrix reference materials are developed which mimic as close as possible the real sample and allow not only the control of the quantification step, but also of operations such as sample preparation as depicted in Figure 1.

Unfortunately a multitude of names is presently used for RMs needed in the analytical process. In dependence on the field of analytical activity, on the awareness of international guidelines, concepts of quality assurance and metrology and even on regional peculiarities one can find different terms for such reference materials. Examples are measurement standard, laboratory standard, reference standard, analytical standard, reference substance, standard material, quality control material, proficiency testing material, laboratory control material, laboratory reference material or calibration material. Part of the terminology confusion seems not only to originate from different traditions of the various analytical/measurement communities, but also from the different understanding of underlying concepts. For instance, the interrelation between the intended use for a reference material in a given measurement procedure and the required minimum material characteristics together with the distributed RM information is often neglected.[1]

Last year the Reference Material Committee of ISO (ISO REMCO) approved new definitions.[2] The term 'reference material' is now characterized as follows: "Material, sufficiently homogeneous and stable with respect to one or more specified properties, which has been established to be fit for its intended use in a measurement process. NOTE 1: RM is a generic term. NOTE 2: Properties can be quantitative or qualitative, e.g., identity of substances or species. NOTE 3: Uses may include the calibration of a measurement system, assessment of a measurement procedure, assigning values to other materials, and quality control. NOTE 4: An RM can only be used for a single purpose in a given measurement".[3]

The first note mentions now explicitly that the expression 'reference material' is an "umbrella term" for various materials which are needed in measurement procedures in addition to the sample to be analyzed. Consequently one could consider the different RM types as members of a family[4] as illustrated in Figure 2.

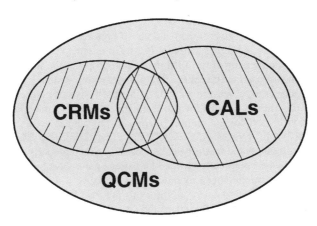

Figure 2 *Graphical representation of the 'Reference Material family' (QCM ... Quality Control Material; CRM ... Certified Reference Material; CAL... calibrant)*

Obviously, all materials possessing the characteristics of adequate homogeneity and stability required for quality control of a given measurement belong to the RM family (grey ellipse in Fig. 2). The ones which are not accompanied by a certificate are often simply called non-certified reference materials. But many other terms such as in-house materials, laboratory control materials or laboratory reference materials are also used. Here the term "Quality Control Material (QCM)" is favored for this subgroup of RMs for which only the material characteristics of homogeneity and stability fit for the intended use are proven. QCMs may support one or more applications from the wide range of both internal and external quality control measures. But they are not sufficiently characterized to be used for method calibration or to provide metrological traceability of a measurement result.

Another subgroup of RMs is formed by the certified reference materials (CRMs) depicted in a special ellipse in Fig. 2. They are now defined as "reference material, characterized by a metrologically valid procedure for one or more specified properties, accompanied by a certificate that provides the value of the specified property, its associated uncertainty, and a statement of metrological traceability. NOTE 1: The concept of value includes qualitative attributes such as identity or sequence. Uncertainties for such attributes may be expressed as probabilities. NOTE 2: Metrologically valid procedures for the production and certification of reference materials are given in, among others, ISO Guides 34 and 35. NOTE 3: ISO Guide 31 gives guidance on the contents of certificates."[3] That means a reference material belongs to this subgroup if in addition to the QCM characteristics a certificate is provided, giving a certified value with its uncertainty and a stated metrological traceability. Further details about these minimum quality characteristics of CRMs are explained in the corresponding ISO Guides.[3,5,6]

The other RM subgroup is composed of the materials used for calibration (indicated by another special ellipse in Fig. 2). They are often denoted as analytical standards, reference standards or simply calibration materials, but not explicitly recognized as "reference materials". Such products, in particular calibrants consisting of pure chemical substances or solutions thereof, are described by some scientists or organisations as having a "higher metrological order" than CRMs. But this misperception originates only from a mixing of classification systems. From the point of necessary material characteristics and provided information, materials for calibration have to be sufficiently homogeneous and stable as to ensure that the assigned property value and its uncertainty are valid for any calibration sample used according to the given specifications. Therefore, they fall under the RM definition given above. Here the term "calibrant (CAL)" is used for such materials, whereas the latest available draft version of the VIM in revision proposes the word 'calibrator' for a "measurement standard used in the calibration of a measuring system". But the term 'calibrant' is much wider used, at least in analytical chemistry, avoids confusion with the person performing the calibration and allows to distinguish between a material (chemical substance) and a device used for calibration (mainly in physics). The necessary additional features of a calibrant in comparison to a QCM are a stated property value with an uncertainty useful for calibration and metrological traceability of the property value. These characteristics are not always completely fulfilled by various materials nowadays used for calibration in different measurement communities or laboratories. But that means only that insufficiently characterized materials are used for this purpose and it does not invalidate the principally required minimum quality characteristics for calibrants, in particular with respect to known uncertainty and traceability of the value used for calibration.

It seems to be acceptable that no formal "certificate", which provides the comprehensive information specified in ISO Guide 31 for CRM certificates,[5] is available for many calibrants used in analytical laboratories. Therefore, strictly speaking only a part of the calibrants can be formally called CRMs and a separate ellipse had to be created in Fig. 2. On the other hand a significant number of existing CRMs is accompanied on their certificates by stated uncertainties of their property values, which make them useful for quality control or other CRM applications, but not for calibration because the uncertainties of CAL values must be included in later calculations. This does not disqualify such materials as CRMs in general (see CRM definition above), but it simply restricts their range of application. A review of current CRM catalogues shows that the stated uncertainties of certified property values for several matrix CRMs, but also for various so-called pure substance CRMs or their solutions, are too large for a useful calibration of measurement instruments or systems. Consequently, only that part of CRMs which provide on their certificates uncertainties of property values fit for calibration of the intended measurement method would belong both to the CRM and the CAL groups as shown in Fig. 2 by the overlapping region.

Sometimes it is believed that only pure chemical compounds are useful for calibration. But many analytical problems are directed to measurands which can only be realized in a molecular environment mimicking the real-world situation (where the targeted chemical species is stabilized by interactions to neighbours) or the analytical process includes operations which are critically influences by such interactions (usually called "matrix effects"). Examples for the latter are many low-energy separation techniques and detection techniques such as solid sampling atomic absorption spectrometry. Consequently well-designed and characterized matrix CRMs are irreplaceable as calibrants for many of the more challenging measurement tasks, especially in molecular analysis.

Typical application areas for reference materials are:

- method development and validation, in particular evaluation of trueness and evaluation of measurement uncertainty
- calibration
- proof of method performance, such as statistical quality control (via control charts etc.), establishing traceable results and qualification of equipment
- proficiency testing, i.e. training and verification of the competence of laboratories

3 REFERENCE MATERIALS AND METROLOGICAL TRACEABILITY

Reviewing the literature and reflecting on presentations and discussions at conferences, other meetings and on university teaching it becomes obvious that the issue of metrological traceability for chemical measurements of samples which are more complex than gases or homogeneous solutions is a topic which is still very much in development and not commonly understood or thoroughly investigated and executed at present. The straightforward approach of traceability chains which can be established for many measurement results in physics, for instance in the case of mass determinations of macroscopic physical objects, is much more difficult to transfer to chemical measurements.

The vast majority of chemical measurement targets are composed of more than one type of atoms with known isotopic composition. Therefore, the realistic expression of common reference points and useful internationally agreed measurement units can represent a significant challenge. This is not only true for the unit of the SI quantity 'amount of substance', the mole, but also for the materialized expression of one kilogram of a specific chemical substance. In that case the purity assessment represents the analytical challenge and becomes very demanding in case of materials composed of macromolecules or many diverse constituents. Consequently the often claimed "traceability to the SI" is difficult to establish in a direct manner. As metrological traceability can also point to other internationally agreed references such as alternative material standards to the kilogram (like the WHO standards), certified reference materials are indispensable for providing common standards for comparisons of measurements from the local to the global scale. In many cases CRMs are basically representing an expression of the "reference quantity" in form of an artefact, namely the specific material status (proteins in frozen human serum, etc.). Therefore, they can be interpreted as 'chemical equivalents' to the famous piece of metals at BIPM defining the kilogram.

Another kind of traceability chains which is very often used in chemical, biochemical and biological measurements points to a method (often called 'reference method'). There exists a huge number of properties each defined via a specific measurement procedure, which is internationally harmonized to a different level. Examples are the so-called total organic carbon (TOC), (total) fat in food and environment samples or Kjeldahl nitrogen and many other parameters. For them international standardization is indispensable, but can ultimately only be realized with the help of common reference materials with sufficiently identical subsamples. As a consequence certified reference materials are representing a crucial tool for establishing metrological traceability of such measurement results and will be essential for that also in the future.

An additional key question for chemical and other species-oriented measurements, which is difficult to be understood with a physically focused background and point of view, is posed by the challenge to define the relevant measurand for the analytical problem of interest. The denotation of this relevant measurand is far from being trivial even in the case of monitoring the biologically active part of calcium in human blood, not to speak about a protein activity in a cell of the liver or somewhere else. Moreover the unit of measurement for such measurands has to be fixed in a useful and achievable manner. This represents a specific challenge if the relevant property does not only depend on the atomic or molecular constitution of the measured target, but also on its chemical or biological activity which is largely dependent on the molecular microenvironment, interactions with solvent and neighbouring molecules as well as physicochemical conditions such as hydrodynamics and temperature near the reaction site. Therefore, modern (bio)chemical measurements have to rely to large extent on common reference systems as further explained in the following.

4 REFERENCE MEASUREMENT SYSTEMS

For establishing harmonized references for measurement results of measurands for which their metrological traceability cannot be straightforwardly established to an independent measurement scale such as the SI, some measurement communities have started to set up so-called reference measurement systems. Their main components are depicted in Figure 3.

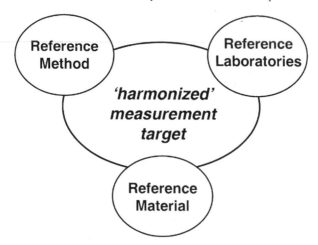

Figure 3 *Components of Reference Measurement Systems*

To define and harmonize a measurement target the respective organisation or network, such as IFCC (International Federation of Clinical Chemistry and Laboratory Medicine) in the field of clinical and laboratory medicine, agrees on a common reference method. This is accompanied by one or more reference materials which allow to install the corresponding reference method in the laboratory of interest and to check regularly the performance of the method. Most of these materials are certified reference materials but there may be also sufficiently qualified reference materials which do not officially come with a full certificate as described in ISO Guide 31. Nevertheless those reference materials could be qualified to be used at the highest metrological level which can be achieved for the time being. An illustrative presentation of corresponding traceability chains for in-vitro diagnostics is published and briefly discussed in ISO 15711.[7]

The reference measurement system is completed by the establishment of reference laboratories which keep the proficiency for the method performance and the production of the most accurate results achievable at present. Usually the reference laboratories are deeply involved in the original development of the reference method and are often participating in the characterisation of the corresponding candidate reference material.

It can be envisaged that reference measurement systems for the sake of achieving international harmonization and comparability of measurement results for properties which cannot be easily or directly linked to the SI will be also formally established at the global scale in more fields than laboratory medicine in the future. They, for instance, already exist in some geographical regions of the world for the area of food analysis. It would be highly desirable to set them up in a systematic manner for environmental monitoring as well, but they require in all cases indeed the availability of specifically tailored reference materials which may limit at present their further spreading.

5 REFERENCE MATERIALS AND ATTACHED INFORMATION

It can often be observed that customers of reference materials do not fully use the potential of their purchased product. One should realize that certified reference materials are coming not only as a physical product in a bottle or ampoule, but that they are accompanied by important information on the material characteristics. CRMs which are fulfilling the requirements of ISO Guides 34 and 35 are delivered with a certificate which contains in addition to the certified value its uncertainty and metrological traceability. Moreover, this certificate should include instructions for use, storage conditions, expiration date of the certificate and either directly or via a link to other documents information about the pretreatment of the material and the procedures of its characterisation. The recommended content for CRM certificates is summarized in ISO Guide 31.[5] Some reference material producers such as IRMM are providing to their customers or even to the public (www.irmm.jrc.be) an even more comprehensive description of the whole CRM preparation and certification process.

It is obvious that one cannot expect from reference materials fabricated many years ago that they are in full compliance with modern concepts and standards of quality assurance and metrology. Moreover various producers and advisory organisations have developed practices and guidelines over the years which may differ because of traditional or area-specific habits of their main customer groups. But there is a recognizable movement towards more harmonized approaches and common understanding in the 'reference material world' in the last few years. This includes a broader appreciation of the factors which are contributing to the total uncertainty budget of a certified value. As it is further explained in the recently published new edition of ISO Guide 35 results from several basic measurements on material characteristics have to be combined for the proper estimation of this uncertainty value.[3] Among others, a number of measurements is needed to assess the stability and homogeneity of a candidate reference material. All of these measurements can only provide results with associated measurement uncertainties. Consequently, such uncertainty contributions have to be properly combined with the measurement uncertainty from the material characterisation (basically the determination of the property value of interest) to obtain the uncertainty of the certified value. There are still many certified reference materials on the market which do not take into account uncertainty contributions from stability and homogeneity measurements. Sometimes this seems not only to reflect limitations in the modern understanding of the metrological design of the procedure for CRM production, but may be also attributed to the goal of achieving lower stated uncertainties on the certificate. There is no doubt that just omitting certain types of measurements, which are actually needed for the proper characterisation of the reference material, is not a scientifically sound way towards low uncertainty values. Further guidance on this topic is provided in the current ISO Guide 35.

It is self-evident that the user of a certified reference material does not have to care about the details of the CRM characterisation, if he can be confident that scientifically sound and internationally accepted approaches have been followed by the reference material producer. But the user should check if this was indeed the case and if proper statements on metrological traceability and uncertainties of the certified values, which include all necessary components, are provided as information together with the material itself. With the increasing implementation of quality management systems and accreditation of testing laboratories, the critical consideration of the confidence in the producer of the reference materials used by a laboratory gains importance as well. There are indications, comparable to developments for the laboratory service of measurements during the last 10 years, for a move towards requests for demonstrated competence of reference material producers through benchmarking against internationally agreed and harmonized criteria rather than accepting 'designated' competence largely based on self-declaration or traditional recognition.

It is certainly not easy and in most cases inefficient for a CRM customer to obtain convincing evidence about the quality management system, the used guidelines and their adequate implementation by a specific reference material producer. Therefore, an appropriate third-party assessment can support also for such activities the building of confidence. For instance, laboratories can look into the database KCDB (Key Comparison Database) of the BIPM (www.bipm.org) for CRMs listed therein. Such reference materials have been produced by signatories or designated laboratories of the CIPM-MRA (Mutual Recognition Arrangement of the International Committee of Weights and Measures) and have been additionally discussed to a certain extent by corresponding groups of the regional metrology organisations. But this database is not intended for use as recommended CRM list and the introduction of CRMs into the KCDB is not based on scientific quality criteria only. Formal requirements include, among others, that only National Metrology Institutes or designated laboratories of the CIPM-MRA can introduce their CRMs and that they had to perform the ultimate measurements used for value assignment in-house. Therefore, one cannot draw any conclusions about the quality of other available reference materials, because the restrictive formal rules of the KCDB exclude the vast majority of reference material producers by themselves. Another way of proving competence is accreditation of reference material producers. After years of controversial debates outside the Asian-Pacific region it is now also recognized at ILAC (International Laboratory Accreditation Co-operation) and CIPM level that ISO Guide 34 is an appropriate accreditation standard for assessing the competence of reference material producers and that it should be applied in combination with ISO 17025 for auditing the quality management system of such producers. There are now also RM producers in Europe and North America accredited according to these international standards and more are in the preparation or even finalization status.

An additional approach to provide confidence in the quality of specific reference materials consists in the creation of CRM brands with stated quality characteristics. Many laboratories know the trademark SRM for certified reference materials coming from NIST in the US. In 1994, a new trademark has been introduced on the CRM market.[8] Based on a co-operation between the Institute for Reference Materials and Measurements of the European Commission's Joint Research Centre, the Federal Institute for Materials Research and Testing (BAM) in Germany and LGC Ltd in the UK the brand ERM® stands for "European Reference Materials". CRMs which carry the ERM trademark are reference materials from the co-operation partners which have been additionally peer-reviewed with respect to their compliance with the principles of ISO Guides 34 and 35. More information on the assessment principles for ERM® CRMs is available on the corresponding website (www.ermcrm.org), where interested parties can also search the ERM catalogue on-line for reference materials of interest. Internet-based search services with information on certificates etc. are also available from other major CRM producers such as IRMM (www.irmm.jrc.be) or NIST (www.nist.gov). The largest database on reference materials worldwide can be found in COMAR (www.comar.bam.de).

Is has been observed that there is still considerable room for improvement in the use of information accompanying the certified reference materials. For instance, further guidance on how to use a CRM within the daily quality control operations of a laboratory is often requested. Besides publications from standardisation bodies like ISO and supporting documents from organisations such as EURACHEM or CITAC (Co-Operation on International Traceability in Analytical Chemistry), several reference material producers are issuing guidance papers CRM users. As a recent example, the ERM co-operation has started a series of so-called "Application Notes" which is available via the ERM webpage. The first one, drafted by IRMM, explains the comparison of measurement results obtained by a laboratory on a CRM with the CRM data provided on the certificate. Such calculations are needed in method validation or verification for assessing the comparability of own measurement data with the certified values. Moreover, it is envisaged that ISO REMCO will revise ISO Guide 33 on the use of reference materials[9] in view of the modern developments and applications of reference materials in chemical, biochemical and biological measurements.

6 OUTLOOK

One of the most difficult tasks for the proper application of quality control measures in measurement laboratories seems to be the selection of a reference material appropriate for the measurement in question. There exists certainly a close correlation between the analytical problem, the nature of the sample to be analysed, the analytical measurement procedure and the reference materials for calibration and quality control (Figure 4).

The selection of the reference materials depends on several characteristics of the analytical procedure such as robustness towards matrix effects. Consequently, the nature of the real-world sample determines the type of matrix which should be represented by the RM in addition to the concentration levels for the measurand (analyte) of interest and possible interfering constituents in the RM. Nowadays there is an increasing use of the term "commutability" to express this similarity in the analytical behaviour of the real-world sample and the chosen reference material. In dependence on the analytical question the metrological level of the analytical procedure has to be defined further. This level is different fo r measurements directed to calibrate a multi-step method or to validate the trueness of a new procedure in comparison to the routine characterisation of products fabricated under specified and long-term applied conditions. Further discussion on this issue is provided in the literature, for instance in ISO 15711 for in vitro diagnostics.[7] Moreover one has to acknowledge that the interrelation between development of a new method and production of a corresponding reference material may be a limiting factor if a new measurement problem has to be tackled. In such cases one can usually only prepare and test a stepwise approach. That means the measurement methods are developed with some kinds of 'test materials', which are not already sufficiently characterized as reference materials, until the methods have achieved a certain level of maturity. Afterwards a full method validation can be achieved with the help of an appropriately designed certified reference material studied by applying the new methods.

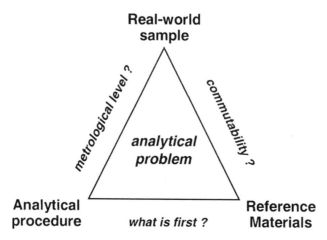

Figure 4 *Some considerations for solving an analytical problem*

In summary, the main purpose of reference materials consists in providing modern tools for quality assurance and control of measurements in different areas. It has been frequently demonstrated and is widely accepted now that appropriate reference materials, which are properly used by the measurement laboratory, support significantly the building of confidence in the corresponding measurement results and consequently their reliability and international comparability.

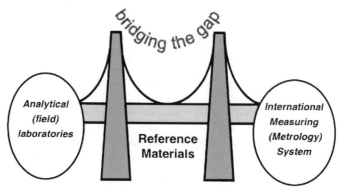

Figure 5 *Reference materials as facilitator*

But it is likewise recognized that the bridging function of reference materials as depicted in Figure 5 can only be fulfilled by materials which are tailored to the needs of the specific measurement task. Therefore, significant efforts are continuously needed also in the future to develop and produce reference materials of appropriate quality, serving the broad range of measurement problems to be tackled. Furthermore, the training of RM users, accreditors and other stakeholders is an important issue as well.

References

1　H. Emons, T.P.J. Linsinger and B.M. Gawlik, *Trends Anal. Chem.*, 2004, **23**, 442.
2　H. Emons, A. Fajgelj, A.M.H. van der Veen and R. Watters, *Accr. Qual. Assur.*, 2006, **10**, 576.
3　International Organization for Standardization, *ISO Guide 35*, 3rd edition, ISO, Geneva, 2005.
4　H. Emons, *Accr. Qual. Assur.*, 2006, **10**, 690.
5　International Organization for Standardization, *ISO Guide 31*, 2nd edition, ISO, Geneva, 2000.
6　International Organization for Standardization, *ISO Guide 34*, 2nd edition, ISO, Geneva, 2000.
7　International Organization for Standardization, *ISO 17511*, ISO, Geneva, 2003.
8　H. Emons, J. Marriott and R. Matschat, *Anal. Bioanal. Chem.*, 2005, **381**, 28.
9　International Organization for Standardization, *ISO Guide 33*, ISO, Geneva, 2000.

THE CSM APPROACH TO THE CALCULATION OF THE UNCERTAINTY IN XRF ANALYSIS OF LOW- AND HIGH-ALLOYED STEELS

E.Celia, and F. Falcioni

Analytical Chemistry Laboratory - Process Chemistry and Environmental Monitorino Section, Centro Sviluppo Materiali S.p.A., Rome

1 INTRODUCTION

The CSM laboratories are, for some tests, accredited by SINAL (the Italian Laboratories Accreditation System) and must meet the specification of ISO/IEC 17025.

This standard states that *"Testing laboratories shall have and shall apply procedures for estimating uncertainty of measurement"*.

The calculation of the uncertainty of XRF analysis of low and high-alloyed steel was made according to the "bottom-up" or "component-by-component" method suggested in the ISO 25 "Guide to the Expression of Uncertainty in Measurements" and in the Eurachem Guide "Quantifying Uncertainty in Analytical Methods".

Repeatability, calibration standards and calibration curves were individuated as the most important contributions to the total uncertainty.

In this work only Cr, Ni, Mo and Mn were considered.

2 EXPERIMENTAL

The repeatability was calculated performing 10 measurements in different days with different operators on two reference samples, as shown in table 1

Table 1

sample	Cr %	Ni %	Mn %	Mo %
QC1				
average	17,4	8,4	1,8	0,30
std dev	0,14	0,13	0,04	0,002
QC8				
average	0,162	0,123	0,707	0,070
std dev	0,002	0,002	0,005	0,001

Being an unknown sample analysed usually only once, it was decided not to consider the standard deviation of the mean ($s/\sqrt{n}$), but only s.

The contribution, as standard uncertainty, of the uncertainty of the certified reference materials used for the calibration was calculated as follows:

If the limits $\pm a$ were given without any confidence level, it was assumed a rectangular distribution with a standard deviation of $a/\sqrt{3}$.

If the limit $\pm a$ were given with a 95% confidence level, the standard deviation considered was $a/1.96$.

Being the calibration curve obtained with more than 15 points, and so with more than 15 reference materials, each of them with its uncertainty, it was decided to consider, for a given concentration of one element, the uncertainty of the reference material with the concentration closest to the value of interest.

The contribution of the calibration curve was calculated considering that in XRF the intensity (I) of the signal of a defined wavelenght is correlated to the element concentration (C)in the sample:

$$I = a + b * C$$

then

$$C = \left(\frac{I - a}{b} \right)$$

The standard errors of a and b were obtained from the linear least square regression, while the standard error of the intensity was derived from instrument resolution information.

All the standard uncertainties were combined and, to obtain the expanded uncertainty, the number of degrees of freedom was calculated by means of the Welch-Satterthwaite equation:

$$\nu_{eff} = \frac{u_c^4(y)}{\sum_{i=1}^{N} \frac{u_i^4(y)}{\nu_i}}$$

Then, the coverage factor was chosen using the Student's t-distribution.

Being the degrees of freedom generally higher than 20, the coverage factors ranged from 2.0 to 2.1.

3 RESULTS

In the following table are shown the uncertainty and the uncertainty budget for the given elements and concentrations

ASTM E2165-01 gives two experimental *Horwitz-like* equations obtained from ISO TC 17 SC 1 and from ASTM Proficiency Test Programs for plain carbon low alloyed and stainless steel.

ISO:　　　　$2s_{95\%} = 0.0303C^{0.6661}$

ASTM:　　　$2s_{95\%} = 0.0384C^{0.58}$

Where s is the interlaboratory standard deviation and C is the analyte concentration in m/m %

In the following graphs (Figure 1 and 2) are plotted the CSM uncertainty values together with those derived from the *Horwitz-like* equations.

Table 2

element	conc	uncertainty	% of contribution		
	%	%	cal. Curve	CRM	repeatability
low-alloyed steel					
Cr	0,16	0,03	48,9	44,1	7,1
Ni	0,12	0,02	73,7	5,3	21,0
Mn	0,71	0,05	20,1	63,9	16,0
Mo	0,070	0,007	37,2	38,3	24,5
high-alloyed steel					
Cr	17,4	0,4	62,5	4,6	32,8
Ni	8,4	0,3	28,0	9,3	62,7
Mn	1,8	0,1	21,4	26,6	52,0
Mo	0,30	0,08	74,3	25,4	0,3

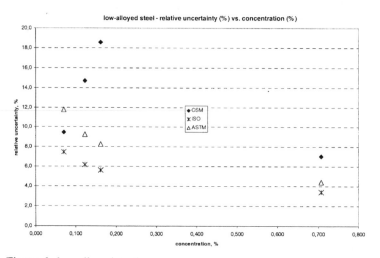

Figure 1 *low alloyed steel*

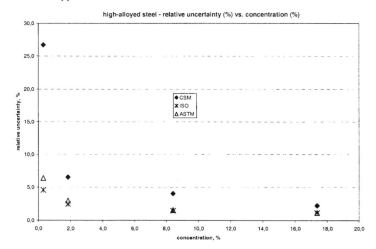

Figure 2 *High alloyed steel*

4 CONCLUSIONS

It can be said that:

The uncertainty values obtained at CSM using the bottom-up model are systematicaly greater than the 2s values derived from the experimental equations of ISO and ASTM proficiency tests.

In the high alloyed steels the trend of the 3 curves is similar (the higher the concentration, the lower the uncertainty), but in the low alloyed steels the CSM points are very scattered.

This could be explained looking at the uncertainty budget.

It is evident that at low concentrations the contributions to the uncertainty of the CRM and of the calibration curve are generally heavier than those at high concentrations.

The uncertainty in these cases depends strongly on the uncertainty of the CRMs used in the calibration.

COMPARISON OF DIFFERENT APPROACHES TO EVALUATE PROFICIENCY TEST DATA

A. Shakhashiro, A. Fajgelj, U. Sansone

International Atomic Energy Agency (IAEA), Agency's Laboratories Seibersdorf and Vienna, P.O.Box 100, A-1400 Vienna, Austria.

1 INTRODUCTION

Proficiency testing (PT) is an essential tool for assessing the competency of analytical laboratories. Proficiency testing involves the regular circulation of test materials for analysis by the participating laboratories, and the subsequent assessment of the resulting data by the organizing body. To this end, several rating systems have been developed for determining a laboratory's performance. The z-scores and u-scores are the most frequently used evaluation approaches, each having its own specifics. By using the z-score the accuracy of participants' results will be evaluated, while using the u-score uncertainties of the participant measurement results and the uncertainty of the assigned PT value are also taken into consideration. Consequently, evaluations of laboratories using the z-score will not necessarily exhibit the same level of performance as in the cases of evaluations using the u-score, where also the measurement uncertainties are considered in the evaluation.

The IAEA Reference Material Group in Seibersdorf, Austria uses a modified u-score evaluation system, thus taking into consideration the trueness and the precision of the reported data and it includes in the evaluation both the total combined uncertainty associated with the target value of proficiency testing samples and the total uncertainty reported by the participating laboratories. A set of data for determination of ^{57}Co in spiked soil material from a proficiency test performed by the IAEA Reference Material Group in Seibersdorf was used to compare different rating systems.

2 VALUE ASSIGNMENT AND STATISTICAL TECHNIQUES USED FOR EVALUATION OF PROFICIENCY TESTING RESULTS

2.1 Assignment of Property Value to PT Materials

The PT provider is responsible for the assignment of the property value of PT samples and the associated uncertainty. The approach in which the property value is assigned influences the claimed uncertainty and relates to the metrological traceability of the assigned value. There are two main different options in establishing the property value. The first one is proactive while the second is retroactive. In the proactive way the PT provider establishes the property value of the PT material/sample before sending it to the participants. The most commonly used approaches are listed below:

2.1.1 Property values assignment based on formulation. In this approach a blank PT material is normally spiked with a standard solution of the analyte(s) of interest, blended or diluted. The property value is calculated from the quantities used, and the associated uncertainty is estimated from the standard uncertainties of these quantities, taking into account also the uncertainties arising from operations such as dilution and weighing. The advantage of this approach is the relatively small uncertainty of assigned property values in comparison to other approaches described below. The metrological traceability of property values assigned to the PT materials can be clearly established through the traceability of the values of the standard solution(s) or constituents used in the preparation. Therefore, PT materials prepared by formulation can be used for the evaluation of the precision of participants' results as well as for their trueness. Furthermore, the false negative and false positive can also be detected. The major drawback of spiked PT materials is that they do not always match the routine samples with respect to physical and chemical characteristics. However, the formulation method is applicable in all cases when non-destructive analytical methods are used and when the total dissolution of the sample is performed.

2.1.2 Property value assignment based on measurement results of a group of expert laboratories. Expert laboratories may use the same or different validated measurement procedures and different methods for characterization of the material. The property value derived according to this approach will be a consensus value. Resulting from the best current measurement practices, results from the expert laboratories are considered as the best estimate of the "true value". Metrological traceability of measurement results of expert laboratories are known and therefore also the metrological traceability of the derived consensus value can be stated. In contrast to the materials prepared by formulation, these materials in most cases match the routine samples with regard to the physical and chemical characteristics.

Different statistical procedures may be applied to derive the consensus value based on the results obtained from a group of expert laboratories, and this might lead to differences in calculated property value and its associated uncertainty.[1,2] In the ISO Guide 35:2006 four statistical approaches for deriving the property value and associated uncertainty are described.[3] We have applied these four approaches on the set of results from an IAEA candidate reference material measurement campaign for determination of ^{57}Co, and the observed differences are presented in Table 1. To allow qualitative comparison between different approaches, the calculated property value and uncertainties were normalized. As expected, it can be seen that the calculated property values are in a good agreement regardless of the approach used, while the associated uncertainties range from 0.8 to 3.6 %.

Consequently, the above described difference in uncertainties might affect the evaluation of the participants performance in a PT when the scoring system takes into account also the uncertainties of the participants' results and/or the uncertainty of the property values. See more details on the u-score in paragraph 2.2.1. It has to be mentioned that characterization of PT materials using a group of expert laboratories might be quite expensive and substantial resources might be required to cover the cost of the characterization campaign.

Table 1. *Normalized Property Values and Associated Uncertainties Obtained Using Different Statistical Approaches.*

Method	Approach	Normalized Property value	u	u%
A–2 ISO Guide 35:2006	Two-way ANOVA	0.993	0.06	3.46
B–6 ISO Guide 35:2006	One-way ANOVA	1.000	0.06	3.55
10.5.2 ISO Guide 35:2006	Mean of mean	1.000	0.06	3.59
10.8.3 ISO Guide 35:2006	Weighted (u) mean	0.993	0.01	0.80
	Robust statistics	1.007	0.013	0.91

2.1.3 Property value assignment based on measurement results from one expert laboratory confirmed by results from two or more expert laboratories. In this case one expert laboratory performs characterization of the PT material in reproducibility conditions and establishes the property value and its uncertainty. Results from other expert laboratories, obtained by the same or different validated methods/procedures, are used for confirmation purpose only. The metrological traceability is established and also the problem of combining measurement uncertainties of different laboratories is avoided. PT materials produced in this way can be used to evaluate trueness and precision.

2.1.4 Property value assignment based on interlaboratory study (retroactively).
The property value is assigned as a consensus value of PT participants and is usually estimated as the mean of the test results after outliers have been rejected. For the PT organizer this is probably the most economic approach for obtaining the characterization data. The shortcoming of this approach is often a lack of control that the PT organizer would have over the quality of the characterization results. Participants' results could be biased due to systematic or random reasons. The use of inadequate methodology or mistakes in the application of the analytical procedure is among the most common sources of errors. Further, calculation of consensus values includes the treatment of extreme values and outliers within the dataset. This can be accomplished by the use of robust statistical methods or outlier detection procedures. The consensus value can then be calculated as, e. g., a robust average, arithmetic mean or weighted mean, etc. Due to the fact that participating laboratories normally use the method/procedure of their choice, and that a number of different calibrants may be used, the metrological traceability of PT materials produced in this way is difficult to describe. PT materials characterized in this way are suitable for assessing the between laboratory performance, but should not be used for evaluation of trueness of participants' results.

2.2 Evaluation of Participants' Performance

2.2.1 z-score, relative bias and ratio. The most common evaluation of PT results is based on the z-score introduced in the IUPAC Harmonized Protocol.[4,5] Participant's result x is converted into a 'z-score' by the equation:

$$z = (x - X) / \sigma \qquad (1)$$

where x is the participant's result, X and σ are the assigned value and the measure of its variability (assigned standard deviation), correspondingly. The model can be used in all cases, regardless of the approach used for the establishment of the PT target value. In an ideal PT scheme, the value of X and σ should be 'fit for purpose', this means they should reflect the requirements of the specific analytical task, including the span of uncertainty that is tolerable in relation to the purpose of the data (results).

Another evaluation of PT results is based on relative bias,

$$\text{Relative bias} = \frac{Value_{Lab} - Value_{Prop}}{Value_{Prop}} \times 100\% \qquad (2)$$

while some PT providers use the ratio between property value and laboratory result for evaluation:

$$\text{Ratio} = \frac{Value_{Lab}}{Value_{Prop}} \times 100\ \% \qquad (3)$$

where $Value_{Lab}$ is the value reported by the laboratory and $Value_{Prop}$ is the property value assigned by the proficiency testing organizer. All three approaches for evaluation of PT results described above are designed for evaluation of bias only. They are very effective for routine proficiency tests especially in regulated areas where standardized measurement procedures/methods are used, and where other method performance indicators, e.g. reproducibility, measurement uncertainty, are assessed collectively during the method validation trials.

2.2.2 u-score. The value of the u-test score is calculated according to the following equation:

$$u_{Score} = \frac{\left| Value_{Prop} - Value_{Lab} \right|}{\sqrt{Unc^2_{Prop} + Unc^2_{Lab}}} \qquad (4)$$

where Unc_{Lab} is the uncertainty associated with the participant's result and Unc_{Prop} is the uncertainty of the assigned property value.

2.2.3 Modified u-score. In many applications in trace elements and radionuclides analysis on non-routine samples and on occasional tests, it is of additional value for participants if their results are evaluated also in respect to the reported measurement uncertainty.[6,7] Based on the more than 40 years experience with the open world-wide laboratory comparison studies, it was decided in the Chemistry Unit of the IAEA Seibersdorf Laboratories to use a modified u-score evaluation, where the trueness and precision of participants' results are evaluated separately.[8]

For trueness[9] evaluation (denoted as T in Table 2) the participants' results are assigned "Acceptable" if:

$$A1 \leq A2 \qquad (5)$$

where:

$$A1 = \left| Value_{Prop} - Value_{Lab} \right| \qquad (6)$$

$$A2 = 2.58 \times \sqrt{Unc_{Prop}^2 + Unc_{Lab}^2} \qquad (7)$$

For evaluation of precision, estimator P (see Table 2) is calculated for each participant, according to the following formula:

$$P = \sqrt{\left(\frac{Unc_{Prop}}{Value_{Prop}}\right)^2 + \left(\frac{Unc_{Lab}}{Value_{Lab}}\right)^2} \times 100\ \% \qquad (8)$$

P directly depends on the measurement uncertainty claimed by the participant. The acceptance limit for precision (ALP) for each analyte respectively is defined for the respective proficiency test in advance, including any adjustment due to the concentration or activity level of the analytes concerned and the complexity of the analytical problem. Participants' results are scored as "Acceptable" for precision when $P \leq ALP$.

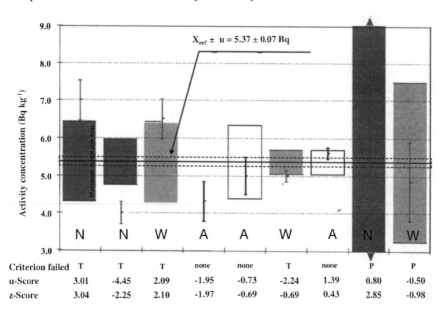

Criterion failed	T	T	T	none	none	T	none	P	P
u-Score	3.01	-4.45	2.09	-1.95	-0.73	-2.24	1.39	0.80	-0.50
z-Score	3.04	-2.25	2.10	-1.97	-0.69	-0.69	0.43	2.85	-0.98

Figure 1 *Graphical presentation of the final evaluation following the IAEA approach, in comparison with u- and z-scores. (A = Acceptable, N = Not Acceptable, W = Warning).*

In the final evaluation, both scores for trueness and precision are combined. A result must obtain an "Acceptable" score in both criteria to be assigned the final score "Acceptable". Obviously, if a score "Not Acceptable" was obtained for both trueness and precision, the final score will also be "Not Acceptable". In cases where either precision or trueness is "Not Acceptable", a further check is applied. The value of bias A1 is compared with the maximum acceptable bias (MAB), which is defined by the IAEA in advance, similarly as

ALP. If A1 ≤ MAB, the final score will be "Warning". "Warning" will mainly reflect two situations. The first situation will be a biased result with a small measurement uncertainty, however still within MAB. The second situation will appear when a result close to the assigned property value is reported, but the associated uncertainty is large. If A1 > MAB, the result will be "Not Acceptable". A graphical presentation of the various situations in the evaluation of results is given in the Figure 1.

3 RESULTS AND DISCUSSION

In Table 2 the results of one of the proficiency tests organized in the frame of the IAEA Network of Analytical Laboratories for the Measurement of Environmental RAdioactivity (ALMERA) are presented. [10] ALMERA is an operational IAEA network comprising 105 laboratories from 66 IAEA Member States. Laboratories are supposed to be ready for emergency situations and should be able to deliver measurement results in a very short time. ALMERA members are from all over the world and comparability of the results is one of the major concerns. Regular proficiency tests covering various matrices and radionuclides are therefore an integral part of ALMERA activity. In this case results from 49 laboratories on the determination of ^{57}Co in a spiked soil were evaluated using five approaches discussed above, namely: relative bias, z-score with σ = 10 % of property value, z-score with standard deviation derived from participants results, z-score with robust standard deviation and the IAEA approach (modified u-score). The comparison of the evaluation results shows that the use of relative bias, z-score with σ = 10 % of property value, z-score with standard deviation derived from participants results and z-score with robust standard deviation give the same number of 'Acceptable" results, actually 82 %. The IAEA approach is shown as more rigorous, with 69 % of results identified as "Acceptable". A similar situation can be observed in the case of "Not Acceptable" scores, where between 12 and 16 % of the results fall into this category respectively. However, the largest difference can be observed with respect to the results obtaining the score "Warning". Using a relative bias, and z-score with σ = 10 % of property value, two results out of 49 (4 %) obtained a score "Warning". 3 results (6 %) were identified when the z-score with standard deviation derived from participants results or z-score with robust standard deviation scores in relative bias were used. In the case of the IAEA approach (a modified u-score) 7 results (14 %) received "Warning".

From this comparison it can easily be concluded that the evaluation approach used by the IAEA is the most rigorous one. However, as mentioned already, the two most common reasons leading to the score "Warning" are under or overestimated measurement uncertainty. The educational component of the proficiency test, including the information on measurement uncertainty is therefore of high importance. It should be pointed out that the predefined acceptance criteria, namely acceptance limit for precision (ALP) and maximum acceptable bias (MAB) reflect the current state of the art in the area of radioactivity measurements and radionuclide determinations and help laboratories to critically assess their way of quantifying and reporting measurement results and associated uncertainties. The experience gained so far with the use of the modified u-score (the IAEA approach) has been very positive.

Table 2: Comparison of IAEA PT scoring approach to z-score and relative bias approaches of ^{57}Co (spiked property value: 17.1 ± 0.44 Bq kg⁻¹).

| | Performance indicators calculations | | | | | | | | Scores in different approaches | | | | |
| Lab. Results | Uncertainty[a] | Relative bias[b] | z-score[c] | z-score[d] | z-score[e] | IAEA[f] Approach | | Relative bias | z-score | z-score | z-score | IAEA Approach |
Bq kg⁻¹	[%]	[%]	[σ = 0.10]	[σ =STD]	[σ =Robust]	T	P	[20, 30]	[σ = 0.10]	[σ =STD]	[σ =Robust]	Final score
15.7	0.96	-8.24	-0.82	-0.74	-0.75	N	A	A	A	A	A	W
17	5.88	-0.64	-0.06	-0.01	-0.01	A	A	A	A	A	A	A
16.9	8.88	-1.23	-0.12	-0.06	-0.07	A	A	A	A	A	A	A
19.77	22.26	15.55	1.55	1.56	1.55	A	N	A	A	A	A	W
18	4.44	5.20	0.52	0.56	0.55	A	A	A	A	A	A	A
18.18	1.05	6.26	0.63	0.66	0.65	A	A	A	A	A	A	A
2.56	4.69	-85.04	-8.50	-8.18	-8.19	N	A	N	N	N	N	N
17.7	6.78	3.45	0.35	0.39	0.38	A	A	A	A	A	A	A
18.1	5.03	5.79	0.58	0.62	0.61	A	A	A	A	A	A	A
18.4	5.98	7.54	0.75	0.79	0.78	A	A	A	A	A	A	A
20.21	3.66	18.12	1.81	1.81	1.80	N	A	A	A	A	A	W
17.9	5.59	4.62	0.46	0.50	0.50	A	A	A	A	A	A	A
39.5	7.09	130.86	13.09	12.73	12.72	N	A	N	N	N	N	N
15.82	3.92	-7.54	-0.75	-0.67	-0.68	A	A	A	A	A	A	A
15.1	5.96	-11.75	-1.17	-1.08	-1.09	A	A	A	A	A	A	A
14.56	11.47	-14.90	-1.49	-1.39	-1.39	A	N	A	A	A	A	W
11.95	3.43	-30.16	-3.02	-2.86	-2.87	N	A	N	N	N	N	N
17.2	1.74	0.53	0.05	0.11	0.10	A	A	A	A	A	A	A
16.9	5.03	-1.23	-0.12	-0.06	-0.07	A	A	A	A	A	A	A
18.9	5.82	10.46	1.05	1.07	1.06	A	A	A	A	A	A	A
10.4	3.27	-39.22	-3.92	-3.74	-3.75	N	A	N	N	N	N	N
18	5.56	5.20	0.52	0.56	0.55	A	A	A	A	A	A	A
18.6	13.44	8.71	0.87	0.90	0.89	A	N	A	W	A	A	W
23.6	6.36	37.93	3.79	3.73	3.72	A	N	N	N	N	N	N
16.6	4.82	-2.98	-0.30	-0.23	-0.24	A	A	A	A	A	A	A
18.6	6.45	8.71	0.87	0.90	0.89	A	A	A	A	A	A	A
16.6	6.02	-2.98	-0.30	-0.23	-0.24	A	A	A	A	A	A	A

		Performance indicators calculations						Scores in different approaches				
Lab. Results	Uncertainty[a]	Relative bias[b]	z-score[c]	z-score[d]	z-score[e]	IAEA[f] Approach		Relative bias	z-score	z-score	z-score	IAEA Approach
Bq kg⁻¹	[%]	[%]	[σ = 0.10]	[σ =STD]	[σ =Robust]	T	P	[20 , 30]	[σ = 0.10]	[σ =STD]	[σ =Robust]	Final score
16.8	5.95	-1.81	-0.18	-0.12	-0.13	A	A	A	A	A	A	A
72.3	0.30	322.57	32.26	31.30	31.28	N	A	N	N	N	N	N
17.5	5.71	2.28	0.23	0.28	0.27	A	A	A	A	A	A	A
16.3	1.53	-4.73	-0.47	-0.40	-0.41	A	A	A	A	A	A	A
24.19	6.74	41.38	4.14	4.07	4.06	N	A	N	N	N	N	N
14.9	5.84	-12.91	-1.29	-1.19	-1.20	A	A	A	A	A	A	A
16.4	3.05	-4.15	-0.41	-0.34	-0.35	A	A	A	A	A	A	A
16.86	2.25	-1.46	-0.15	-0.08	-0.09	A	A	A	A	A	A	A
14.93	6.16	-12.74	-1.27	-1.18	-1.18	A	A	A	A	A	A	A
14.7	7.48	-14.08	-1.41	-1.31	-1.32	A	A	A	A	A	A	A
14.3	5.59	-16.42	-1.64	-1.53	-1.54	A	A	A	A	A	A	W
18.42	5.37	7.66	0.77	0.80	0.79	A	A	A	A	A	A	A
17.3	4.05	1.11	0.11	0.16	0.16	A	A	A	A	A	A	A
14.86	8.75	-13.15	-1.31	-1.22	-1.22	A	A	A	A	A	A	A
18.8	2.66	9.88	0.99	1.01	1.01	A	A	A	A	A	A	W
14.9	4.03	-12.91	-1.29	-1.19	-1.20	A	A	A	A	A	A	A
17.1	8.19	-0.06	-0.01	0.05	0.04	A	A	A	A	A	A	A
20.75	7.18	21.28	2.13	2.12	2.11	A	A	W	W	W	W	W
12.27	9.78	-28.29	-2.83	-2.68	-2.69	N	N	W	W	W	W	N
16.8	7.98	-1.81	-0.18	-0.12	-0.13	A	A	A	A	A	A	A
18.1	6.63	5.79	0.58	0.62	0.61	A	A	A	A	A	A	N
16.9	5.33	-1.23	-0.12	-0.06	-0.07	A	A	A	A	A	A	A
						Acceptable scores per approach		40 (82%)	40 (82%)	40(82%)	40 (82%)	34 (69%)
						Warning scores per approach		2 (4%)	2 (4%)	3 (6%)	3 (6%)	7 (14%)
						Not Acceptable scores per approach		7 (14%)	7 (14%)	6 (12%)	6 (12%)	8 (16%)
						Total of scores		49	49	49	49	49

Legend for Table 2: a) Combined uncertainty of laboratory results at 1σ; b) Calculated as per formula (2); c) z-score calculated with target $\sigma = 10$ % of the property value; d) z-score with standard deviation derived from participants' results; e) z-score with robust standard deviation and mean; f) calculated as per formulas (5 to 8), A = Acceptable; N = Not Acceptable; W = Warning.

References

1   EURACHEM Guide 'Selection, use and interpretation of proficiency testing (PT) schemes by laboratories', EURACHEM, 1st edition, 2000.

2   D. L. Duewer, A Robust Approach for the Determination of CCQM Key Comparison Values and Uncertainties, Paper presented at the 10th meeting of the CIPM Consultative Committee for Amount of Substance - Metrology in Chemistry, Sevres, France, April 2004.

3   ISO Guide 35, Reference materials — General and Statistical Principles for Certification, International Organization for Standardization, Geneva, 2006.

4   M. Thompson, S.R.L. Ellison, R. Wood, The International Harmonized Protocol for the Proficiency Testing of Analytical Chemistry Laboratories, Pure Appl. Chem., 2006, **78**, 145.

5   ISO Guide 43, Proficiency testing by interlaboratory comparisons, International Organization for Standardization, Geneva 1993.

6   Guide to the Expression of Uncertainty in Measurement, International Organization for Standardization, Geneva, 1995.

7   Quantifying Uncertainty in Nuclear Analytical Measurements, TECDOC-1401, International Atomic Energy Agency, Vienna, 2004.

8   C. J. Brookes, I. G. Betteley, and S. M. Loxton, Fundamentals of Mathematics and Statistics, Wiley, UK, 1979.

9   ISO 5725 (E), "Accuracy (trueness and precision) of measurement methods and results", International Organization for Standardization, Geneva, 1994.

10   IAEA/AL/152 Final report on the Proficiency Test of the Analytical Laboratories for the Measurement of Environmental Radioactivity (ALMERA) Network, International Atomic Energy Agency, Vienna, 2005.

DISTRIBUTION OF PROFICIENCY TESTING RESULTS AND THEIR COMPARABILIY

I.Kuselman

The National Physical Laboratory of Israel (INPL), Danciger "A" Bldg, Givat Ram, Jerusalem 91904, Israel

1   INTRODUCTION

To develop a proficiency testing (PT) scheme, a PT provider uses the knowledge of analytical methods and of reference materials (RM) that can be applied in the experiment as the test/sample units. The experiment (interlaboratory comparison or intercomparison[1]) consists of simultaneous analysis of the same RM samples in laboratories participating in PT. Therefore, determination of metrological properties of this material is one of the first steps of any PT. The properties are: the assigned/certified value C_{cert} of the analyte concentration in RM, the standard uncertainty σ_{cert} of this value (including uncertainty components arising from RM inhomogeneity and instability) and traceability of this value. International documents[1-4] describe five methods of determining RM properties for PT. From a metrological point of view and for the sake of simplicity it is preferable for a PT provider to have a certified reference material (CRM) which can be distributed among the laboratories participating in PT. In this case, C_{cert}, as well as its traceability and σ_{cert} value are already determined by the CRM producer and reported in the CRM certificate[5,6] (σ_{cert} equals approximately to half the expanded uncertainty of C_{cert} given in the certificate). If CRM is expensive or not available, it can be helpful to use an in-house reference material (IHRM) or a spike with traceable property values determined during their development[7,8]. In cases when the number of PT participants (N) is large enough, the assigned value of RM sent to laboratories participants in PT is accepted after completing the experiment, as the consensus value of expert laboratories (out of the PT participants), or as the consensus value of all the N participants[2].

The performance of a laboratory-PT participant is assessed by the difference between the result it obtained and the assigned value, compared to the PT standard deviation σ_{PT}. The latter one can be equal to either the combined uncertainty of both the laboratory result and of the assigned value, or to a standard deviation of the PT results (when N is large enough), or to a target standard deviation of the analysis[1-3, 9-12]. Since ISO 17025[13] and other standards for laboratory accreditation define the performance as an important element of the laboratory technical competence, PT results are assessed individually for each PT participant. However, comparability/equivalence of the PT results is also an important group performance characteristic of laboratories participating in PT, which should be assessed.

One of the principal tools for achieving comparability is traceability, since results obtained by different laboratories in different countries and at different times are comparable through their relation to the same references, which are internationally agreed and recognized measurement standards[14]. CRM, IHRM and spikes with traceable property values applied in PT (further generally referred to as RM) are a kind of measurement standard. Therefore, to assess comparability, one should determine how PT results are distributed and how the parameters of this distribution are close to the parameters of the RM (C_{cert} and σ_{cert}) used for the PT.

A criterion of comparability of PT results in situations when distributions are normal, is proposed in ref.[15]. Recently, a test has been developed that allows a PT provider and/or an accreditation body to assess comparability of PT results when their distribution differs from a normal one and cannot be normalized[16]. The purpose of the present paper is to discuss the hypotheses necessary for development of the comparability criteria.

2   RELATIONSHIPS BETWEEN THE DISTRIBUTION OF RM ASSIGNED VALUE DATA AND THE DISTRIBUTION OF PT RESULTS

Data used for calculation of the RM assigned value, and the measurement/analysis results of the laboratories participating in PT can be considered as independent random events. Therefore, the relation between them can be characterized by the common area P under the density function curves for both RM data and for PT results. The P value is the probability of joint events and, therefore, the probability of obtained PT results belonging to the population of RM data. Taking into account the traceability of the RM data, the larger the P value, the higher comparability of the PT results.

For the sake of simplicity, both distributions are assumed to be normal, with parameters C_{cert}, σ_{cert} and C_{PT}, σ_{PT}, as shown in Fig. 1. The figure refers to a simulated example of aluminum determination in coal fly ashes using standard reference material SRM 2690 (NIST, USA) with $C_{cert} = 12.35$ % and $\sigma_{cert} = 0.14$ % by weight[15]. It can be seen that probability P shown in Fig.1 by the shaded area tends to zero when the difference between C_{cert} and C_{PT} is significantly larger than both σ_{cert} and σ_{PT} (Fig. 1a). The closer C_{PT} to C_{cert}, the higher the P value is. When C_{PT} and C_{cert} coincide (Fig.1b), P achieves max value at the given standard deviations σ_{cert} and σ_{PT}. In the general case (Fig. 1c), the values of the analytical results C_1 and C_2 corresponding to the crossing points of the density function curves should be found to calculate P.

Since both density functions, f_{cert} of RM data and f_{PT} of PT results are equal at these C values, one can write

$$f_{PT} = \frac{1}{\sigma_{PT}\sqrt{2\pi}} e^{-(C-C_{PT})^2/2\sigma_{PT}^2} = \frac{1}{\sigma_{cert}\sqrt{2\pi}} e^{-(C-C_{cert})^2/2\sigma_{cert}^2} = f_{cert} \qquad (1)$$

After transformation of expression (1) into a quadratic equation, C_1 and C_2 are calculated as its roots[17]:

$$C_1, C_2 = \frac{(\sigma_{cert}^2 C_{PT} - \sigma_{PT}^2 C_{cert}) \pm \sigma_{cert}\sigma_{PT}\sqrt{\rho}}{\sigma_{cert}^2 - \sigma_{PT}^2}, \qquad (2)$$

where

$$\rho = (C_{cert} - C_{PT})^2 + 2(\sigma_{PT}^2 - \sigma_{cert}^2)\ln\frac{\sigma_{PT}}{\sigma_{cert}}. \qquad (3)$$

When C_1 and C_2 are known, the probability calculation is convenient by the next formula:

$$P = \int_{-\infty}^{C_1} f_{cert} dC + \int_{C_1}^{C_2} f_{PT} dC + \int_{C_2}^{+\infty} f_{cert} = 1 + \phi\left(\frac{C_1 - C_{cert}}{\sigma_{cert}}\right) + \phi\left(\frac{C_2 - C_{PT}}{\sigma_{PT}}\right) - \phi\left(\frac{C_1 - C_{PT}}{\sigma_{PT}}\right)$$

$$- \phi\left(\frac{C_2 - C_{cert}}{\sigma_{cert}}\right) \tag{4}$$

where ϕ stands for normalized normal distribution.

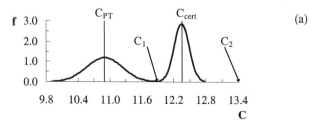

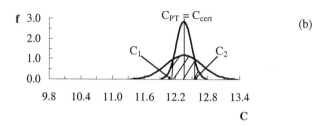

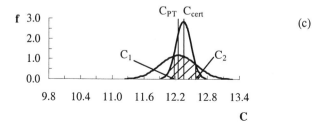

Figure 1 *Density functions f of PT results and of RM data C (aluminum in fly ashes, % by weight) at $C_{cert} = 12.35$, $\sigma_{cert} = 0.14$ and $\sigma_{PT} = 0.34$, when $C_{PT} = 10.9$ (Fig. 1a), $C_{PT} = 12.35$ (Fig. 1b), and $C_{PT} = 12.25$ (Fig. 1c). Values C_1 and C_2 are the analytical results corresponding to the crossing points of the density function curves.*

For example, calculations by formulas (2)-(4) in the cases shown in Fig. 1 yield the following values: (a) $C_1 = 11.90$, $C_2 = 13.41$ and P=0.00; (b) $C_1 = 12.15$, $C_2 = 12.55$ and P=0.60; (c) $C_1 = 12.16$, $C_2 = 12.58$ and P=0.58.

The closer σ_{PT} to σ_{cert}, the higher P value is at the given C_{PT} and C_{cert}. Theoretically, the situation is possible when $\sigma_{cert} = \sigma_{PT} = \sigma$, i.e. the dispersion of the PT results and the dispersion of the results used by the RM producer (or by the PT provider) for C_{cert} evaluation, are on the same level. In such a case, there is only one intersection point of the density function curves (Fig. 2a), while the second point tends to infinity. Moreover, it follows from equation (1) that at $\sigma_{cert} = \sigma_{PT} = \sigma$ the crossing point is $C_1 = (C_{cert} + C_{PT})/2$. Then, probability P from formula (4) is

$$P = 2\left[1 - \phi\left(\frac{|C_{cert} - C_{PT}|}{2\sigma}\right)\right].$$ (5)

For example, in conditions shown in Fig. 2a, $C_1 = (12.35 + 12.00)/2 = 12.18$ and $P = 0.21$.

According to formula (5), the max value $P=1$ can be achieved at $\sigma_{cert} = \sigma_{PT}$ and $C_{cert} = C_{PT}$, when $\phi(0) = 0.5$. Thus, both crossing points are absent, and the distributions coincide (Fig. 2b).

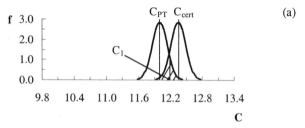

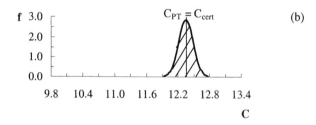

Figure 2 *Density functions f at* $\sigma_{PT} = \sigma_{cert} = 0.14\%$, *when* $C_{PT}=12.00$ *and* $C_{cert}=12.35$ *(Fig. 2a), and* $C_{PT}=C_{cert}=12.35$ *(Fig. 2b).*

This is the only case when one can definitely conclude that the comparability is satisfactory. In other words, the condition of the highest comparability is when both the means of the distributions and their standard deviations are equal. Any other values of distribution parameters lead to $P<1$. In particular, even at $C_{PT} = C_{cert}$ and $\sigma_{PT} > \sigma_{cert}$, it is impossible to assess the comparability unambiguously, in opposite to the case when all the values of the distribution parameters are equal. This conclusion seems odd, since the common rule states that the smaller uncertainty σ_{cert} of RM certified value C_{cert}, the better. However, in an extreme case, when $\sigma_{cert} \to 0$ and $\sigma_{cert}/\sigma_{PT} = \gamma \to 0$, the P value also tends to zero. It means only that the certified RM property value in this situation is practically non-random, unlike the PT results. In this aspect, PT results do not match RM certification data. In such a case (for $\gamma < 0.3$) the uncertainty of C_{cert} value can be ignored and the discussion of its distribution becomes unnecessary.

Calculation of P for more complicated (non-normal) theoretical distributions may be performed numerically, if the analytical mathematical tools are not available. Of course, information on the distributions of both PT results and RM data is limited by experimental statistical sample sizes. Therefore, the P value can adequately characterize the comparability only as much as the goodness-of-fit empirical and theoretical distributions is high. However, the P value is of practical importance since it allows one to choose a suitable null hypothesis for a sample criterion of a "yes-no" type for comparability assessment of relatively small (not infinite) number of PT results.

3 DISCUSSION OF NULL HYPOTHESES FOR COMPARABILITY ASSESSMENT

The null hypothesis (H_{01}) states that the comparability is satisfactory if the bias of PT results from the RM certified value is negligible in comparison with the combined uncertainty of the bias:

$$H_{01}: \left| C_{cert} - C_{PT} \right| \le 0.3 \left(\sigma_{cert}^2 + \sigma_{pT}^2 \right)^{1/2}, \tag{6}$$

where coefficient 0.3 is used according to the known metrological rule defining one standard deviation insignificant in comparison with another one when the former does not exceed 1/3 of the latter (i.e. the first variance is smaller than the second one by an order).

Substituting γ into H_{01} by formula (6) allows to limit the bias in σ_{cert} units:

$$H_{01}: \left| C_{cert} - C_{PT} \right| \le 0.3 \sigma_{cert} \left(1 + 1/\gamma^2 \right)^{1/2}. \tag{7}$$

Probability P corresponding to the null hypothesis can be calculated based on formula (4) transformed into the following:

$$P \ge 1 + \phi \left(\frac{\sigma_{cert}(C_{cert} - C_{PT}) - \sigma_{PT}\sqrt{\rho}}{\sigma_{PT}^2 - \sigma_{cert}^2} \right) + \phi \left(\frac{\sigma_{PT}(C_{cert} - C_{PT}) + \sigma_{cert}\sqrt{\rho}}{\sigma_{PT}^2 - \sigma_{cert}^2} \right) -$$

$$\phi \left(\frac{\sigma_{PT}(C_{cert} - C_{PT}) - \sigma_{cert}\sqrt{\rho}}{\sigma_{PT}^2 - \sigma_{cert}^2} \right) - \phi \left(\frac{\sigma_{cert}(C_{cert} - C_{PT}) + \sigma_{PT}\sqrt{\rho}}{\sigma_{PT}^2 - \sigma_{cert}^2} \right). \tag{8}$$

The calculation requires substituting ratio γ and bias $C_{cert} - C_{PT}$ from the right side of expression (7) into formula (3) for ρ, and into formula (8) for P. For example, H_{01} at ratio $\gamma = 0.4$ corresponds to probability $P \ge 0.56$, as in Fig. 1c, and at ratio $\gamma = 0.7$ it corresponds to $P \ge 0.77$. When $\sigma_{cert} = \sigma_{PT}$ and $\gamma = 1.0$, the probability values are $P \ge 0.83$ by formula (5).

Another null hypothesis H_{02} requires negligibility of the bias in comparison with the sum of the standard deviations:

$$H_{02}: \left| C_{cert} - C_{PT} \right| \le 0.3 \left(\sigma_{cert} + \sigma_{PT} \right) \tag{9}$$

Hypothesis H_{02} corresponds to: probability $P \ge 0.54$ at ratio $\gamma = 0.4$, $P \ge 0.63$ at $\gamma = 0.7$, and $P \ge 0.76$ at $\gamma = 1.0$. Apparently, hypothesis H_{01} is preferable as it provides higher probability P for the same γ values.

The increase of the numerical coefficient in the right-hand side of hypotheses H_{01} and H_{02} results in the distribution divergence and a decrease in probability P value. For example, the condition $\left| C_{cert} - C_{PT} \right| > 2 \left(\sigma_{cert} + \sigma_{PT} \right)$ leads to $P \to 0$, as in Fig.1a. In this case, the comparability of the PT results under consideration is highly unlike. The same is true for the requirement $\left| C_{cert} - C_{PT} \right| > 2 \left(\sigma_{cert}^2 + \sigma_{pT}^2 \right)^{1/2}$. Therefore, the traditional two-sigma criterion based on the hypothesis

$$H_{03}: \left| C_{cert} - C_{PT} \right| \le U_{0.95} \left(\sigma_{cert}^2 + \sigma_{pT}^2 \right)^{1/2} \tag{10}$$

is unacceptable here, as $U_{0.95} = 1.96$ is the percentile of normal distribution at the level of confidence 0.95. Moreover, a statement like $|C_{cert} - C_{PT}| = 2(\sigma_{cert}^2 + \sigma_{pT}^2)^{1/2}$ can be considered only as a hypothesis alternative to the null hypothesis (6).

4 HYPOTHESES AND A CRITERION FOR ASSESSMENT OF COMPARABILITY OF PT RESULTS BEING A NORMAL DISTRIBUTION SAMPLE

The null hypothesis established in ref.[15] for assessment of comparability by a "yes-no"-type criterion is based on the assumption that if the bias $C_{cert} - C_{PT}$ exceeds σ_{cert}, only by a value which is insignificant in comparison with random interlaboratory errors. In this case, the null hypothesis has the following form:

$$H_{04}: |C_{cert} - C_{PT}| \leq [\sigma_{cert}^2 + (0.3\sigma_{PT})^2]^{1/2}. \tag{11}$$

By hypothesis H_{04}, the probability P of considering the PT results as belonging to the population of RM data is $P \geq 0.53$ for $\gamma \geq 0.4$, when the right-hand side of expression (11) reaches the value of $1.25\sigma_{cert}$[17].

The alternative hypotheses H_1 assume that the bias exceeds σ_{cert} significantly:

$$H_{11}: |C_{cert} - C_{PT}| \leq 2.0[\sigma_{cert}^2 + (0.3\sigma_{PT})^2]^{1/2}, \tag{12}$$

$$H_{12}: |C_{cert} - C_{PT}| \leq 2.1[\sigma_{cert}^2 + (0.3\sigma_{PT})^2]^{1/2}, \tag{13}$$

etc.

The criterion for not rejecting H_{04} for a population sample of size N, i.e. for results of N laboratories participating in the PT, is

$$|C_{cert} - C_{PT/av}| + t_{1-\alpha/2} S_{PT} / \sqrt{N} \leq [\sigma_{cert}^2 + (0.3\sigma_{PT})^2]^{1/2}, \tag{14}$$

where $C_{PT/av}$ and S_{PT} are the sample estimates of C_{PT} and σ_{PT} calculated from the same N results as the sample average and standard deviation, correspondingly; the left-hand side of the expression represents the upper limit of the confidence interval for the bias $C_{cert} - C_{PT}$; $t_{1-\alpha/2}$ is the percentile of the Student's distribution for the number of degrees of freedom N-1; the $1-\alpha/2$ value is the probability of the bias not exceeding the upper limit of its confidence interval.

By substituting the ratio γ and $S_{PT}/\sigma_{PT} = [\chi_{\alpha/2}^2/(N-1)]^{1/2}$, where χ^2 is the percentile of chi-square distribution for the number of degrees of freedom N-1, into formula (14), the following transformation of the criterion is obtained:

$$|C_{cert} - C_{PT/av}| / S_{PT} \leq \left\{ \frac{(N-1)}{\chi_{\alpha/2}^2}(0.09 + \gamma^2) \right\}^{1/2} - \frac{t_{1-\alpha/2}}{\sqrt{N}}. \tag{15}$$

Table 1 gives the numerical values for the right-hand side of the criterion at $\alpha = 0.05$. These values are the norms for the bias of the average PT result from the analyte concentration certified in RM (in S_{PT} units). Value γ is set based on the requirements to the analysis, taking into account σ_{PT} value that is equal either to the standard analytical/measurement uncertainty or to the target standard deviation calculated using the Horwitz curve[2] or another database.

For the above mentioned example described in ref.[15], SRM 2690 (NIST, USA) of coal fly ashes has the certified value of aluminum concentration $C_{cert} = 12.35\%$ and $\sigma_{cert} = 0.14\%$ by weight, while σ_{PT} calculated based on the information contained in the ASTM standard[18] is 0.38% by weight, and $\gamma = 0.14/0.38 = 0.4$. Comparability of 15 simulated PT results (N=15, $C_{PT/av}$ =12.30%, S_{PT} = 0.34%) is assessed as satisfactory, since $|C_{cert} - C_{PT}| = |12.35 - 12.30| = 0.05 < 0.23\, S_{PT} = 0.23 \cdot 0.34 = 0.08$ % by weight, where 0.23 is the bias norm from Table 1. The same is true for the comparability of 30 simulated results (N=30, $C_{PT/av}$ =12.38%, S_{PT} = 0.35%; the bias norm from Table 1 is 0.30), since $|C_{cert} - C_{PT}| = |12.35 - 12.38| = 0.03 < 0.30\, S_{PT} = 0.30 \cdot 0.35 = 0.11$ %.

Reliability in such comparability assessment is determined by the probabilities of not rejecting the null hypothesis H_0 when it is true, and rejecting it when it is false (i.e. when the alternative hypothesis H_1 is true). Criterion (15) does not allow rejecting hypothesis H_{04} with probability 1-α/2 when it is true.

Table 1 *The bias norms in S_{PT} units by criterion (15)*

γ	N						
	5	10	15	20	30	40	50
0.4	0.20	0.20	0.23	0.26	0.30	0.32	0.34
0.7	0.95	0.68	0.65	0.64	0.65	0.66	0.67
1.0	1.76	1.19	1.09	1.06	1.03	1.02	1.02

Probability of an error of type I by this criterion (to reject the H_{04} hypothesis when it is true) is α/2. Probability of rejecting H_{04}, when it is false, i.e. when the alternative hypotheses H_1 are actually true (the criterion power - CP) is:

$$CP = \phi\left\{\frac{t_{\alpha/2} + \lambda}{\left[1 + t_{1-\alpha/2}^2 / 2(N-1)\right]^{1/2}}\right\}, \qquad (16)$$

where

$$\lambda = \frac{|C_{cert} - C_{PT}| - \sigma_{PT}(0.09 + \gamma^2)^{1/2}}{\sigma_{PT}/\sqrt{N}}. \qquad (17)$$

The value of the deviation parameter λ is calculated by substituting the bias $C_{cert} - C_{PT}$ in equation (17) for its value corresponding to the alternative hypothesis.

For example, for hypothesis H_{11} by formula (12) the substitution is $2.0[\sigma_{cert}^2 + (0.3\sigma_{PT})^2]^{1/2}$. Therefore, for H_{11} it is $\lambda = [(0.09 + \gamma^2)N]^{1/2}$, and for H_{12} by formula (13) it is $\lambda = 1.1[(0.09 + \gamma^2)N]^{1/2}$. The probability of an error of type II (not rejecting the H_0 when it is false) equals to β = 1-CP. Thus, the reliability of the comparability assessment using the hypotheses H_{04} against H_{11} for the PT scheme for aluminum determination in coal fly ashes (where γ = 0.4) can be characterized by 1) probability 1- α/2 = 0.975 of the correct assessment of comparability as successful (i.e. not rejecting the null hypothesis H_{04} when it is true) for any number N of the laboratories participating in PT, and by 2) probability CP = 0.42 of correct assessment of comparability as unsuccessful (i.e. rejecting H_{04} when the alternative hypothesis H_{11} is true) for N = 15, and probability CP = 0.75 for N = 30 results. Probability α/2 of a type I error is 0.025 for any N, while probability β of a type II error is 0.58 for N = 15, and it is 0.25 for N = 30, etc.

The power of criterion (15) is high enough (CP > 0.5) for a number of PT participants N $\geq$ 20. However, it is known that reliability of the assessment and probabilities of erroneous decisions based on such a criterion depends on the statement of the alternative hypotheses. In particular, CP is increasing as λ values increase from alternative hypothesis H_{11} to hypothesis H_{12}, since the latter is more "remote" from the null hypothesis H_{04}. Thus, detecting greater bias is simpler: for H_{04} the bias is $[\sigma_{cert}^2 + (0.3\sigma_{PT})^2]^{1/2}$, for H_{11} it is $2.0[\sigma_{cert}^2 + (0.3\sigma_{PT})^2]^{1/2}$ and for H_{12} it is $2.1[\sigma_{cert}^2 + (0.3\sigma_{PT})^2]^{1/2}$.

The role of the ratio γ of the population standard deviations σ_{cert} and σ_{PT} is more complicated. Formally, the larger the γ, the larger the λ and the criterion power. However, $\gamma > 1$ ($\sigma_{cert} > \sigma_{PT}$) has no metrological sense even when IHRM or a spike is used, since it means that either less accurate methods were applied for the material preparation and study than those used by laboratories participating in the PT, or the material is not suitable due to its inhomogeneity and/or instability. On the other hand, if possible to neglect the uncertainty of the RM certification ($\sigma_{cert} \ll \sigma_{PT}$ and $\gamma < 0.3$), the criterion (15) is reduced to a simpler one. The null hypothesis H_{04} is transformed here into hypothesis stating that the bias of PT results from the RM certified value is negligible in comparison with the interlaboratory standard deviation, as $|C_{cert} - C_{PT}| \leq 0.3\sigma_{PT}$. The corresponding criterion is less powerful and requires $N = 50 \div 100$ for achieving the operational characteristics of criterion (15). When both σ_{cert} and σ_{PT} are values of the same order, i.e. when $0.4 \leq \gamma \leq 1$, the null hypothesis H_{01} can be also applied. However, this hypothesis leads to a criterion stricter than (15) which might be more appropriate for comparability assessment of key comparison results.

From the chemical / metrological point of view, the discussion of results comparability should be restricted by the definition of the analyte and the matrix for which C_{cert} and σ_{cert} are quantified. Therefore, the adequacy of RM used is of high importance. On the other hand, if the RM prepared for PT is not certified and a consensus value (the average or the median of the PT results) is used instead of C_{cert}, traceability of this value is questionable. Thus, comparability of PT results cannot be assessed as "tested once, accepted everywhere". In such cases, especially when the number of participants is limited, a local comparability, i.e. among the participants only, is tested based on known requirements to the method's reproducibility[15].

5 HYPOTHESES AND A NON-PARAMETRIC TEST FOR COMPARABILITY ASSESSMENT OF PT RESULTS WITH UNKNOWN DISTRIBUTION

In the case of unknown distributions differing from the normal one, the median is more robust than the average, i.e. better reproduced in the repeated experiments, being less sensitive to extreme results/outliers. Therefore, the null hypothesis assuming here that the bias of PT results exceeds σ_{cert} by a value which is insignificant in comparison with random interlaboratory errors, has the following form:

$$H_{05}: |C_{cert} - M_{PT}| \leq [\sigma_{cert}^2 + (0.3\sigma_{PT})^2]^{1/2} = \Delta, \qquad (18)$$

where M_{PT} is the median of PT results of hypothetically infinite number N of participants, i.e. the population median.

If $M_{PT} \geq C_{cert}$, the null hypothesis H_{05} implies that probability P_e of an event when a result C_i of the i-th PT-participating laboratory exceeds the value $C_{cert} + \Delta$, is $P_e\{C_i > C_{cert} + \Delta\} \leq \frac{1}{2}$ according to the median definition. If $M_{PT} < C_{cert}$, the probability of C_i yielding the value $C_{cert} - \Delta$ is also $P_e\{C_i < C_{cert} - \Delta\} \leq \frac{1}{2}$. The alternative hypotheses assume that the bias exceeds σ_{cert} significantly and probabilities of the events described above are $P_e > \frac{1}{2}$:

$$H_{13}: |C_{cert} - M_{PT}| > \Delta, \qquad (19)$$

where Δ is the same as in expression (18). For example,

$$H_{14}: \left| C_{cert} - M_{PT} \right| = 2.0\Delta. \tag{20}$$

Probabilities P_e of the events according to the alternative hypothesis H_{14} at normal distribution depending on the permissible bias Δ (in σ_{PT} units) at different γ values are shown in Table 2.

Table 2 *Probability P_e according to alternative hypothesis H_{14}*

γ	Δ/σ_{PT}	P_e
0.4	0.50	0.69
0.7	0.75	0.77
1.0	1.04	0.85

Since the population median is unknown in practice, and results of N laboratories participating in PT form a N-size statistical sample from the population, hypothesis H_{05} is not rejected when the upper limit of the median confidence interval does not exceed $C_{cert}+\Delta$, or the lower limit does not yield $C_{cert} - \Delta$. The limits can be evaluated based on the simplest non-parametric sign test[16]. According to this test, the number N_+ of results $C_i > C_{cert} + \Delta$ or the number N_- of results $C_i < C_{cert} - \Delta$ should not exceed the critical value A (the bias norm) in order not to reject H_{05}. The A values are available in known statistical handbooks, for example, in ref.[19]. For $N = 5 \div 50$ PT participants and levels of confidence 0.975 ($\alpha/2 = 1-0.975 = 0.025$) and 0.95 ($\alpha/2 = 0.05$), these values are shown in Table 3. The A value for fewer than six participants at $\alpha/2 = 0.025$ cannot be determined, and therefore, is not presented in Table 3 for $N = 5$.

The test does not allow rejecting hypothesis H_{05} with a probability of $1-\alpha/2$, when it is true. Probability of an error of type I by this test (to reject the H_{05} hypothesis when it is true) is $\alpha/2$. Probability TP of rejecting the null hypothesis when it is false, i.e. when the alternative hypothesis is actually true (the test power), is tabulated in ref.[19]. The probability of a type II error (not rejecting H_{05} when it is false) equals $\beta = 1-TP$. The operational characteristics of the test (TP and β) at alternative hypothesis H_{14}, as well as $\alpha = 0.05$, different γ values and different numbers N of PT participants are presented in ref.[16].

For example, regarding SRM 2690 applicable for PT of aluminum determination in coal fly ashes, a statistical sample of N=50 PT results is simulated with a bimodal unknown distribution[16]. Taking into account $C_{cert} = 12.35\%$, $\sigma_{cert} = 0.14\%$, $\sigma_{PT} = 0.38\%$ by weight, and $\gamma = 0.14/0.38 = 0.4$, one can calculate $\Delta = 0.50 \cdot 0.38 = 0.19\%$ (Table 2), $C_{cert} + \Delta = 12.54\%$ and $C_{cert} - \Delta = 12.16\%$. There are simulated $N_+ = 15$ results $C_j > 12.54\%$, $N_- = 12$ results $C_j < 12.16\%$, and $N - N_+ - N_- = 23$ values in the range $C_{cert} \pm \Delta$. The sample median found is $C_{25} = C_{26} = 12.49 > C_{cert} = 12.35\%$ and $N_+ > N_-$. However, N_+ is lower than the critical value $A = 17$ at $\alpha/2 = 0.025$ and N=50 (Table 3). Therefore, null hypothesis H_{05} concerning successful comparability of the results is not rejected.

Reliability of the assessment using hypotheses H_{05} against H_{14} for this case can be characterized by: 1) probability $1- \alpha/2 = 0.975$ of correct assessment of comparability as successful (not rejecting the null hypothesis when it is true) for any number $N \geq 6$ of the PT participants, and 2) probability $TP = 0.73$ of correct assessment of comparability of $N = 50$ PT results as unsuccessful (rejecting H_{05} when alternative hypothesis H_{14} is true).

Table 3 *The bias norms A by the sign test*

$\alpha/2$	N						
	5	10	15	20	30	40	50
0.025	-	1	3	5	9	13	17
0.05	0	1	3	5	10	14	18

Probability $\alpha/2$ of a type I error is 0.025 for any $N \geq 6$, while probability β of type II error is 0.27 for N = 50.

Since the sign test critical A values are determined for $N \geq 4 \div 8$ depending on probabilities α, and the test power is calculated also only for $N \geq 6 \div 8$, the proposed comparability assessment cannot be performed for a smaller sample size. The power efficiency of the sign test in relation to the t-test (ratio of the sizes N of statistical samples from normal populations allowing the same power) is from 0.96 for N = 5 to 0.64 for infinite N. For example, practically the same power (0.73 and 0.75) was achieved in the sign test of comparability of PT results for aluminum determination in coal fly ashes at N = 50 discussed above, and in the t-test for the same purpose at N = 30 in the previous paragraph 3. The power efficiency here is approximately of 30/50 = 0.6. On the other hand, when information about the distribution of PT results is limited by small sizes of statistical samples of experimental data $(N < 50)$, it is a problem to evaluate the goodness-of-fit empirical and theoretical/normal distributions, a decrease of the t-test power and the corresponding decrease of reliability of the comparability assessment caused by deviation of the empirical distribution from the normal one.

Of course, reliability of the assessment and probabilities of erroneous decisions based on the sign test, as any other statistical criterion or test, depends on how the hypotheses have been stated. In particular, if $\gamma < 0.3$ and the null hypothesis is $\left|C_{cert} - M_{PT}\right| \leq 0.3\sigma_{PT}$, the same power requires a bigger sample size, i.e. N = 50 ÷100. Another null hypothesis: $\left|C_{cert} - M_{PT}\right| \leq 0.3\left(\sigma_{cert}^2 + \sigma_{pT}^2\right)^{1/2}$, like H_{01}, also leads to a test that is stricter than the one described above.

The role of RM adequacy, of the ratio γ and of an alternative hypothesis (its "distance" from the null hypothesis) is similar to the one shown in paragraph 3.

6 CONCLUSIONS

1. Relationships are discussed between 1) the distribution of analytical results obtained in PT and 2) the distribution of the assigned value of RM used in PT as the test/sample units. It is shown that different null hypotheses assuming different relationships between these distributions can be used for assessment of comparability of PT results.

2. At normal distributions, comparability of PT results is assessed based on a "yes-no"-type criterion for testing the null hypothesis concerning insignificance of the bias of the results mean from the RM traceable value.

3. In cases when the distributions differ from the normal ones, the sign test of the hypothesis concerning insignificance of the bias of the results median from the RM traceable value is helpful.

4. The reliability of the assessment is discussed in terms of probabilities of type I and type II errors in decisions concerning rejection or non-rejection of the null hypothesis.

5. A combination of chemical/metrological and statistical knowledge is necessary for careful statement of the null hypothesis, since different hypotheses can lead to different decisions on comparability of the measurement/analysis results obtained in the same PT scheme.

References

1 ISO/IEC Guide 43. *Proficiency Testing by Interlaboratory Comparisons. Part 1: Development and Operation of Proficiency Testing Schemes. Part 2: Selection and Use of Proficiency Testing Schemes by Laboratory Accreditation Bodies*, ISO, Geneva, 1997.

2 ISO/DIS Standard 13528 (Provisional Version). *Statistical Methods for Use in Proficiency Testing by Interlaboratory Comparisons*, ISO, Geneva, 2002.

3 ISO/AOACI/IUPAC. *The International Harmonized Protocol for the Proficiency Testing of (Chemical) Analytical Laboratories,* ISO, Geneva, 2005.

4 ILAC G13. *Guidelines for the Requirements for the Competence of Providers of Proficiency Testing Schemes*, ILAC, 2000.

5 ISO Guide 35. *Certification of Reference Materials. General and Statistical Principles*, 3d Edn., ISO, Geneva, 2003.

6 ILAC–G12: 2000. *Guidelines for the Requirements for the Competence of Reference Material Producers*, ILAC, 2000.

7 I.Kuselman and M.Pavlichenko. *Accred. Qual. Accur.,* 2004, **9**,399.

8 I.Kuselman. *J. of Metrology Society of India,* 2004, **19/4**, 245.

9 I.Kuselman, I.Papadakis, W.Wegscheider. *Accred. Qual. Assur.,* 2001, **6**, 78.

10 D.Kisets. *Accred. Qual. Assur.,* 2006, **10**, 461.

11 P.Armishaw, B.King, R.G Millar. *Accred. Qual. Assur.,* 2003, **8**,184.

12 E.Kardash-Strochkova, Ya.I. Tur'yan, I.Kuselman, N.Brodsky. *Accred. Qual. Assur.,* 2002, **7**, 250.

13 ISO/IEC Standard 17025. *General Requirements for the Competence of Testing and Calibration Laboratories*, ISO, Geneva, 1999.

14 EURACHEM/CITAC Guide. *Traceability in Chemical Measurement. A Guide to Achieving Comparable Results in Chemical Measurement,* 2003.

15 I.Kuselman. *Accred. Qual. Assur.,* 2006, **10**, 466.

16 I.Kuselman. Non-parametric assessment of comparability of analytical results obtained in proficiency testing based on a metrological approach. *Accred. Qual. Assur.,* 2006, **11**, in press.

17 I.Kuselman. *Talanta*, 1993, **40, 1**.

18 ASTM Standard D 2795 -74. *Standard Methods of Analysis of Coal and Coke Ash*, 1974.

19 D.B.Owen. *Handbook of Statistical Tables*. Addison-Wesley Publishing Company, Inc., London, 1962.

INTER-LABORATORY COMPARISON: THE APAT APPROACH

P. de Zorzi, S. Balzamo, S. Barbizzi, S. Gaudino, A. Pati, S. Rosamilia, M. Belli

Agenzia per la Protezione dell'Ambiente e per i Servizi Tecnici (APAT),
Servizio Laboratori Misure ed Attività di Campo.
Via di Castel Romano, 100,
00128 Roma, Italy

1 INTRODUCTION

Around one hundred laboratories, belonging to the Italian Regional and Provincial Environmental Protection Agencies (ARPAs/APPAs) are daily involved in analytical activities for monitoring the Italian environment. The activities are mainly directed at: i) supporting the environmental policy-makers and ii) verifying the compliance with the statuary prescriptions and requirements (i.e. contaminated site clean-up, industrial discharges). In general monitoring requires the development of common and transparent environmental data set, to be used by public and private stakeholders.

It is now widely recognized the need for laboratories of an appropriate quality assurance/quality control system (QA/QC) to demonstrate their ability to produce consistent and reliable data. Amongst the actions required for QA/QC, there is the need to demonstrate that the laboratory analytical systems are under statistical control, the analytical methods are validated; the results are "fit-for-purpose", also through the participation in proficiency tests and inter-laboratory exercises.

Proficiency testing is becoming an integral feature of laboratory accreditation and the results generated in proficiency testing are used for the purpose of a continuing assessment of the technical competence of laboratories. With the advent of "mutual recognition", there is, also in Italy, an increased demand of proficiency testing schemes, that will provide an interpretation and assessment of results, which are transparent to the participating laboratory and its "customer".

In this framework, the Italian National Environmental Protection Agency (APAT), plays an important role in assisting the Regional/Provincial laboratories in improving the quality of their analytical measurements. This is accomplished through the provision of matrix reference materials and through the evaluation of measurement performance by the organization of proficiency tests, laboratories and sampling intercomparison exercises. The matrix reference materials produced by APAT in recent years were soil, sediments, compost, water, wastes, and the analytes of interest were metals and organic compounds (for solid), metals, pesticides and anions/cations (for liquid matrices). The paper presents and discusses the main outcomes and issues arising from the proficiency tests and sampling intercomparison exercises organized by APAT in the last four years, in terms of definition of the performance criteria, determination of the assigned values and assessment of the laboratory performance.

2 THE ITALIAN CONTEXT

The environmental monitoring in Italy is performed by more than one hundred laboratories belonging to 21 Regional/Provincial Environmental Protection Agencies. The activities are addressed to different analytical fields: chemical, biological and physical. The laboratories have a long tradition in participating in proficiency testing schemes and intercomparison exercises. A study carried out in 2003[1] showed that more than 90 % of the chemical and biological laboratories is regularly participating in proficiency tests (PTs) and inter-laboratory exercises, while the percentage decreases for the laboratories performing physical measurements (<30%).

The PT providers are not limited to the Italian institutions and for chemical measurements, the main providers are: UNICHIM, an Italian private organization federated with the Italian National Standardization Body; QUASIMEME, an European quality assurance program for marine environmental monitoring; APAT, the Italian environmental protection agency, and ISS, the technical and scientific branch of the Italian National Health Service. About 1300 test samples are annually analyzed and during 2003, in the framework of chemical inter-laboratory comparison exercises, 63 % of them were provided by the above mentioned institutions.

The matrices used in the inter-laboratory exercises and the measurands determined are linked both to the most frequently analyzed matrices and to the availability of reference materials. In this frame, the determination of metals in water is the most requested exercise.

Table 1 *Interlaboratory intercomparison exercise organized by APAT*

Chemical				Ecotoxicological				Sampling			
Matrix	**Analyte**	**PT**	$N°$ **Lab**	**Matrix**	**Analyte**	**PT**	$N°$ **Lab**	**Matrix**	**Analyte**	**PT**	$N°$ **Lab**
Sediment	Metals	APAT IC001	56	Water	Daphnia magna	Ecotox . 1	41	Soil	Metals	APAT IC003	14
Compost	Metals	APAT IC002	51	Water	Daphnia magna	Ecotox 2	29				
Water	Metals	APAT-IC004	68	Water	Daphnia magna	APAT IC006	52				
Soil	Metals	APAT IC004	59								
Water	Anions Cations	APAT IC005	27								
Water	Pesticides	APAT SC01	18								
Water	Anions Cations	APAT IC007									

APAT started the organization of inter-laboratory comparison exercises in 2002. The primary "end users" of the APAT programme in this field are the Italian regional laboratories belonging to the 21 Regional/Provincial Environmental Protection Agencies. The objective is to support the laboratories to improve their analytical quality assurance and quality control practices. As no-profit public institution, the support provided by APAT is free of charge for the regional environmental laboratories. The need of reference materials in a wide range of analyte/matrix combinations required to APAT the establishment of a laboratory for reference material production. The reference materials produced by APAT are strictly connected to the routine analytical activities performed by the regional environmental laboratories in Italy, and the inter-laboratory comparison exercises organized reflect the main requests and issues coming from the same agencies. In **Table 1** a list of the inter-laboratory exercises carried out from 2002 to January 2006 is shown, sorted by field of analysis.

The participation to the inter-laboratory exercises organized by APAT shows a positive trend through the years. The number of participating laboratories in the different exercises is highly influenced by the objective proposed. Not all ARPA/APPA laboratories have the same level of competence for all environmental matrices. Metals and ions in water are determined in all laboratories, while organic compounds in soil and waste are determined in some specialized laboratories. The target in a medium-term is to assure the participation in the APAT inter-laboratory comparison exercises of 70% of the regional environmental protection laboratories. Up to today, on average, the participation is ranging between 30 and 50%.

Among the inter-laboratory comparison exercises carried out in the recent years, the soil sampling exercise represents a new challenge. Most of the regional environmental agencies recognize the relevance of sampling in environmental monitoring activities and the need for a harmonized approach in the Italian context. The availability of an agricultural reference site, previously characterized by APAT in terms of spatial distribution of trace elements within the framework of an international project[2,3], enabled the performing of soil sampling intercomparisons. These exercises involved different regional laboratories to test their own sampling strategies and techniques. The same reference site was used by the IAEA in 2005, in the frame of a collaboration with APAT, to perform a similar soil sampling intercomparison among a restricted group of analytical laboratories of the IAEA ALMERA network (Analytical Laboratories for the Measurements of the Environmental Radioactivity)[4].

3 PT AND INTERLABORATORY COMPARISON SCHEME ADOPTED BY APAT

The success of an intercomparison exercise depends on many factors, involving technical and management aspects. APAT established a general system suitable to manage each intercomparison exercise, whatever its objective, the matrix and the analyte of interest (**Figure 1**). The APAT scheme foresees that the coordination of the inter-comparison is assured by a Project Manager (DP), while the role of technical consultants, in accordance with ISO Guide 43-1[5], is covered by a Permanent Advisory Group (GTP). GTP is composed by laboratory experts, quality managers and data users coming from the regional environmental protection agencies. GTP members are requested to contribute to the definition of the inter-laboratory exercise's scheme, to submit proposals for new inter-laboratory exercises, to assure the continuous link between APAT and the regional laboratories, to comment and review the draft reports before the plenary meeting with the laboratories. The details of the scheme adopted during each exercise are reported in a protocol prepared before the starting of each run and sent to the laboratories with the reference material and the forms to report to APAT the laboratory results. Key points of the protocol are:

- the definition of the assigned value of the measurands;
- the fitness for purpose of the reference material;
- the scoring system used to evaluate the analytical performance of the participating laboratories.

These aspects have already been tackled at international level and some guidelines and/or international rules are available to harmonize the approaches, such as ISO Guide 43-1[5], ISO 13528[6] and the harmonized protocol for proficiency testing of analytical laboratories, recently revised by IUPAC[7].

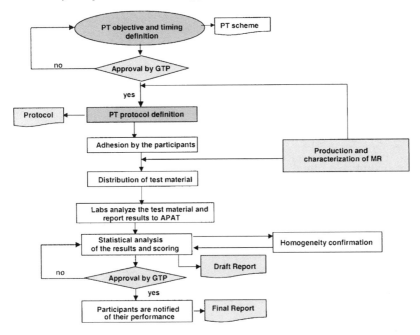

Figure 1 : *General PT process"*

The protocol prepared by APAT in agreement with the GTP reports the method used to assign the property value to the reference material. In the proficiency testing organized until now, the property value was assigned as:

- consensus value of expert laboratories (using an operationally defined analytical method);
- consensus value from the participants reported values;
- values obtained by reference laboratories.

Homogeneity and stability studies of the analytes of interest are carried out in APAT laboratory before the distribution of the material following the indication of the ISO Guide 35[8]. The protocol reports the information on the homogeneity tests carried out and on the stability test assuring their validity for the period of the interlaboratory comparison. In addition, the protocol reports the statistical method be used to assure the fitness-for-purpose of homogeneity between the RM units distributed to the participants. The method adopted is suggested by ISO 13528 and compares the variation between bottles (S_S) with the inter-laboratory variation ($\hat{\sigma}$). As stated by ISO 13528 (Annex B), the RM units may be considered to be adequately homogeneous if:

$$s_s/\hat{\sigma} < 0{,}3 \tag{1}$$

If this criterion is not met, the between-bottles standard deviation is combined with the proficiency testing standard deviation:

$$\hat{\sigma} = \sqrt{\hat{\sigma}_1^2 + s_s} \tag{2}$$

where $\hat{\sigma}_1$ is the standard deviation for proficiency testing. **Table 2** reports an example of the homogeneity confirmation.

Table 2 *Example of homogeneity confirmation in accordance with ISO 13528 (Annex B). Inter-laboratory comparison APAT-IC004 "Metals in water"[9]*

Element	$s_s/\hat{\sigma}$
Cd	0.03
Cr	0.12
Cu	0.05
Hg	0.01
Ni	0.04
Sb	0.14
Zn	0.24

The protocol gives also details on the statistical analysis used for the assessment of the inter-laboratory comparison results and the scoring system adopted.

Figure 2 reports one example of result assessment. Laboratory results are reported grouped by the analytical method used. In **Figure 2** the values obtained with different digestion methods are compared with the consensus value obtained, using robust statistics[6] on mean values given by expert laboratories with only one digestion method. **Figure 2** shows that there is not difference between digestion methods for lead mass fraction determination in soil. The grouping of results on the basis of the analytical methods can give information on systematic underestimation/overestimation of an analytical method and consequently highlights the need of future harmonization.

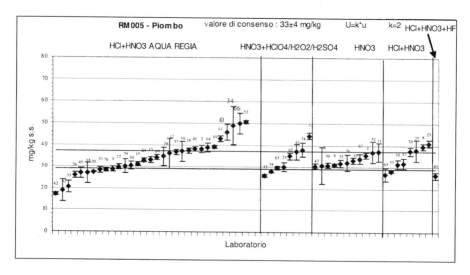

Figure 2: *Lead mass fraction mean values sorted by digestion method and ordered by increasing values - APAT-IC004 "Metals in agricultural soil"[10]*

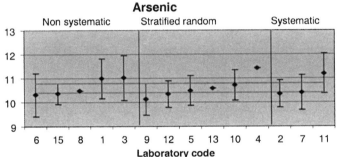

Figure 3 : *Arsenic mass fraction mean values (mg/kg) sorted by sampling strategies and ordered by increasing values - APAT-IC003 "Soil sampling strategies intercomparison in agricultural area"[10]*

Figure 3 reports the results obtained in an interlaboratory comparison for soil sampling. In this case, the objective of the exercise was to assess the metal content mean values in an agricultural area. The results obtained with different sampling strategies are compared with the mean values obtained during the reference sampling[2,3].

The scoring system mainly used is the z-score, defined by the equation:

$$z = \frac{X_{LAB} - X_{RIF}}{\hat{\sigma}}$$

(3)

where:

- X_{LAB} = laboratory's result;
- X_{RIF} = assigned value for the measurand[6];
- $\hat{\sigma}$ = robust standard deviation for proficiency testing[6].

The z-score is calculated on a statistical basis and its assessment is versus a fixed scale of judgment (satisfactory, questionable and not acceptable or requiring action). **Figure 4** reports an example of z-score plotting.

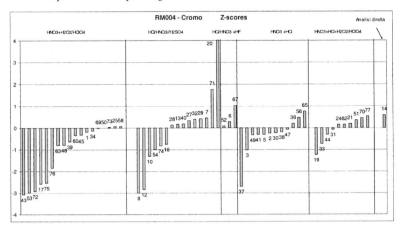

Figure 4 : *Chromium z-score graph, sorted by digestion method and ordered by increasing values - APAT-IC004 "Metals in agricultural soil"[9]*

4 PERSPECTIVES AND CONCLUSION

As above reported, the evaluation approach mainly used in APAT until now is z-score. This approach uses the robust standard deviation ($\hat{\sigma}$) calculated on the basis of all laboratory results. The standard deviation is strongly influenced by the characteristics of the participating laboratories. Furthermore this approach does not allow to monitor the improvement of the network of the laboratories in a period of time, because the standard deviation depends strongly on the population of the laboratories, on the type of the analyte and on its content in the matrix. The IUPAC Harmonised Protocol for the Proficiency Testing of Analytical Chemistry Laboratory[7] gives useful suggestions and recommendations on the evaluation approaches. The identification of a standard deviation σ_p stated as fitness for purpose for each monitoring activities could give a target that all the network of environmental laboratory should reach. An exemplification of the effects due to different way to estimate the standard deviation for the proficiency testing is reported in **Figure 5.** A comparison between *z-score* assessment using both $\hat{\sigma}$, calculated on the participant's data, and σ_p, fixed equal to 15 % of the assigned value, is shown. The comparison is based on analytical data from the same set of laboratories (Hg measurements) in order to avoid any distorting effect due to an heterogeneous data set.

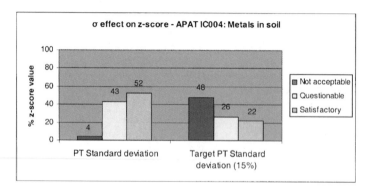

Figure 5 : *Comparison between z-score - APAT-IC004 "Metals in agricultural soil"*[9]

As the standard deviation is calculated in different ways, the ratio between "satisfactory" and "not acceptable" *z-score* may change substantially. In APAT-IC004[9] the final performance assessment for the laboratories was basically positive: more than 50% of the participants had "satisfactory" *z-score*, and only 4% showed *z-score* requiring remedial actions. But adopting a target or fitness-for-purpose standard deviation for the proficiency testing (15%), the conclusion might be far from positive. This behavior could be due to the low content of mercury in the reference material (0,18 ± 0,03 mg/kg d.w.). In conclusion the definition of a target standard deviation needs an agreement between data users and laboratories and a strong consensus across the Country. For this reason GTP and data users will be strongly involved in the definition of the target values for each monitoring activities.

Furthermore, in the next year, a program to support the comparability of analytical data of the regional laboratories, during the waste characterization activities will start. To this end APAT has launched a project[11] in cooperation with the Centro Sviluppo Materiali S.p.A (CSM) for the production of matrix reference materials, similar to the test samples measured in the regional laboratories (Solid Refuse Fuel (SRF), fly ash and bottom ash deriving from incineration of Municipal Solid Waste (MSW))

References

1   S. Barbizzi, M. de Zorzi, R. Mufato, G. Sartori, G. Stocchero, Circuiti d'Interconfronto e Materiali di Riferimento Certificati in matrice nei laboratory italiani del Sistema delle Agenzie per l'Ambiente nell'anno 2003, APAT-Servizio Metrologia Ambientale, 2005. ISBN 88-448-0156-6.

2   P. de Zorzi, M. Belli, S. Barbizzi, S. Menegon, A. Deluisa, *Accreditation and Quality Assurance*, 2002, **7**(5), 182.

3   S. Barbizzi, P. de Zorzi, M. Belli, A. Pati, U. Sansone, L. Stellato, M. Barbina, A. Deluisa, S. Menegon, V. Coletti, *Environmental Pollution*, 2004, **127**, 131.

4   Report on the 2nd ALMERA network coordination meeting & the ALMERA soil sampling intercomparison exercise (IAEA/SIE/01), 2006, IAEA/AL/164

5   ISO Guide 43-1, Proficiency testing by interlaboratory comparison. Part 1, International Standardization Organization, Geneve, Switzerland 1997.

6   ISO 13528 Statistical methods for use in proficiency testing by interlaboratory comparisons, International Standardization Organization, Geneve, Switzerland 2005

7   M. Thompson, S.R. Ellison, R. Wood, *Pure and Applied Chemistry*, 2006, **78** (1), 145.

8   ISO Guide 35, Certification of reference materials – general and statstical principles, International Standardization Organization, Geneve, Switzerland 2006

9   M. Belli, G. Sartori, S. Barbizzi, P. de Zorzi, R. Mufato, G. Stocchero, S. Balzamo, D. Centioli, C. Galas, S. Gaudino, T. Guagnini, H. Muntau, A. Pati, C. Ravaioli, S. Rosamilia, G. Sentina, Rapporto conclusivo interconfronto APAT-IC004 – Metalli in acqua sotterranea e suolo in publication

10 M. Belli, G. Sartori, S. Barbizzi, P. de Zorzi, R. Mufato, G. Stocchero, S. Balzamo, D. Centioli, S. Gaudino, H. Muntau, A. Pati, C. Ravaioli, S. Rosamilia, Rapporto conclusivo interconfronto APAT-IC003 - Strategie di campionamento di suoli in un area agricola in publication

11 S.Balzamo, M.Belli (2005) Proceedings of Sardinia 2005, 507-508

EXTERNAL QUALITY ASSESSMENT SCHEMES IN BIOMEDICAL FIELD: FROM DEFINITION OF TARGETS TO END USERS

M. Patriarca, I. Altieri, M. Castelli, F. Chiodo, A. Semeraro and A. Menditto

Department of Food Safety and Veterinary Public Health
Istituto Superiore di Sanità, Rome, Italy

1 INTRODUCTION

Continuous development of scientific knowledge over the last century has brought changes in the field of public health (including veterinary public health) with dramatic improvement in the prevention, diagnosis, prognosis and treatment of human pathologies. A large part of these improvements is based on the increased ability to measure quantities that are related to human and animal health. To support the growing needs of care and prevention, measurements are carried out in various fields: from fundamental research and development, where measurements support the understanding of physio-pathological processes aimed to the development of new diagnostic and treatment methods, to the implementation of current regulations, based on compliance with stated concentrations, aimed to limit human and animal exposure to potentially toxic agents, to clinical care, where measurement results are often an essential part of the decision process toward diagnosis and treatment of pathologies (Table 1).

Table 1. *Fields of public health sciences in which measurements have a key role*

Research and development
• Physio-pathology
• New diagnostic tools
• New drugs

Prevention
• Human biomonitoring (occupational exposure, environmental exposure)
• Environmental biomonitoring
• Epidemiology (transitional studies, field studies)

Control
• Foods, beverages
• Therapeutic drugs, including blood products
• Drugs of abuse and doping substances

Clinical medicine and veterinary public health
• Prevention, diagnosis, prognosis and treatment of human disease

Table 2. *Examples of analytes and matrices frequently subjected to measurement in clinical laboratory medicine*

Analytes	Matrices
Electrolytes	Blood
Metabolites and substrates	Serum
Enzymes	Urine
Proteins	Saliva
Hormones	Cerebrospinal fluid
Coagulation factors	Breast milk
Nucleic acids	Hair
Vitamins	Nails
Trace elements	Semen
Drugs/Residues	Tissues
Contaminants/Metabolites	Expired air

At large, measurements relevant to public health include determinations of major constituents and potential contaminants in food, water and environmental samples, carried out for preventive purposes, and the assessments of drugs and blood products before, and sometime even after, their release on the market. In clinical laboratory medicine only, 400 to 700 types of quantities are measured routinely in different matrices (Table 2), consisting of an enormous variety of molecules of different size, structure, action and origin, at concentrations ranging from a few ng/l (e.g. trace elements, contaminants) to g/l (major constituents, e.g. proteins).

Meaningful measurements are therefore essential for risk assessment, prevention, diagnosis, treatment and monitoring of disease in humans and animals. Inadequate or incorrect analytical measurements may have consequences for patients, clinicians, the health care system and the society. Assuring the reliability and fitness-for-purpose of these measurements is a major challenge facing the biomedical world.

2 TRACEABILITY OF MEASUREMENTS IN PUBLIC HEALTH SCIENCES

2.1 General concepts

Reliable results are expected to be accurate, specific and comparable among laboratories over time and space. Metrologists have long established the principle of traceability of a measurement result to the SI-system to ensure comparable results. Traceability is defined as the "property of the result of a measurement or the value of a standard whereby it can be related to stated references, usually national or international standards through an unbroken chain of comparisons all having stated uncertainties".[1] In metrological terms, comparability can only be achieved if results are traceable to the same reference.

In analytical chemistry and in laboratory medicine a traceability chain to the SI can be established by comparing the laboratory result with certified reference materials (CRM) which value derives, in turn, from the comparison with CRMs of higher order, linked to or representing the mole unit. Such CRMs are equivalent to metrological primary or secondary standards for the quantity "concentration" of a given substance. The strength of the traceability chain of an individual result is expressed in terms of measurement uncertainty. A graphical presentation of these concepts is given in the international standard ISO 17511.[2] However, several problems hamper the practical realization of such chains and, at present, metrological traceability can be established for only 20-30 of the routine measurements carried out in the field of laboratory medicine.[2] Comparison of test results obtained from External Quality Assessment Schemes (EQAS) have highlighted for several measurands/matrices significant biases. A special need for standardisation has been outlined for specific classes of analytes (such as nucleic acids and hormones). Appropriate CRMs in suitable matrices are not always available, e.g. because of the lack of primary/reference methods for their certification, the difficulty of obtaining the analyte in a pure form, inadequate stability of the analyte and/or the matrix. A clear definition of the measurand, linking it to an SI unit, is not always possible because of the complexity of biological systems, where molecular activity, structure and sequence may be more important characteristics than mass or number of moles or else the measurement of the combined effect of a mixture of molecules with similar/antagonist action, such as in the determination of serum total antioxidant status, provides more biologically or clinically useful information than the quantification of individual components. For many biological substances of crucial diagnostic importance, measurements throughout the world are referred to specific World Health Organization (WHO) International Standards, i.e. large batches of human samples or purified biological factors, to which a value is assigned by consensus by means of several analytical techniques and on which the relevant International Unit is based. In some cases, the analyte is defined by the method itself and only results obtained with the same agreed procedure can be compared.

2.2 International initiatives for comparability and traceability in the biomedical field

The issues of comparability of analytical data relevant to public health has led to the development of initiatives to improve the comparability of data across countries. A key-issue of this process is to understand the nature of such measurements and how their reliability can be improved. For this reason, the General Conference on Weights and Measures - an intergovernmental organisation created in 1875 according to a diplomatic treaty to assure the comparability of measurements in all technical and scientific fields - recommended in 1999 that initiatives should be taken by the Bureau International des Poids et Mesures and the National Metrological Institutes for the development of metrology in chemistry and the biotechnologies and for a wider application of the SI units in human health. Following this advice, a new working group of the Consultative Committee on Amount of Substance and Metrology in Chemistry was established in 2001, to explore the level of comparability that can be achieved in all fields of analysis involving biological macromolecules.

At the same time, networks of reference laboratories in some critical sectors (e.g. air monitoring and food control) have been established throughout the European Union, to warrant that legislation is applied uniformly and univocally, thus allowing the removal of barriers to trade and free movements of goods and ensuring the same level of protection of public health. Such laboratories are requested to qualify by achieving accreditation according to the international standard ISO/IEC 17025[3] (first issued in 1999, revised edition 2005), which requires laboratories to give evidence of the traceability of their results and estimate the uncertainty of their measurements.

In the field of laboratory medicine, the European Directive 98/79 EC[4] on *in vitro* diagnostic medical devices (IVDMD) states the rules to be followed by manufacturers of such products to be used within the European Union. In particular it is requested that "traceability of values assigned to calibrators and/or control materials must be assured through available reference measurement procedures and/or available reference materials of a higher order". Because of the worldwide impact of this European legislation for manufacturers and clinical laboratories, international, regional and national collaboration has been sought among intergovernmental, professional, IVDMD manufacturers' and EQAS providers' organisations, aiming to establish a metrologically sound reference system in laboratory medicine.[5] Specific standards and guidelines have been issued (ISO TC212 and CEN TC140) covering requirements for the competence of medical[6] and reference measurement laboratories[7] and other aspects related to the traceability of IVDMD.[2,8] The Joint Committee on Traceability for Laboratory Medicine[9] (JCTLM) was established in 2002, based on a collaboration agreement between CIPM, ILAC and IFCC, with the support of other relevant organisations and institutions (Advanced Medical Technology Association, AdvaMed; American Association for Clinical Chemistry, AACC; Bureau National de Métrologie, BNM; Centers for Disease Control and Prevention, CDC; Chemical Science and Technology Laboratory of the NIST, NIST CSTL; Clinical and Laboratory Standards Institute, NCCLS; European Committee For External Quality Assurance Programmes in Laboratory Medicine, EQALM; European Diagnostic Manufacturers Association, EDMA; HECTEF Standard Reference Center Foundation Ltd, HECTEF Foundation; Institute for Reference Materials and Measurements, IRMM; Korea Research Institute of Standards and Science, KRISS; Laboratory of the Government Chemist, LGC; National Institute for Biological Standards and Control, NIBSC; Paul-Ehrlich-Institut, PEI; Langen;Physikalisch-Technische Bundesanstalt, PTB; Reference Institute of Bioanalysis, RfB)

The aim of the JCTLM is "to support world-wide comparability, reliability and equivalence of measurement results in laboratory medicine, for the purpose of improving health care and facilitating national and international trade in *in vitro* diagnostic devices". This shall be accomplished by means of the development of international conventional reference systems including reference materials certified by reference measurement procedures applied by reference measurement laboratories, at least for selected analytes, prioritised according to recognised medical needs. In this context, networks of expert laboratories, by using the best internationally recognised analytical procedures and a metrologically sound approach, will assign values to reference materials for well-defined measurands. These will be used to establish, through a chain of comparisons, the traceability and uncertainties of routine methods. In addition to metrological principles, criteria such as clinical usefulness, clinical needs, biological variability and disease associated variations, will be considered, when the allowable analytical biases for diagnostic purposes are defined. It is envisaged that the JCTLM initiatives will result in harmonisation and/or standardisation of procedures used in medical laboratories, will achieve true and world wide comparable results and will have an impact on clinical decision criteria. To this aim, calls for the selection of Reference Methods, CRMs and Reference Laboratories for key analytes have been organised and the results are available on the BIPM website. Further work is in progress to expand the procedure to other analytes.

These initiatives will establish the principles of traceability in laboratory medicine and, with time, expand the number of analytes for which traceability to SI can be demonstrated. In the meantime, routine laboratories can demonstrate at least equivalence of measurements by participation in appropriate interlaboratory comparisons.

3 ROLE OF EXTERNAL QUALITY ASSESSMENT SCHEMES IN LABORATORY MEDICINE

EQAS are a particular type of interlaboratory comparisons, other types of interlaboratory comparisons being: method standardisation; certification of reference materials and key-comparisons. A definition of interlaboratory comparison is the following: "organisation, performance and evaluation of test on the same or similar test items by two or more laboratories in accordance with pre-determined conditions".[10]

A more extensive definition of EQA is given by WHO as: "system of objectively checking laboratory results by means of an external agency. The checking is necessarily retrospective, and the comparison of a given laboratory's performance on a certain day with that of other laboratories cannot be notified to the laboratory until some time later. This comparison will not therefore have any influence on the tested laboratory's output on the day of the test. The main object of EQA is not to bring about day to day consistency but to establish between-laboratory comparability".[11] According to ISO/REMCO, EQA refers to "a system of objectively checking laboratory results by means of an external agency, including comparison of a laboratory's results at intervals with those of other laboratories, the main objective being the establishment of trueness".[12]

The main objective of EQAS is to assess objectively the quality of the measurements performed by participants. In addition EQAS have a key role in laboratory medicine since by them it is possible an independent assessment of equivalence among laboratories' results. EQAS and Proficiency Testing (PT) are not synonyms. While EQAS focus mainly on education and continuous improvement, PT focuses on the evaluation of (satisfactory) performances for authorization purposes. EQAS are especially valuable in developing or specialised fields (e.g. where IVDMD are not yet available). In addition, under stated conditions, EQAS can demonstrate traceability/comparability of measurements. The role of EQAS in the chain of traceability in the field of laboratory medicine is outlined in Fig 1.

The history of EQAS/PT in quality assessment of clinical laboratories dates back to 1946.[13] After its introduction in clinical chemistry a dramatic improvement of analytical performance was observed. EQAS were subsequently extended to other fields of laboratory medicine (haematology, immunology and several non-quantitative fields such as microbiology, histo-and cytopathology). A specific legislation was developed in the USA since 1967 (Clinical Laboratory Improvement Act, 1967) and then amended (Clinical Laboratory Improvement Amendments, 1988[14]). Later, similar provisions were developed in European countries. In Italy participation in EQAS is mandatory for all clinical laboratories operating within the national health service.[15] Participation in EQAS is recommended in both ISO/IEC 17025[3] and ISO 15189.[6]

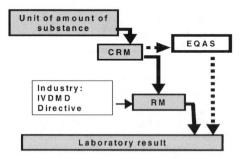

Figure 1 *Role of EQAS in laboratory medicine*

4 EQAS: BASIC ASPECTS AND GENERAL REQUIREMENTS

4.1 General aspects

The organisation of EQAS/PT is described in several authoritative documents[10,16-18] some of which recently issued.[19,20] Some of the fundamental aspects of EQAS/PT will be discussed below, to highlight the different role of EQAS/PT with respect to CRMs and internal quality control, as part of the quality management system.

EQAS/PT samples should be real or commutable, generally based, in the field of laboratory medicine, on pools of native materials (blood, urine, etc.), in order to provide reliable information on the performance of the laboratory on patients' samples of the same nature. Animal fluids can be used in studies of exogenous substances. The range of matrices/concentrations should cover the scope of the assay. Since participants should not be able to identify samples, several batches are prepared frequently and used within a short period of time. Because of this less stringent requirement on stability, EQAS/PT samples undergo lesser treatments and are closer to real samples than CRMs. The frequency of the scheme should be appropriate to the assay/test, possibly not less than two weeks and not more than four months apart. This is to avoid, on one hand, confusion or misuse of the EQAS/PT as internal quality control and, on the other, the developing and worsening of problems undetected by other quality control procedures. EQAS results cannot be used to reject or accept analytical series, as the outcome is often not known for several weeks after the test. Procedures for data analysis should include: a) methods to identify deviations from normal distribution, b) methods to identify outliers or robust statistics, since deviations from normality may indicate problems with the samples and occasional outlier and blunders create artifactually wide values of the dispersion of results. Performance evaluation should be based on criteria that are fit for the purpose of the assay and, whenever possible, should take into account the uncertainty of the control materials and the uncertainty reported by the laboratory as part of the assessment.

4.2 End-users of EQAS in the biomedical field

The organisation of EQAS in the biomedical field should take into account to some extent the different needs of the several categories of end-users of EQAS data. EQAS direct customers, i.e. the participating laboratories, should be granted a fair and independent evaluation of their performance. In addition, they may wish to use the information from participation in EQAS as part of method validation and uncertainty evaluation. Physicians and patients need to rely on the information provided by the laboratory in order to make potentially crucial decisions. They may have increased confidence in results provided by laboratories demonstrating satisfactory performance in appropriate EQAS, i.e. EQAS covering the range of matrices and concentrations of interest and which target of performance reflect clinical needs. Accreditation Bodies may wish to use the results of laboratories' participation in EQAS as part of the objective evidence for the assessment of competence. The issue of whether satisfactory performance in EQAS could justify reduction of the frequency or extent of surveillance visits to accredited laboratories has been debated. Particular attention has been given, to the extent of a specific standard being issued,[8] to the use of EQAS results in the assessment of IVDMDs, therefore including IVDMD manufacturers among the possible end-users of EQAS, expecting a fair and independent evaluation of IVDMD performance. In this context, the specific requirements for EQAS include: classification of laboratories in method groups, with adequate numbers; ability to distinguish between the method own faults and laboratories' failure of performance; values of the control materials traceable to higher metrological order procedures; assigned values determined by accredited Reference Measurement Laboratories; implementation of a quality management system and (possibly) accreditation of the EQAS Provider.

5 KEY-ISSUES FOR EQAS DEVELOPMENT

5.1 General considerations

Over the last few years, accreditation of laboratories to support customers' confidence and mutual recognition of analytical data has gained increasing importance. As a consequence, more and more attention is now paid to the organization of EQAS/PT, which play a major role in assuring equivalence, if not metrological comparability, of analytical data, especially in fields were demonstrating traceability to SI is more difficult or not yet possible. Accordingly, the issue has been risen of formal accreditation of EQAS/PT and the need for specific standards. New documents have been published recently addressing the practice of EQAS/PT and in particular statistical procedures and methods of performance evaluation.[19,20] Whereas a comprehensive discussion of such issues is out of the scope of this paper, selected examples of key-issues relevant to the future development of EQAS/PT will be given. In particular, the following issues will be addressed: 5.1.1) the need for organisers to document the uncertainty of the assigned values and provide such information to their customers; 5.1.2) the need for collaboration among EQAS/PT organisers and other interested parties to establish equivalence of judgement and evaluation criteria in EQAS/PT within the same scope and 5.1.3) appropriate use of EQAS/PT data for method validation and evaluation of measurement uncertainty.

In this context, previous experience from the Italian national EQAS for lead in blood (MeTos)[21] and collaborative work with both the EU JRC Institute for Reference Materials and Measurements (IRMM) project International Measurement Evaluation Programme (IMEP)[22] and the Network of Organisers of EQAS in Occupational and Environmental Laboratory Medicine (Network of EQAS in OELM[23]) will be exploited.

    5.1.1 Uncertainty of assigned values. Methods for value assignment in EQAS/PT as given in ISO Guide 43-1,[10] in hierarchical order, are listed in Table 3 together with the respective method for uncertainty estimate, as given in ISO 13528.[19] This information is particularly valuable as it may affect the judgement of laboratory performance, and, to this aim, ISO 13528 recommend the uncertainty of the assigned values to be checked against the performance standard deviation and be ≤ 0.3. Of the listed methods, formulation is only applicable when exogenous substances, not subjected to changes in a biological environment, are to be tested. In the biomedical field, this only applies to some drugs and persistent contaminants. However, in this last case, the pollution of the general environment makes it difficult to obtain really appropriate blank matrices. At present, the assignment of values to EQAS/PT control samples by means of traceable, reference methods is largely unpractical and sometimes impossible, for both technical and economical reasons.

    Control samples with values assigned by metrologically valid procedures are used within the framework of IRMM-IMEP, a metrological interlaboratory comparison scheme whose aim is to enable the benchmarking of laboratory performance, in support of the elimination of trade or border crossing barriers, to support the EU policies and the establishment of

Table 3 *Methods for value assignment in EQAS/PT (hierarchical order) with their respective method for uncertainty estimate.*

Methods for value assignment	Methods for uncertainty estimated
Formulation	Uncertainty of preparation
Reference method in a single laboratory	Uncertainty of the measurement
Average from results of expert laboratories	Combined uncertainty from individual reported uncertainties or standard deviation of the mean of the results
Consensus value	Estimated uncertainty from robust statistics according to ISO 13528

a chemical measurement infrastructure within the EU, based on the national measurement systems. Within this framework, a relevant initiative was undertaken by IRMM in the field of clinical chemistry with the organisation of an evaluation programme, IMEP-17,[24-26] devoted to unveil the actual status of comparability among routine laboratories measurements all over the world, with regard to trace and minor constituents in serum. Two fresh frozen human serum samples were distributed to 1037 routine laboratories from 35 countries in geographic areas on all continents. The assigned values, unknown to participants, were certified by traceable reference methods (Table 4) and their uncertainty estimated.

Table 4 *Analytes included in IMEP-17 and methods used for the certification of values (modified from 24)*

Component	Applied primary/reference methods*
Ca	ID-ICP-MS, ID-TIMS
Cl	ID-ICP-MS, coulometry, titrimetry
Cu	ID-ICP-MS, ID-TIMS
Fe	ID-TIMS
K	ID-ICP-MS, FAES, ion chromatography
Li	ID-ICP-MS, ID-TIMS
Mg	ID-ICP-MS, ion chromatography
Na	Gravimetry, FAES
Se	NAA, ID-ICP-MS
Zn	ID-ICP-MS, ID-TIMS
Glucose	ID-GC-MS
Cholesterol	ID-GC-MS
Creatinine	ID-GC-MS
Urea	ID-GC-MS
Uric acid	ID-GC-MS
Thyroxine (T4)	ID-LC-MS
Albumin	RID
IgG	RID
Amylase	IFCC enzymatic primary reference method, 37 °C
γ-Glutamyltransferase (γ-GT)	IFCC enzymatic primary reference method, 37 °C

* ID=isotope dilution; ICP=inductively coupled plasma; MS=mass spectrometry; TI=thermal ionisation; FAES=flame atomic emission spectrometry; NAA= neutron activation analysis; GC=gas chromatography; LC=liquid chromatography

Regional co-ordinators selected participants in their region, distributed the samples and related information, collected results and acted as the local contact point for any problem. In Italy the co-ordinator of the program was the Istituto Superiore di Sanità and the selection of laboratories was designed to be representative of the whole country.

The exercise provided a reliable picture of the quality of routine measurements for most of the analytes included in the study and highlighted technical problems related to the use of the same method/instrument/conventional units within one area which undermine the comparability of results. In a few cases (albumin, IgG, amylase and chloride) the uncertainty of the certified values was still too high compared with expected quality specifications (QS) based on biological variation.

Few EQAS/PT can demonstrate the traceability of values assigned to the control samples and most use consensus values from the participants. These may be unreliable, if there is no real consensus amongst the participants, or biased by the general use of techniques based on the same (faulty) principle. However, it is recognised that, for tests where sufficient experience exists, consensus values are usually very close to those provided by other, more reliable, techniques (i.e. formulation, analysis by reference laboratories or consensus of expert laboratories). Although better estimates of the assigned values and their uncertainties are sometimes possible and even desirable, this is not always practicable or even effective. An important issue, however, is the provision of information on the uncertainty of the assigned values, which assessment may be less straightforward than with other assignment procedures.

A method for such task is outlined in ISO 13528,[19] using robust statistics --robust average (X) and standard deviation (s*)-- calculated by means of an iterative procedure (Algorythm A), also reported in ISO 5725-5.[27] The standard uncertainty of the assigned value X is estimated as:

$$u_x = \frac{1.23 \times s^*}{\sqrt{p.}}$$

where p is the number of participating laboratories.

This procedure was applied to assess the uncertainty of the EQAS samples distributed over the last two years as part of the Italian national EQAS for lead in blood. The results (Fig. 2) indicate that the uncertainty of the assigned values was small compared with both the scheme QS for total error allowable (TEa) (AL-ITEQAS) and the observed robust standard deviation among participants' results. Similar considerations may apply in general, as the spread of data observed in different EQAS for lead in blood[28] were similar (Table 5) across the EU, with the exception of Denmark, where only few selected laboratories perform such assay.

5.1.2 Equivalence of judgement and evaluation criteria.   At present, EQAS/PT schemes are mainly organised at a national level, with exception for less frequent, new or specialised tests. Since each scheme is organised in its own way, laboratories taking part in different EQAS/PT schemes for the same assay may be judged differently. This potential problem has been examined within the framework of projects supported by the European Commission in different analytical fields. A study involving laboratories who analysed the same blood samples for their lead content showed that an individual laboratory's performance could be evaluated as unsatisfactory by one scheme but acceptable by another.[29]

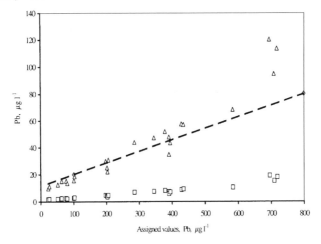

Figure 2 *Uncertainties of the assigned values (□) compared with the observed robust standard deviations of the participants' results (Δ) and the limits for acceptable peformance (dashed line) defined by the organisers of MeTos for Pb in blood over the chosen concentration range for 24 samples distributed within two years.*

Table 5 *Trimmed CV% observed in EQAS for lead in blood (Modified from A. Taylor et al.[28])*

	Concentration, µg/l		
Country	**100**	**400**	**700**
Great Britain	12.6	10.6	10.3
Germany	17.0	10.6	9.8
Italy	21.2	14.2	11.1
Belgium	16.9	16.0	14.0
France	16.3	15.9	14.3
Spain		12	12
Netherlands	14	13	13
Denmark	7.0	7.0	

The same conclusion has since been reached in similar projects which examined EQAS for haematology[30] and for analyses in water, foods, soil and occupational hygiene.[31] It is apparent that the possibility of inconsistency in performance assessment is a general problem. The comparison of statistical performance evaluation protocols, applied in nine EQAS in OELM, suggested that such inconsistencies arose mainly from the adoption of different criteria for performance evaluation.[32] Even when z-scores,[10] or variants that involve the measurement uncertainty associated with the assigned value and/or the participant's result,[19] are used for performance scoring, different results would be obtained if different criteria are chosen to define the performance standard deviation σ (σ_p). Examples may include: i) state of the art (e.g. the SD of all results reported on a sample), ii) stated multiples of the SD of results given by a group of expert laboratories, iii) limits set by the organisers (such as a simple percentage of the assigned concentration, a value that will include/exclude a given proportion of the participants) and iv) limits set by law. Recent guidance recommends using a value that is deemed to produce a performance score to demonstrate whether a laboratory provides results that are fit for the purpose for which the measurement is being carried out.[20] However, unless organisers of EQAS within the same technical sector collaborate and agree on the 'σ_p' value, the performance scores will still fail to provide comparable information and it may be difficult to establish mutual recognition agreements among accreditation bodies and other organisations for the use of such data.

Within the Network of EQAS in OELM, collaborative work has been undertaken to introduce harmonisation in a way that will avoid imposition of a procedure from one scheme onto others and will also ensure that assessments demonstrate when analytical data from participants are fit for purpose. This approach is based on the application of common QS, developed according to an established hierarchy within the scientific medical community,[33] to give equivalence of assessment among EQAS.

The first recommendations for the definition of QS for analytical tests in laboratory medicine were published in 1963 and were based on state-of-the-art.[34] In 1968, Barnett introduced the concept of medical significance of the variation of clinical tests[35] and later in 1970, pioneer work started to compare biological and analytical variability of common serum constituents in relation to the physiological and medical implications.[36] This was followed by further developments both in the theoretical principles [37-40] and application to specific analytes.[41-46]

At present, a hierarchy of methodologies for the definition of QS in laboratory medicine has been agreed [Consensus agreement, Stockholm, 1999[33]] which consider as most relevant the effect of analytical performance on clinical outcomes (in specified clinical settings) and clinical decisions (based on previous knowledge of the biological variability and opinions of clinicians). Published professional recommendations, limits set by law, regulatory organisations or EQAS organisers and state-of-the-art, assessed from results of relevant EQAS or scientific literature on analytical methodologies are the following decreasing steps of this ladder. Whereas little information is available on the effect of analytical variability on clinical outcomes, a simple and scientifically sound model has been developed to derive QS from information on the components of biological variability, which is becoming increasingly available. Therefore, such approach is the most likely to be exploited.

Desirable QS for analytical imprecision (CVa), bias (B) and total error allowable (TEa) can be derived from data on intra- and inter-individual biological variability according to the following formulas:[40]

$$CVa \% = 0.50*CVintra$$
$$Bias\,(\%) = 0.25*(CV^2intra + CV^2inter)^{1/2}$$
$$TEa \% < Bias \% + z*CVa\%$$
$$z = 1.65\,(95\%)$$

Similar formulas can be used to derive minimal and optimal QS.

A database was set up and is regularly updated by Ricós et al.,[47] which currently includes analytical QS (for CVa, B, and TEa) for 289 analytes, derived from published data on within and between-subject biological variation components in healthy subjects. Selected examples are given in Table 6.

Table 6 *Selected examples of quality specifications for imprecision (CVa), bias (B) and total allowable error (TEa) based on the intra (CVintra) and inter (CV inter) individual biological variation. Data from the database at* www.westgard.com, *last updated in 2006.*[47]

Analyte*	CV_{intra}	CV_{inter}	Cva	Bias	$TEa_{0.05}$
Albumin(s)	3.1	4.2	1.6	1.3	3.9
CA 125 (s)	13.6	46.5	6.8	12.1	23.3
Calcium (s)	1.9	2.8	1.0	0.8	2.4
Cholesterol (s)	6.0	14.9	3.0	4.0	9.0
Copper (s)	4.9	13.6	2.5	3.6	7.7
Creatinine (U, 24h)	24.0	24.5	12.0	8.6	28.4
Factor VIII coagulation (s)	4.8	19.1	2.4	4.9	8.9
Free thyroxine (s)	7.6	12.2	3.8	3.6	9.9
γ-Glutamyltransferase (s)	13.8	41.0	6.9	10.8	22.2
Prostatic specific antigen (s)	18.1	72.4	9.1	18.7	33.6
Retinol (s)	14.8	18.3	7.4	5.9	18.1
Vanilmandelic acid (U, 24h)	22.2	47.0	11.1	13.0	31.3
Zinc (s)	9.3	9.4	4.7	3.3	11.0

* s= serum, U= urine

According to this model and taking into account factors such as effect of analytical error on possible outcomes and current analytical capabilities, the Members of the Network of EQAS in OELM have agreed on common QS for TEa (95%) for lead in blood,[28] aluminium,[28] copper, selenium and zinc in serum.[48] In order to calculate z-scores in the usual way, a performance standard deviation equal to half of the QS will be used.

In this way, EQAS organisers may continue to use well established scoring procedures which are familiar to participants, but performances for similar concentrations, can be compared across schemes using z-scores based on the agreed common QS, provided that enough confidence can be placed in the quality of specimens distributed within EQAS/PT and the assigned values.

Although the practical application of these QS is still to be fully implemented, it is evident that, when in place, assessment of performance will immediately demonstrate whether results from participants can be regarded as fit for the purpose for which they are undertaken and therefore also provide univocal comparability among laboratory performance.

Similar collaboration among the EQAS/PT scheme organisers of other sectors is recommended to provide the dual objective of using a z-score that clearly demonstrates when a laboratory is achieving results that are fit for purpose and allowing for comparison of performance from one scheme to another.

5.1.3 Use of EQAS/PT data in method validation and the evaluation of uncertainty of measurement. Participation in EQAS/PT is recommended in ISO standards for the accreditation of laboratories.[3,6,7] Data from EQAS/PT provide independent information on the performance of the laboratory on samples as close as possible to real ones, processed in the same way. In addition to other evaluations and experiments, such information can have an important role in method validation and in the evaluation of the uncertainty of measurement, in particular because it allows the assessment of the performance of the method over the entire range of concentrations and matrices within its scope.

In theory, provided that systematic deviations from the assigned values and any other sources of uncertainty are negligible or otherwise taken into account, the results obtained by a laboratory over regular participation in EQAS/PT could give information on the uncertainty of its measurements for a given test (as covered by EQAS/PT). The scheme should fulfil conditions such as: full coverage of the routine analytical range, traceability and small uncertainty of the assigned values (compared to the spread of the results).

As organisers of the Italian national EQAS in OELM, we carried out a preliminary study to assess how such information could be used by the participants and how different EQAS/PT features (frequency of trials and number of samples) may affect a laboratory estimate of its uncertainty.[49] Such information can be helpful to improve EQAS/PT organisation and , for a given test: a) assess the state of the art of the uncertainty of current measurement procedures, b) identify needs for improvement of analytical methodologies and c) set targets for acceptable uncertainty values. We used data from 14 participants in the EQAS for lead in blood, which had shown consistent good performances over two years (8 trials, 24 samples, including 7 unknown duplicates distributed in two separate occasions) without evidence of significant bias. For each laboratory, the difference between the results provided and the assigned values was calculated and the relative SD of the differences was taken as the best estimate of the uncertainty of their results. The contribution of imprecision to total uncertainty was evaluated from the differences between the results provided for the unknown duplicates. From these data we estimated uncertainty values for the determination of lead in blood by routine laboratories ranging from 3% to 23.5%, which were consistent with previous reports in the literature about single laboratory evaluation of uncertainty.[50,51]

6 CONCLUSIONS

EQAS/PT is a powerful tool to demonstrate equivalence of measurements, if not their metrological comparability, and to promote education and improvement of laboratory practice. In the biomedical field, with more than 700 types of quantities being analysed everyday and new tests being continuously under development, EQAS also provide independent information on the state of the art of routine analyses and the performances of IVDMD, also allowing continuous monitoring of their quality. Work is in place to improve the scientific background and everyday practice of EQAS to fulfil the (expressed or implicit) needs of end-users, including promoting the understanding and application of new metrological concepts by participants.

Acknowledgments
The Authors are grateful to the Members of the Network of EQAS Organisers in OELM, for their fruitful and continuous collaboration, to the EURACHEM Proficiency Testing Working Group for the useful discussions on the role and future of EQAS/PT, to Maria Belli, Head of the Metrological Service at APAT, Rome, Italy and to Ales Fajgelj IAEA for the organization of the Workshop.
The financial support from the Italian Ministry of Health, Project "Analysis of risk connected with the presence of residues in food of animal origin - SARA", 2003-2006 is gratefully acknowledged.

References
1   International Vocabulary of Basic and General Terms in Metrology (1993). ISO, Geneva, Switzerland.

2   ISO 17511: 2003 In vitro diagnostic medical devices -- Measurement of quantities in biological samples -- Metrological traceability of values assigned to calibrators and control materials. ISO, Geneva, Switzerland.

3   ISO/IEC 17025: 2005 General requirements for the competence of testing and calibration laboratories. ISO, Geneva, Switzerland.

4   Directive 98/79/EC of the European Parliament and of the Council of 27 October 1998 on in vitro diagnostic medical devices. Official Journal L 331, 07/12/1998 P. 0001 - 0037.

5   MM. Müller, *Clin. Chem.*, 2000, **46**, 1907.

6   ISO 15189: 2003 Medical laboratories -- Particular requirements for quality and competence. ISO, Geneva, Switzerland.

7   ISO 15195: 2003 Laboratory medicine -- Requirements for reference measurement laboratories. ISO, Geneva, Switzerland.

8   EN 14136: 2004 Use of external quality assessment schemes in the assessment of the performance of in vitro diagnostic examination procedures. CEN, Brussels, Belgium.

9   www.bipm.org

10   ISO/IEC Guide 43: 1997 Proficiency testing by interlaboratory comparisons - Part 1: Development and operation of proficiency testing schemes. ISO, Geneva, Switzerland.

11   External Assessment of Health Laboratories, Report on a WHO Working Group EURO Reports and Studies No 36, WHO, 1981.

12   ISO/REMCO N. 231 Harmonized proficiency testing protocol, 1991. ISO, Geneva, Switzerland.

13   W.P. Belk and F.W. Sunderman, *Am J Clin Pathol*, 1947, **17**, 853.

14   US Department of Health and Human Services. Medicare, Medicaid and the CLIA programs; regulations for implementing the Clinical Laboratory Improvement Amendments of 1988 (CLIA). Final rule. Federal Register 1992, 57, 149.

15   D.P.R. 14 gennaio 1997 Approvazione dell'atto di indirizzo e coordinamento alle regioni e alle province autonome di Trento e di Bolzano, in materia di requisiti strutturali, tecnologici ed organizzativi minimi per l'esercizio delle attività sanitarie da parte delle strutture pubbliche e private, G.U. 20/02/1997, n. 42, S.O.

16   ILAC G13: 2000 Guidelines for the Requirements for the Competence of Providers of Proficiency Testing Schemes. ILAC, Rhodes, Australia.

17   M. Thompson, R. Wood, *Pure Appl. Chem.*, 1993, **65**, 2123.

18   IFCC/EMD/C-AQ. Guidelines for the Requirements for the Competence of EQAP organizers in medical laboratories. http://www.ifcc.org/division/EMD/Documents/EQAP_version _3-2002.pdf

19   ISO 13528: 2005. Statistical methods for use in proficiency testing by interlaboratory comparisons. ISO, Geneva, Switzerland.

20   M. Thompson, S.L.R. Ellison and R Wood, *Pure Appl. Chem.*, 2006, **78**, 145.

21   M. Patriarca, F. Chiodo, M. Castelli, F. Corsetti, A. Menditto, *Microchemical J.*, 2005, **79**, 337.

22   www.imep.ws

23   www.occupational-environmental-laboratory.com

24   U. Örnemark, A. Uldall, L. Van Nevel, Y. Aregbe and P.D.P. Taylor, IMEP-17 Trace and Minor constituents in Human Serum. Certification. Report EUR 20243 EN. IRMM GE/R/IM/36/01, Retieseweg, Geel, Belgium, 2002.

25   L. Van Nevel, U. Örnemark, P. Smeyers, C. Harper and P.D.P. Taylor, IMEP-17 Trace and Minor Constituents in Human Serum. EUR 20657 EN. Report to Participants, Part 1: International comparability. IRMM GE/R/IM/42/02, Retieseweg, Geel, Belgium, revised 2003.

26 U. Örnemark, L. Van Nevel, P. Smeyers, C. Harper and P.D.P. Taylor, IMEP-17 Trace and Minor Constituents in Human Serum. EUR 20694 EN. Report to Participants, Part 2: Methodology and quality specifications. IRMM GE/R/IM/04/03, Retieseweg, Geel, Belgium, revised 2003.

27 ISO 5725-5:1998 Accuracy (trueness and precision) of measurement methods and results - Part 5: Alternative methods for the determination of the precision of a standard measurement method. ISO, Geneva, Switzerland.

28   A. Taylor, J. Angerer, F. Claeys, J. Kristiansen, O. Mazarrasa, A. Menditto, M. Patriarca, A. Pineau, I. Schoeters, C. Sykes, S. Valkonen, C. Weykamp, *Clin. Chem.* 2002, **48**, 2000.

29 J.M. Christensen, E. Olsen, *Ann. Ist. Super. Sanità*, 1996, **32**, 285.

30 M. Van Blerk, J.C. Libeer, *EQANews*, 2003, **14**, 62.

31 N. Boley, presented at the 5th EURACHEM Workshop "Proficiency Testing in analytical chemistry, microbiology and laboratory medicine. Current practice and future directions", Portorož, Slovenija 26-27 September 2005. Book of abstracts.

32   A. Taylor, M. Patriarca, A. Menditto, G. Morisi, *Ann. Ist. Super. Sanità*, 1996, **32**, 295.

33 D. Kenny, *Scand. J. Clin. Lab. Invest.*, 1999, **59**, 585.

34 D.B. Tonks, *Clin. Chem.*, 1963, **9**, 217.

35 R.N. Barnett, *Am. J. Clin. Pathol.*, 1968, **50**, 671.

36 E. Cotlove, E.K. Harris and G. Williams, *Clin. Chem.*, 1970, **16**, 1028.

37 C.G. Fraser, *Clin. Chem.*, 1987, **33**, 1298.

38 C.G. Fraser, *Arch. Pathol. Lab. Med.*, 1988, **112**, 404.

39 C.G. Fraser, P.H. Petersen, C. Ricos, R. Haeckel, *Eur. J. Clin. Chem. Clin. Biochem.*, 1992, **5**, 311.

40 C.G. Fraser, *Scand. J. Clin. Lab. Invest.*, 1999, **59**, 487.

41 P.H. Petersen, C.G. Fraser, L. Jorgensen, I. Brandslund, M. Stahl, E. Gowans, J.C. Libeer, C. Ricos, *Ann. Clin. Biochem.*, 2002, **39**, 543.

42 M.C. Browing, R.P. Ford, S.J. Callaghan, C.G. Fraser, *Clin. Chem.*, 1986, **32**, 962.

43 C.G. Fraser, *Am. J. Clin. Pathol.*, 1987, **88**, 667.

44 R.P. Ford, P.E. Mitchell, C.G. Fraser, *Clin. Chem.*, 1988, **34**, 1733.

45 M. Plebani, A. Giacomini, L. Beghi, M. De Paoli, G. Roveroni, F. Galeotti, A. Corsini, C.G. Fraser, *Anticancer Res.*, 1996, **16**, 2249.

46 S.M. Ross, C.G. Fraser, *Ann. Clin. Biochem.*, 1998, **35**, 80.

47 C. Ricós, J.V. García-Lario, V. Alvarez, F. Cava, M. Domenech, A. Hernández, C.V. Jiménez, J. Minchinela, C. Perich, M. Simón, C. Biosca. Biological variation database, and quality specifications for imprecision, bias and total error . The 2006 update. www.westgard.com

48   A. Taylor, J. Angerer, J. Arnaud, F. Claeys, R.L. Jones, O. Mazarrasa, E. Mairiaux, A. Menditto, P.J. Parsons, M. Patriarca, A. Pineau, S. Valkonen, J.I. Weber, C. Weykamp, *Accred. Qual. Assur.* 2006, DOI 10.1007/s00769-006-0118-8.

49 M. Patriarca, F. Chiodo, M. Castelli, A. Menditto, presented at the 5th EURACHEM Workshop "Proficiency Testing in analytical chemistry, microbiology and laboratory medicine. Current practice and future directions", Portorož, Slovenija, (25) 26-27/09/2005. Book of abstracts.

50 M. Patriarca, M. Castelli, F. Corsetti, A. Menditto, *Clin. Chem.*, 2004, **50**, 1396.

51 J. Kristiansen, J. Molin Christensen, J.L. Nielsen, *Mikrochim Acta*, 1996, **123**, 2419.

MEASUREMENT UNCERTAINTY AND ITS ROLE IN PROFICIENCY TESTING SCHEME-CASE STUDY ILC WASTE WATER

Magda Cotman, Andreja Drolc, Milenko Roš,

National Institute of Chemistry, Hajdrihova 19, SI-1000 Ljubljana, Slovenia, Tel:+386 1 4760238, Fax:+386 1 4760300, E-mail: magda.cotman@ki.si,

1 INTRODUCTION

The concept of comparability (equivalence) of measurement results is increasingly important since it allows minimizing technical barriers in trade, to improve environmental monitoring and cut down expenditures for international cooperation. Nowadays it is almost a mandatory requirement that a testing laboratory has to demonstrate to its clients or accreditation body its measurement capability through participation in external quality assurance programmes such ac proficiency testing (PT). Proficiency testing is use of interlaboratory comparison (ILC) for the determination of laboratory testing or measurement performance [1]. Firstly, a PT scheme acts in the capacity of policeman when the outcome of the PT is used by accreditation bodies to monitor laboratory competence in specific measurement procedure. Secondly, for the laboratories that are not yet competent in the test, a PT programme provides an opportunity for improvement since the PT programmes reports are usually accompanied by an analysis of results and recommendation for improving the quality of measurements. For the proficiency testing scheme this "teaching" function is at least as important as the monitoring component. Learning from mistakes, investigating problems, implementing corrective actions and then observing whether this is successful in the next PT represents what many in our profession would regard as a proper and effective use of PT [2].

2 METHODS AND RESULTS

2. 1 General description of ILC

To assess the quality of waste water analyses in Slovenia a PT was carried out in accordance with ISO Guide 43-1 [1] and ILAC Requirements [3]. The PT scheme named ILC-Waste Water (MP-Odpadne vode) was organized by a project group at the National Institute of Chemistry (NIC), namely the Laboratory for Chemistry, Biology and Technology of Waters, which was responsible for designing the scheme, the preparation and validation of test materials, the production and distribution of instructions and test materials to participating laboratories, the collection and statistical analyses of the data obtained from the test and feedback of the results to the participants. Important decisions about performance criteria, determination of the reference value and participation fee were made by a Technical Committee which included representatives of the Ministry of Environment and Spatial Planning, the Slovenian Accreditation Agency, the Faculty of Chemistry and Chemical Technology of the University of Ljubljana, representatives of the participants and the organizers. Homogeneity and stability studies on the samples were performed for most parameters at the NIC Laboratory for Chemistry, Biology and Technology of Waters, which has a quality system accredited according to ISO 17025. For the homogeneity and stability study for metals, we cooperated with the laboratories which are very experienced laboratories in these fields of measurements and have been involved in laboratory quality control systems. The PT scheme was organized in two languages (Slovene and English). This was important since it facilitated better cooperation with small field laboratories. Laboratories, operating essentially on national level, may have no need to require from their staff that they can cope with documents in another language than the national one. A way out can be to provide documentation in different languages, but this remedy is a considerable logistic and linguistic burden.

We had two distributions yearly. Details of the organization of the scheme were published elsewhere [4]. Each PT scheme involved around fifty laboratories, mostly from Slovenia. In the field of waste water analyses there is a lack of certified reference materials (CRMs) due to the problem of the instability of samples. Reference water materials are stable for only a short period of time. Thus it is very important for such laboratories to cooperate in interlaboratory comparisons. Samples used in ILC have matrix effects such that participants would typically meet in connection with their routine work. The samples had two different concentration levels of analyte. Samples were prepared from the inflow to a waste water treatment plant, additionally filtered through a special filter device to remove particles. Some samples were spiked with an additional amount of analyte if necessary to adjust the concentration level. According to the literature and from our own experience some samples were acidified to increase the stability of the samples [5]. Before we dispatched the samples we also made a homogeneity study. We randomly took an appropriate number of samples to check in-bottle and between-bottle homogeneity. The samples were analysed in triplicate. From the calculated between-bottle and within-bottle deviations, it was shown that the prepared samples were homogeneous enough to be used in the PT.

Table 1 *Characteristic of the samples used in ILC-Waste Water 11*

Sample identification	Matrix	Analyte	Concentration level, $mg\ L^{-1}$
A1	Deionised water	F^-, PO_4^{3-}-P	0,5-5
		Cl^-, SO_4^{2-}	0,5-50
A2	Municipal waste	F^-, PO_4^{3-}-P	5-50
		Cl^-, SO_4^{2}	50-500
S1	Deionised water	COD	30–100
S2	Municipal waste	COD	100 – 2000

2.2 Measurement uncertainty

Measurement uncertainty can be estimated by identifying all possible sources of uncertainty associated with a method, quantifying uncertainty components and calculating total uncertainty by combining the individual uncertainty components following appropriate mathematical rules. For the proficiency testing scheme "teaching" function is at least as important as the monitoring component so the need to adapt statistical and evaluation procedures of current PT schemes to include uncertainty data are most important. The issue of whether PT should include uncertainty when most participants do not use or understand it was considered. Knowledge of assigned value uncertainty was seen as important, so scheme organisers should be encouraged to include the uncertainty of the assigned value if possible. This was seen as important for education as well as assessment, particularly with increasing demand for uncertainty estimates. In any case, the provider needs to know the reliability of an assigned value for proper assessment and interpretation of results [6].

In our PT scheme the assigned value and its standard uncertainty was determined in different ways depends on the number of participants. When the number was large enough (more than twenty), the assigned value can be calculated from analytical results of the laboratories as a consensus value (as the mean or median of the results). The choice which results were used for calculating consensus value is responsibility of organizer. The value of PT material was taken as the mean or median of all participants, who analyzed the parameter with a standard method, without outliers and other unsuitable results. An unsuitable result is defined a result which is not made with a standard procedure. In waste water analyses for some parameters (e.g. Kjeldahl nitrogen, oil and grease, chemical oxygen demand (COD)) results dependent on the method, so the organizer which is responsible for assigned value use only results performed by standardized methods. Moreover the organizers used only results made in the proper time period, because same parameters (e.g. nitrite, nitrate, mercury, BOD_5) are unstable during short time especially the samples are not properly storage.

The standard uncertainty of assigned value is calculated according to ISO 13528 [7] as follows:

$$u_x = 1,25 x \frac{s^*}{\sqrt{p}} \qquad (1)$$

u_x standard uncertainty of assigned value, s^* is the robust standard deviation, p is number of results.

If the number of participants measured defined parameter are less then twenty we use certified reference material (CRM). Assigned value and calculation of uncertainty are made by producer. CRMs have assigned value traceable to SI, such materials are stable and homogenous. The limitation of this approach is that it can be expensive to provide every participants and the facts that adequate matrix CRMs are not available, due to instability of samples.

Uncertainty data from participants was generally thought to be a useful addition. This was seen as important initially as a test of understanding, a check on feasibility, and a method of promoting education in the concepts. However, there remain concerns about the methodology to use, both because there is a need for consistency in a given scheme and because it is not clear that existing approaches are sufficiently reliable. From the beginning of organizing PT schemes the report forms included data on measurement uncertainty. If the laboratory had no procedure for estimating its measurements uncertainty these data could be omitted. We examined the number of participants that had reported measurement uncertainty for their results in successive PT schemes. The participants paid more attention to procedures for calculating measurement uncertainty as the schemes proceeded, so every year more participants reported measurement uncertainty. Measurement uncertainty estimation seems to be still in the development stage. Some participants calculated uncertainties according to the Guides for Quantifying Measurement Uncertainty (GUM)

[8], another participants calculated their uncertainty with data from validation procedure. Some participants reported standard uncertainty (u_{lab}) another participants expanded measurement uncertainty ($U_{lab}=k\ u_{lab}$, k=2). In the future more and more attention should be paid to unification procedure.

2.3 Performance criteria

Among various statistic [1, 7] Z scores and E_n or ζ (Zeta) are being expressed in symbols of as follows:

$$Z = (x - X)/S_{t\,arg} \tag{2}$$

$$E_n = (x - X)/\sqrt{U_{lab}^2 + U_{ref}^2} \tag{3}$$

$$\zeta = (x - X)/\sqrt{u_{lab}^2 + u_{ref}^2} \tag{4}$$

where x and X are the results of participating laboratory and the assigned value respectively; S_{targ} is the estimated or measure of variability. U_{lab} and U_{ref} are the expanded uncertainty of the result of participating and reference laboratory respectively. ζ- scores differ from E_n numbers by using standard uncertainties u_{lab} and u_{ref} rather than expanded uncertainties U_{lab} and U_{ref}. The usage of performance indicators is specified as Z≤2 is satisfactory; 2<Z<3 is questionable; Z≥3 is unsatisfactory. ζ- scores may be used instead of Z-scores and shall be interpreted in the same way as Z-scores using the same critical values of 2 and 3. In the contrast to these critical values, it is common to use critical values 1,0 and 1,5 with E_n numbers. This is because E_n numbers are calculated using expanded uncertainties in the denominator instead of standard deviations. When the expanded uncertainties are calculated using a coverage factor of 2,0 a critical value of 1,0 for E_n is equivalent to the critical value 2,0 used with Z scores.

2.4 Evaluation of the results of the ILC

Around 35 samples A1 and A2 for determination of anions and 39 samples S1 and S2 for determination COD were distributed among the participants in the ILC-Waste Water 11. A final number of results were presented in Table 2. The mean and median values from results of the participants were calculated. In the Table 2, these statistical values, as well as maximum and minimum values and number of results for every sample are presented.

Table 2 *Statistical values grouped by types of samples*

Sample	Analyte	Mean, mg L^{-1}	Median, mg L^{-1}	Minimum	Maximum	p
A1	F$^-$	1,76	1,76	1,55	2,20	21
A1	PO$_4$$^{3-}$-P	3,13	3,06	1,20	7,50	33
A1	Cl$^-$	13,2	13,1	10,3	16,0	32
A1	SO$_4$$^{2-}$	24,8	25,1	17,8	37,8	31
A2	F$^-$	22,1	20,9	16,7	50,6	20
A2	PO$_4$$^{3-}$-P	28,8	28,7	14,3	70,3	33
A2	Cl$^-$	176	176	149	202	32
A2	SO$_4$$^{2-}$	265	266	191	346	32
S1	COD	51,8	50,0	42,3	92,9	39
S2	COD	827	828	758	890	39

The next steps are calculation of the assigned values and its uncertainties according to Eqs.1. Standard uncertainty of assigned value depends on the robust standard deviation and the number of results. The assigned value was taken as the mean value of all participants, who analysed the parameter with the standard method, without outliers, and other unsuitable results. An unsuitable result was defined as a result made in proper time, which was not made with procedure and not made in proper time period. The results of calculations are presented in Table 3.

Table 3 *Assigned values and its uncertainties*

Sample	Analyte	X_{ref}, mg L^{-1}	u_{ref}, mg L^{-1}	u_{ref}, %	S_{targ}, mg L^{-1}
A1	F$^-$	1,74	0,025	1,4	0,087
A1	PO$_4$$^{3-}$-P	2,99	0,02	0,7	0,149
A1	Cl$^-$	13,2	0,119	0,9	0,66
A1	SO$_4$$^{2-}$	24,3	0,44	1,8	1,21
A2	F$^-$	20,7	0,312	1,5	1,03
A2	PO$_4$$^{3-}$-P	27,9	0,25	0,9	1,39
A2	Cl$^-$	176	0,865	0,5	8,8
A2	SO$_4$$^{2-}$	263	2,11	0,8	13,1
S1	COD	50,9	0,579	1,1	2,54
S2	COD	826	4,6	0,6	20,6

The assigned values were used for performance criteria. The internationally recommended Z- score was used as the performance criteria. For each results x_i, the individual Z-score was calculated according to Eqs. 2. The target performance criteria were known participants in advance and were determined according to analytical method and the concentration level of analyte. Target performance criteria for different measurands are listed in Table 3. In addition ζ- scores may be used as the performance criteria. For each results x_i, the individual ζ-score was calculated according to Eqs. 4 and shall be interpreted in the same way as Z-scores using the same critical values of 2 and 3. As an example results of determination COD in sample S2 (waste water) are presented.

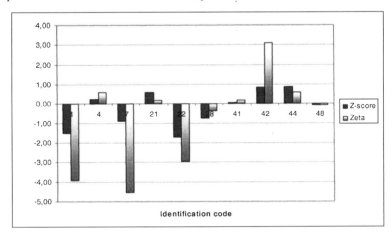

Figure 1: *Z-scores and ζ-scores for determination COD in sample S2*

3 CONCLUSIONS

The quality control of chemical measurements tests is currently a major priority in many countries. Only accurate results allow valid conclusions to be drawn about the harmful properties of different types of waste water and the risk related to biota. The quest of quality is a long-term process, which requires adherence to rules such as the use of control charts, reference materials, certified reference materials and participation in interlaboratory comparisons. In order to verify the quality of such tests the National Institute of Chemistry organized the ILC-Waste Water.

References
1   ISO/IEC Guide 43-1. ISO, Geneve, Switzerland, 1996, 1.
2   Boly, N., *Accred. Qual. Assurance*, 2004, **9**, 633
3   ILAC-G13, ILAC:www.ILCA.org, 2000
4   Cotman, M., Drolc, A. and Roš, M., *Accred. Qual. Assurance*, 2003, **8**, 156
5   ISO 5667, Part 3. ISO, Geneve, Switzerland 1994
6   Brookman, B., Papadakis, I., Squirrell, A., et.al., *Accred. Qual. Assurance*, 2004, **9**, 635
7   ISO 13528, ISO, Geneve, Switzerland, 2005
8   Guide to the Expression of Uncertainty in Measurement, BIPM, IEC, IFCC, ISO, IUPAC, IUPAP, OIML, ISO, Geneve, 1995.

THE PROFICIENCY TESTING OF LABORATORIES: A FIRST APPROACH TO IMPLEMENT BAYESIAN METHODS IN THE ASSESSMENT OF PERFORMANCE

R. Núñez-Lagos[1], M. Barrera[2], M.L. Romero[2]

[1] Dep. Física Teórica, Fac. Ciencias, C/ Pedro Cerbuna 12, E-50.009 Zaragoza, Spain
[2] CIEMAT, Dep. Medio Ambiente, Av. Complutense 22, E-28.040 Madrid, Spain

1 INTRODUCTION

Periodical intercomparison exercises have been organised by CIEMAT and CSN among laboratories providing data for environmental monitoring programmes in Spain. Since 1.985 different statistical methodologies have been applied for evaluation, with the result that the recommendations of the IUPAC protocol, the z-score, has shown to be the best working method to assess laboratory performance[1].

The last revision of the IUPAC International protocol for proficiency testing of laboratories[2], recommends the assessment of laboratory performance against established criteria based on fitness for purpose criteria (the z-score).

$$z = \frac{(b-a)}{\sigma_{ffp}} \qquad (1)$$

where a = The best estimate of the measurand, b = The participant's result, σ_{ffp} = The fitness-for-purpose based "standard deviation for proficiency assessment".

This method provides objective means to assess the accuracy of laboratory results and assumes that participants perform in a manner consistent with the scheme's criteria, thus it does not take into account the participants' reported uncertainties. However a more precise assessment of performance should require a better estimation of the quality of the measurement (i.e.: when a participant has a fitness for purpose requirement inconsistent with that of the scheme).

On the other hand, the u-score method of performance assessment takes into account the uncertainty of the laboratory result, but unreliable uncertainties reported by participants can disguise the scoring.

$$u = \frac{|b-a|}{\sqrt{\sigma^2 + \tau^2}} \qquad (2)$$

σ = uncertainty on the assigned value τ = uncertainty on the laboratory result

Although experience shows that uncertainty estimates are often incorrectly evaluated by laboratories, the Bayesian statistical methods allow the inclusion of such information by means of probability distribution functions, pdf.

An attempt to include this information of the laboratories in the performance assessment has been initiated. A pdf is assigned both to the reference value and to the laboratory results and their respective uncertainties and the intrinsic difference[3] between the functions is then computed and evaluated.

The proposed approach has been applied to a real case, the proficiency test organized by the Spanish Nuclear Regulatory Council (CSN) and CIEMAT in Spain in 2004, among 40 environmental radioactivity laboratories, where the test material was an aqueous synthetic solution containing certified amounts of radionuclides. The results are analyzed and compared to those obtained by the z-score method.

2 METHODOLOGY

To establish an evaluation method that will consider, objectively, the laboratory result and its uncertainty, a methodology has been developed. Bayesian methods [4,5,6,7,8] make use of the concept of intrinsic discrepancy, δ, a general measure of the divergence between two probability distributions. The intrinsic discrepancy between two distributions of the random vector $x \in X$, $p(x)$, $q(x)$, is defined as:

$$\delta(p,q) = Min\left\{ \int_X p(x)\ln\left(\frac{p(x)}{q(x)}\right)dx, \int_X q(x)\ln\left(\frac{q(x)}{p(x)}\right)dx \right\} \qquad (3)$$

The intrinsic discrepancy is symmetric, non negative and is zero if, and only if, both distributions are identical.

When one of the probability distributions is supposed to be true, say for instance $p(x)$, then the intrinsic discrepancy of the distribution $q(x)$ respect to the true probability distribution is simply

$$\delta(p,q) = \int_X p(x)\ln\left(\frac{p(x)}{q(x)}\right)dx \qquad (4)$$

In other words the intrinsic discrepancy is the expected log-likelihood ratio in favour of the true distribution. The application of this concept to the field of intercomparison exercises requires that the corresponding probability functions be assigned to the reference value and to the laboratory results. It is possible to associate, as a first approximation, normal distributions both to the reference $p(x|a,\sigma)$ and to the laboratory results $q(x|b,\tau)$. The expected value of the x variable being respectively the results a and b and their widths σ and τ.

$$\text{Reference value: } a \pm \sigma \rightarrow p(x\,|\,a,\sigma) = \frac{1}{\sqrt{2\pi}\sigma}e^{-\frac{(x-a)^2}{2\sigma^2}} \qquad (5)$$

$$\text{Laboratory result: } b \pm \tau \rightarrow q(x\,|\,b,\tau) = \frac{1}{\sqrt{2\pi}\tau}e^{-\frac{(x-b)^2}{2\tau^2}} \qquad (6)$$

By substituting in (3) we obtain the intrinsic discrepancy of the associated probability distributions to the reference and laboratory as:

$$\delta = min\left[\ln\frac{\tau}{\sigma} - \frac{1}{2} + \frac{\sigma^2}{2\tau^2} + \frac{(a-b)^2}{2\tau^2}, -\ln\frac{\sigma}{\tau} - \frac{1}{2} + a\frac{\tau^2}{2\sigma^2} + \frac{(b-a)^2}{2\sigma^2} \right] = \delta(a,\sigma,b,\tau) \qquad (7)$$

The discrepancy, δ, is then a function of reference values a, σ, and laboratory results b, τ. It is positive or zero and symmetric under the exchange of reference and laboratory $\sigma \Leftrightarrow \tau$, $a \Leftrightarrow b$.

In our case we can consider that the reference distribution is the true distribution and we then measure the different laboratory results with respect to the reference. In this case the intrinsic discrepancy is simply:

$$\delta(a,\sigma,b,\tau) = \ln\frac{\tau}{\sigma} - \frac{1}{2} + \frac{\sigma^2}{2\tau^2} + \frac{(a-b)^2}{2\tau^2} \qquad (8)$$

To express δ in a more simple and generic way, we shall use σ_{ffp} for the width of the reference probability distribution and we define the following parameters: τ^*, normalized laboratory uncertainty, $\tau^* = \dfrac{\tau}{\sigma_{ffp}}$ and the usual z-score $z = (b-a)/\sigma_{ffp}$. Thus the discrepancy $\delta = \delta(z, \tau^*)$, becomes a function of two variables, z, that measures the deviation of the laboratory result in units of σ_{ffp} and, τ^*, that performs the measure of the laboratory uncertainty in the same units. The discrepancy, δ, becomes:

$$\delta(z,\tau^*) = \ln\tau^* - \frac{1}{2} + \frac{1}{2\tau^{*2}} + \frac{z^2}{2\tau^{*2}} \qquad (9)$$

The intrinsic discrepancy (9) has a minimum at $\tau^* = \sqrt{1+z^2}$ (only the positive values can be accepted). For τ^* values below the minimum, δ grows very rapidly. This translates, in this variables, the fact that the probability that the true value, a, be within the interval $b \pm \tau$ decreases as τ decreases. For τ^* values above the minimum δ grows more slowly and for very large values of τ^* grows as lnt^*. Therefore, when the uncertainty of the laboratory is very high, the intrinsic discrepancy depends very little of the value of z, i.e. on the laboratory result, but such extreme case is out of the real world.

This can be seen easily in Figures 1 and 2 where the intrinsic difference is plotted against $|z|$ and τ^*. Figure 1 that presents $\delta(p,q)$ versus z for different constant τ^* values show the expected parabolas with a minimum at $z=0$. Figure 2 that represents the intrinsic discrepancy $\delta(p,q)$ versus the uncertainty τ^* for different constant z-score values is a very interesting curve that validates the proposed methodology.. The generic shape of this curve shows a minimum for $\tau^* = \sqrt{1+z^2}$ and it can be seen that the intrinsic discrepancy grows rapidly for uncertainty values $\tau^* < \sqrt{1+z^2}$ and more slowly for $\tau^* > \sqrt{1+z^2}$. The existence of a minimum for τ^* is a very good behaviour as small and large laboratory uncertainties respect the reference one, are both penalized in contrast with the behaviour of the u-score. The only case with a zero discrepancy is reached when both, z is zero, and then

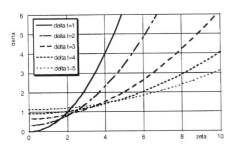

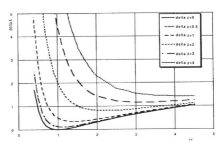

Figure 1 *The intrinsic difference $\delta(p,q)$ as a function of the z-score for different t* values*

Figure 2 *The intrinsic difference $\delta(p,q)$ as a function of t* for different z-score*

the laboratory matches the reference result, and also $\tau^* = 1$, that is, the laboratory results are identical to the reference. The intrinsic discrepancy is zero only when both density functions are identical.

3 APPLICATION TO A REAL CASE

The results from a proficiency test organized by the Spanish Nuclear Regulatory Council (CSN) and CIEMAT in Spain in 2004 have been utilized to validate the proposed methodology[9]. The exercise was organized among 39 environmental radioactivity laboratories, and the test material was an aqueous synthetic solution containing certified amounts of radionuclides. The determination of ^{137}Cs activity has been studied. The assigned value, a, was 498,8 Bq/m^3, its uncertainty percentage was 2 % and the relative σ_{ffp} was 8%. Table 1 shows the activity values and uncertainties informed by the laboratories, and the corresponding values of z, τ^* and δ. The last two columns show the laboratory numbers ordered by $|z|$ and δ values. The difference showed by this two evaluation methodologies is due to the fact that intrinsic discrepancy, δ, depends, not only on z, but also on the uncertainty τ^*. Some laboratories with very good z-score values have bad δ values, as for example the best z-score laboratory, lab number 22, is classified as the 17th by the intrinsic discrepancy δ and the best δ laboratory, lab number 2, is he 10th by its z-score.

Table 1. *Cs-137 data corresponding to the intercomparison exercise CSN/CIEMAT-2004, ordered by laboratory numbers. Last two columns give the laboratory numbers ordered by $|z|$ and the intrinsic discrepancy values δ. (Reference activity a= 498.8 Bq/m^3, $\sigma = 2$ %, $\sigma_{ffp} = 8$ %)*

Lab num	b Bq/m3	τ Bq/m3	z	τ*	δ	Lab |z|	Lab δ	Lab num	b Bq/m3	τ Bq/m3	z	τ*	δ	Lab |z|	Lab δ
1	515,0	48,0	0,41	1,20	0,09	25	2	24	547,2	33,5	1,21	0,84	1,08	40	37
2	486,0	36,0	0,32	0,90	0,07	22	19	25	500,0	73,0	0,03	1,83	0,25	3	35
3	460,0	20,0	0,97	0,50	2,68	30	1	26	472,0	100,0	0,67	2,51	0,53	5	15
4	354,0	80,0	3,63	2,01	1,96	34	32	27	495,6	59,0	0,08	1,48	0,12	35	16
5	539,0	19,0	1,01	0,48	3,20	27	27	28	477,3	21,9	0,54	0,55	1,04	15	39
9	486,0	206,0	0,32	5,16	1,16	19	34	29	493,7	23,2	0,13	0,58	0,46	24	28
10	448,0	109,0	1,27	2,73	0,68	29	25	30	496,0	114,0	0,07	2,86	0,61	10	24
12	550,0	60,0	1,28	1,50	0,49	32	38	31	600,0	52,0	2,54	1,30	1,95	12	33
13	620,0	60,0	3,04	1,50	2,17	17	17	32	492,0	56,0	0,17	1,40	0,10	18	9
14	560,0	200,0	1,53	5,01	1,18	2	20	33	610,6	134,3	2,80	3,37	1,10	14	14
15	545,4	36,1	1,17	0,90	0,85	9	29	34	502,0	64,0	0,08	1,60	0,17	37	18
16	413,0	98,0	2,15	2,46	0,86	36	12	35	542,6	34,5	1,10	0,86	0,83	39	31
17	506,0	25,0	0,18	0,63	0,35	21	26	36	485,0	10,0	0,35	0,25	7,03	16	4
18	555,0	35,0	1,41	0,88	1,31	1	21	37	570,0	88,0	1,78	2,21	0,72	31	13
19	503,0	54,2	0,11	1,36	0,08	20	40	38	472,0	77,0	0,67	1,93	0,35	33	3
20	516,0	86,0	0,43	2,16	0,40	28	30	39	577,0	61,0	1,96	1,53	0,96	13	5
21	514,0	24,4	0,38	0,61	0,54	38	22	40	535,0	35,0	0,91	0,88	0,55	4	36
22	500,00	21,0	0,03	0,53	0,66	26	10								

Figure 3 represents δ versus z, for different τ values. The theoretic curve $\delta(z)$ previously described in Figure 1 can be seen, but the shape is not so well defined, due to the experimental variability of the τ^* values, covering different ranges. As it was deduced, the discrepancy grows rapidly with respect to z for low uncertainty values. The biggest discrepancies are reached for high z, and for intermediate z values combined with small uncertainty values. This last situation occurs for laboratory results, with $z \approx 1$, but with $\tau^* < 1$ such as labs 17 and 3.

Figure 4 represents δ versus τ^* for z values covering different ranges. The theoretical curves of Figure 2 can be seen as well as that the smaller discrepancies correspond to low z values and uncertainties consistent with z. On the other side, bigger discrepancies occur for high z values or small τ^* values.

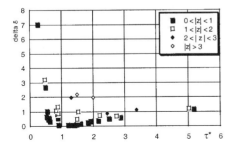

Figure 3 *δ values versus z-score, for CSN/CIEMAT-2004 intercomparison Cs-137 data. δ as a function of t* for z-score values*

Figure 4 *δ values versus z-score, for CSN/CIEMAT-2004 intercomparison Cs-137 data. δ as a function of t* for different different z-score values*

4 CONCLUSIONS

The methodology described in this work could be used as an adequate tool to evaluate the results of proficiency testing among analytical laboratories, as it takes into account the laboratory values and its uncertainties in a suitable way. The basis of the method is to assign probability distributions to the laboratory results and to the reference value and then evaluate the difference of such distributions by computing their Intrinsic Discrepancy as a measure of such deviation.

The method not only evaluates the deviation of the laboratory value with respect of the reference, but also evaluates the quality of the determination, avoiding the partiality of u-score, which gives advantage to the values with bigger uncertainties.

The validity of this first approach has been tested trough the application to a real proficiency test. The future development of the method would involve the study of the classification intervals, the possibility of using non-normal distributions for the laboratory and reference values and the evaluation of the optimum fitness for purpose standard deviation σ_{ffp}.

Acknowledgements

One of us R. N-L wishes to thank Ciemat for financial support.

References

1   M.L.Romero, L.Ramos. *Radioprotección* 2000, *8*, 26
2   Thompson, M., Stephen, L.R.Ellison, Wood, R. *Pure and Applied Chemistry* 2006, *78*, 145
3   Bernardo J. M. (ed). *Bayesian Methods in the Sciences*. Real Academia de Ciencias Madrid 1999.

4   Bernardo, J. M. and Rueda, R. *International Statistical Review* 2002, **70**, 351

5   Berry, D.A. *Statistics: A Bayesian Perspective.* Wadsworth Publishing Company; Belmont, California 1996:

6   Iman,R.L., Conover, W.J. *A modern approach to statistics.* John Wiley and Sons. New York 1983

7   Strom, D.J., *Introduction to Bayesian Statistics.* PNNL-SA-31527. Health Physics Society Annual Meeting, June 27-July 1, 1999. Pacific Northwest National Laboratory Richland, Washington 1999

8   Bayes, T. *The Philosophical Transactions* 1763, **53**, 370; Reproduced in. Biometrika 1958, **45**, 293

9   Romero M.L., Barrera M. *Resultados del ejercicio Interlaboratorios de Radiactividad Ambiental CSN/CIEMAT-04 (Solución Acuosa)* Informes Técnicos CIEMAT (nº 1047). Ciemat, Madrid 2004

THE E_n, U_{score}, AND ACCURACY PARAMETERS - A TOPIC TO DEBATE

Em. Cincu, L. Cazan, V. Manu

Horia Hulubei National Institute for Nuclear Physics and Engineering {IFIN-HH},
Bucharest - Magurele, POB MG-6, Str. Atomistilor No. 407, Code 077125, Romania

1 INTRODUCTION

Under the quality requirements of the Standards in force[1,2], laboratories must participate in International ILC (Inter-Laboratory Comparison) or PT (Proficiency Tests) exercises in order to check their performance and prove that results reported in their Bulletins of Analysis are valid. Taking into account the quite old content of the Standard ISO/IEC Guide 43-1:1997[2] we suggest adding some prescriptions by which metrology and routine laboratories might have different limits of the 'accepted' values characterizing their performance, as the results of routine laboratories will always include reference data provided by metrology or other authorized institutions (e.g. CRMs certificates).

To this end, we propose expanding the restrictive range of accepted values of the performance criterion E_n as established in the Standard[2] to include the 5 classes of values specific to the $U_{test/score}$ parameter[3], usually accepted by IAEA and NPL in evaluating results of routine laboratories within ILC /PT exercises, because their mathematical expressions are identical and because it becomes possible to cover by an unique performance parameter all the range of values provided by metrology and routine laboratories. The advantages and disadvantages of E_n and $U_{test/score}$ are discussed.

The influence of the experimental uncertainty u_{exp} stated by a laboratory participant in an ILC/PT exercise on the performance parameter E_n is studied in correlation with variation of the E_n numerator, the expression of which is shown to be the closest to the definition of the Accuracy parameter given in the metrology literature[4]; the experimental data used for illustration were obtained at our laboratory in applications of the NAA technique[5].

We also present in short the way we calculate the overall/combined uncertainty at our Laboratory; an illustration is given for the case of determining element concentration by INAA using the relative Standardization method based on a CRM external standard.

2 PERFORMANCE EVALUATION CRITERIA

2.1 Evaluation of Results using the E_n Parameter

Under the ISO Guide Standard 43-1:1997[2], the definition of the parameter E_n used in evaluating results of ILC /PT-participant Laboratories has the expression:

$$E_n = \frac{\left|X_{ref} - X_{exp}\right|}{\sqrt{\left(u_{ref}\right)^2 + \left(u_{exp}\right)^2}} \tag{1}$$

where u_{ref}, u_{exp} are the relative overall /combined uncertainties associated with the reference value and with the experimental result reported by the analyst/ Laboratory, respectively. The 'accepted' values of E_n include those results the values of which are ≤ 1; any other results are considered out of the acceptance range.

2.2 Evaluation of Results using the U_{test} Parameter

The $U_{test/score}$ parameter[3], which is usually employed by IAEA and NPL for interpreting results of the analysts/ laboratories participating in ILC/PT exercises, has a similar definition, namely:

$$U_{test/score} = \frac{\left|X_{ref} - X_{exp}\right|}{\sqrt{\left(u_{ref}\right)^2 + \left(u_{exp}\right)^2}} \tag{2}$$

where the meanings of the parameters are identical to those given above.

The results of the analyst /Laboratory are rated by the way they fit in one of the 5 evaluation classes established by Brookes et al[3], we named e_1 ...e_5, as follows:

e_1/ If $u < 1.64$,   the values do not differ significantly,
e_2/ If $1.64 < u < 1.96$, the values probably do not differ significantly but more data are
                                 required to confirm this,
e_3/ If $1.96 < u < 2.58$, one cannot say whether there is a significant difference without further
                                 data,
e_4/ If $2.58 < u < 3.29$, the values probably differ significantly but more data are required to
                                 confirm this,
e_5/ If $u > 3.29$,   the values differ significantly.

2.2.1 Comments: The class e_1/ may be considered the class of 'excellent' results, the class e_2/ - the class of 'very good' results, e_3/ - the class of 'good' results and e_4/ - the class of 'acceptable' results; the last one – e_5/ - includes the 'out-of-range' results.

The 5 classes of evaluation associated with the U_{score} parameter are adequate to the routine laboratories, as they allow a quite large variation in performance from 'excellent' to acceptable'; its disadvantage is the symbol U (also used as **u**), which is similar to the symbol of the uncertainty parameters it includes. For this reason, the symbol **U** or **u** is not the most suitable to be used as such.

By comparison, the E_n parameter is adequate for characterizing metrology laboratories, just because it is more restrictive.

If the range of accepted results associated with the E_n parameter were considered a class e_0 corresponding to values ≤ 1 specific of metrology performance, and e0 ... e5 were put together, then a unique criterion for performance evaluation could be established, that would cover all performance aspects of both metrology and routine laboratories, as necessary in practice.

2.3 Accuracy

The definition of "Accuracy," according to the International Vocabulary of Basic and ....
Terms in Metrology (1995)[4] is "...**closeness of the agreement between the result of a measurement and a true value of the measurand.**" Such a definition clearly suggests a

deviation (difference) from the 'true value of the measurand' as a quantification measure, in contrast with the expression the IAEA indicated some years ago in its kind efforts to help first-time participants in ILC exercises, namely:

Accuracy: result passes if $\left|\text{Value}_{ref} - \text{Value}_{Analyst}\right| \leq$ **3.29** x $\sqrt{\text{Unc}_{ref}^2 + \text{Unc}_{Analyst}^2}$ (3)

2.3.1 Comments. Obviously, the right-hand side of eq. (3) corresponds to the maximum acceptance limit according to the U_{score} expression given by Brookes et al[3].

Taking into account that the unknown 'true' value may be expressed in practice by the 'reference' value, and that in the above definition the "closeness....between the result and the true value of the measurand..." is given by the difference between the analyst/laboratory result X_{exp} (eq.1, 2), or Value$_{Analyst}$ (eq.3) and the reference quantity X_{ref}, or Value$_{ref}$, respectively, we suggest expressing "Accuracy" (absolute value) just by that difference, as it is the closest to the meaning/concept stated in the metrology literature[4]. More clearly, the absolute 'Accuracy' value would be given by the numerator of eqs.(1, 2), or by the left part of the eq.(3), respectively, and the relative value would be given by dividing the absolute

value to X_{ref} (eq.1, 2), or Value$_{Analyst}$ (eq.3), respectively. So, we suggest defining:

   Accuracy$_{relative\ value}$ = $|X_{ref} - X_{exp}| / X_{ref}$ (4)

which corresponds to the 'relative difference' usually determined in comparison exercises.

The 'Accuracy' parameter is frequently used nowadays as a necessary **measure** of a method, technique or equipment performance, and in this sense **a quantitative expression** /definition is strongly required. The definition we above proposed has the advantage to give a mathematical expression to the quantity closest to the 'Accuracy concept'; in addition, it will permit to definitely *avoid the frequent confusion between 'Accuracy'* and *'Precision'*.

This short discussion aims to point out that the performance and accuracy parameters need specific definitions /formulae best expressing each concept, at the same time ensuring a general, unique frame for interpreting the experimental results, without confusions and superposition of definitions of the same quantities.

3. DISCUSSION

3.1. Influence of the u_{exp} Variation on the Evaluation Parameter E_n

Taking into account the E_n definition (eq. 1), it is obvious that if the experimental uncertainty is improved /lowered, the denominator should decrease and consequently the E_n value should increase.

Is it possible that this effect clash with the performance improvement concept? In other words, is it possible that the improved values top the critical limit of E_n?

Such an unexpected situation occurred to us as we determined element concentrations by INAA (Instrumental Neutron Activation Analysis) in a sample of $\underline{S}$tainless $\underline{S}$teel AISI 316 (**SS**) using several relative standardization methods and 'improved' the initial nuclear decay data some time later by updated values given in the literature.

That work[5] presented in 2004 at the MTAA 11 Conference (Guilford, UK) aimed at identifying the most favorable standardization procedure in terms of minimum relative overall uncertainty. That standardization procedure was found by comparing the relative overall uncertainties of the elements concentration values in the **SS** matrix calculated using

several Standardization procedures but an unique set of experimental data, and the E_n score, in each case. The Stainless Steel AISI 316 material had been certified by IRMM.
Table 1 contains our results on **Fe concentration** in the **SS** material (an excerpt from Tables 2-3 of the complete work[5] presented at MTAA 11) as basis of discussion .

Table 1 *Effect of improving the P_γ uncertainty on the evaluation parameter E_n*

Certified Reference Standard (IRMM)	Fe concentration values determined by INAA in a Stainless Steel sample using 4 variants of Standardization methods[5], based on:							
	a) External Standard (CRM AISI 316)		**b) Internal Reference Standard** { C_{Cr}^{IRMM} }, using:					
Stainless Seel AISI 316			Effective cross sections, calculated with data (σ_{th}, I_0) from the literature				$k_{0,Au}$ values (u_{rel-c})	
	Variant 1		Variant 2a[ref 5]		Variant 2b[ref 5]		Variant 3[ref 5]	
C_{Fe}^{ref} (%) (u_{ref})	C_{Fe}^{x} (%) (u_{exp})	E_n	C_{Fe}^{x} (%) (u_{exp})	E_n	C_{Fe}^{x} (%) (u_{exp})	E_n	C_{Fe}^{x} (%) (u_{exp})	E_n
67.0 (2.5 %)	68.16 (3.70 %)	0.26	71.5 (7.50 %)	0.57	77.63 (10,6) (9.0 %)	0.98 initial* 1.14 corrected **	69.32 (4.39 %)	0.46

* Initial value of the P_γ (absolute γ-ray emission probability) uncertainty: **3,36 %**
** Corrected value of the P_γ uncertainty: **0,37 %**

The Fe concentration was determined by three relative standardization procedures based on external and internal standards. In the first case (Variant 1), we determined the Fe concentration in a SS sample taking it as 'unknown' and using another piece of the same material as 'standard', simultaneously irradiated and identically analyzed. In the case of the Internal Standardization procedure (Variants 2, 3), the **Cr** concentration certified by IRMM was used as internal reference standard; three variants were employed, based on effective cross sections calculated using two sets of dàta from the literature (Variant 2a and Variant 2b), and on the k0 parameters (Variant 3).

In the initial paper (May 2004), all the E_n values were under the critical limit or very close (case of Fe concentration determined by Variant 2b). In October 2005, during the revision for publication, the P_γ data were replaced by updated values found in the recent literature[6]; the corrected value was $E_{n-cor}= 1.14$, which is beyond the critical limit of accepted results.

That effect is explainable, as the relative uncertainty of **Fe** concentration (C_{Fe}) varied from 10.6% to 9% in the denominator of E_n due to the variation of the P_γ uncertainty from 3.36% to 0.37% for the γ- ray of energy $E_\gamma = 1099.26$ keV of ^{59}Fe. unexpected

However, that unexpected result raised questions about the influence of the relative experimental uncertainty u_{exp} on E_n :
• How does the experimental uncertainty variation influence the E_n parameter?
• How does that variation depend on the E_n numerator, which corresponds to the absolute value of 'Accuracy' according to eq. (4)?

In order to understand such effects, we conducted a short study using the 4 situations from **Table 1** as starting points for **Table 2**, while keeping constant the E_n numerator. i.e. C_{Fe}^{ref} , and C_{Fe}^{x} (with x corresponding to the Variants 1, 2a, 2b, 3), and u_{ref} (C_{Fe}^{ref}); the experimental uncertainty u_{exp}(C_{Fe}^{x}) was varied in each case by decreasing (i.e.) improving its value, or by increasing it so as to top the critical limit of E_n.

For each case /Variant (1)...(4), the <u>Rel</u>ative <u>D</u>ifference (**Rdif**) parameter (*corresponding to the relative Accuracy, in our previous definition*) was calculated as follows:

$$\text{Rdif} (\%) = 100 \times |C_{Fe}^{ref} - C_{Fe}^{x}| / C_{Fe}^{ref} \qquad (5)$$

The calculated **Rdif** values indicated in the headers of the 4 columns of **Table 2,** were: 1.73% for Variant 1; 6.72% for Variant 2; 15.87% for Variant 3, and 3.46% for Variant 4. Starting from the initial u_{exp} values given in each column, their numerical values were lowered/improved in cases (1, 2, 4), and increased in case 3.

Table 2 *Influence of the experimental uncertainty variation on the performance Parameter E_n*

CONCENTRATIONS VALUES KEPT CONSTANT		C_{Fe}^{ref} = 67,0 % and u_{ref} = 2.5 %						
		Variant **1:** C_{Fe}^{1} = 68.16 (%)		Variant **2a** C_{Fe}^{2a} : 71.5 (%)		Variant **2b** C_{Fe}^{2b} : 77.63 (%)		Variant **3** C_{Fe}^{3} : 69.32 (%)
		Rdif*$_1$ = 1.73 %		**Rdif*$_{2a}$** = 6.72 %		**Rdif*$_{2b}$** = 15.87 %		**Rdif*$_3$** = 3.46 %
V		$u_{rel\text{-}Fe}$ (%)	E_n	$u_{rel\text{-}exp}$ (%)	E_n	$u_{rel\text{-}exp}$ (%)	E_n	$u_{rel\text{-}exp}$ (%)	E_n
A R I A		3.70	0.26	7.50	0.57	9.0 10.6	1.14 corrected 0.98initial**	4.39	0.46
B		3.20	0.29	6.00	0.69	10.0	1.03	4.00	0.49
L E		2.50	0.33	5.40	0.76	10.3 ➡	1.00	3.50	0.54
	U	2.00	0.36	3.8 ➡	0.99	12.0	0.87	3.00	0.59
	N	1.50	0.40	3.4	1.07	20.0	0.53	2.50	0.66
	C	1.00	0.43	3.00	1.15	50.0	0.21	2.00	0.72
	E R	0.50	0.45	2.00	1.40	80.0	0.13	1.50	0.8
	T A	0.10	0.46	1.00	1.67	90.0	0.118	0.50	0.92
	I N	0.05	0.46	0.50	1.77	95.0	0.112	0.10	0.93
	T Y	0.01	<u>0.46</u> <u>limit</u>	0.10	<u>1.80</u> <u>limit</u>	98.0	0.11	0.01	<u>0.93</u> <u>limit</u>

* **Rdif** (Relative Difference) = $100 \times |C_{Fe}^{ref} - C_{Fe}^{x}| / C_{Fe}^{ref}$

** Initial P_γ value for $E\gamma$:

3.1.1 Comments. Case **(1)** where **Rdif** had the smallest value, 1.73%: by improving/ the experimental uncertainty from 3.7% to 0.5%, the corresponding E_n values varied from 0.26 to 0.46, which appeared to be the limit; in that case the critical limit of E_n was not attained. Case **(4)**, where **Rdif** was 3.46%: the variation of u_{exp} from 4.39% to 0.1% induced a variation in E_n from 0.49% to 0.93%. The limit 0.99 would have been attained if **Rdif** = 3.50%. Case **(2)**, where **Rdif** was 6.72%: the improvement in u_{exp} from 7.50% (the starting point) to the lowest possible value 0.5% determined a variation in E_n from 0.57 to 1.80 (limit); the critical limit of E_n was topped at u_{exp} = 3.80%, an 'improved' value by comparison with the starting point (7.50%).

Case **(3)** where **Rdif** = 15.87%: the situation was different, because in order to top the critical value of E_n, it was necessary to increase u_{exp}. So, the variation from 9.0% to 98%

determined an E_n variation from 1.14 (starting point) to 0.11. The critical limit of E_n was exceeded when u_{exp} was **10.3%**.

The conclusion is that topping /attaining the critical limit of E_n strongly depends on the value of the E_n numerator, i.e. on the difference between the reference value and the result experimentally determined of the measurand, meaning on the absolute 'Accuracy'.

When that difference (Accuracy) has a low value a good performance/ E_n value is obtained but as the E_n numerator increases an improvement in the experimental performance (by reducing u_{exp}) can determine over passing of the accepted limit.

A more detailed investigation of these aspects would be useful, as well as a corrected definition formula of the performance criterion (E_n and /or U_{score}) so that an improvement in the experimental performance be adequately reflected in the E_n numerical value.

3.2. The relationship between performance and experimental uncertainty

The relation between performance and experimental uncertainty deserves attention as some problems occur sometimes in practice. **Table 3** illustrates such a case, by presenting a selection of results concerning the ^{40}K activity determined in the IAEA Soil -375 sample by the Laboratory 'Z' within the framework of an ILC exercise we launched in 2005.

Table 3 *Contradiction between performant results and experimental uncertainty. Case of the ^{40}K radionuclide activity determined in the IAEA Soil -375 sample within the framework of an ILC exercise (2005)*
Laboratory Z (Selection)

Radio-nuclide	IAEA Certificate (updated values)		Experimental Results Laboratory Z		E_n (U_{test})
	C_A/31.10.2005 (Bq/kg)	$u_{rel\text{ -ref}}$ (1k) (%)	C_A/31.10.2005 (Bq/kg)	u_{exp} (1k) (%)	score
^{40}K	424	1,77	441	57	0,31

At Lab. 'Z' the γ-ray background is high, especially in the energy range of 1460.8 keV, which is typical for the ^{40}K specific gamma -ray.

The significance of the E_n value (0.31) generally means 'excellent' results, but that value was obtained with a high relative experimental uncertainty: 57%. The question, which requires further discussion, is whether such an experimental result really qualifies as an excellent performance (?)

A possible solution in such cases would be to use two performance criteria for interpreting the analyst/laboratory performance: the criterion (En and/or $U_{test/score}$) and another criterion by which the experimental uncertainty associated to the evaluated physical quantity be appreciated as function of its correspondence to one of several possible classes of values, as (we suggest): 5%, 10%, 20%, and over 20%.

3.3 Evaluation of the Overall/Combined Uncertainty at the 'ACTIVA-N' Laboratory

The operating procedure[7] applied at the 'ACTIVA-N' Laboratory for establishing the relative overall/combined uncertainty that should be attached to the physical quantity under investigation (radionuclide activity, or element concentration determined by INAA), generally denoted by **X**, indicates to fill out a spreadsheet typical for each case with: the numerical values of the uncertainty components that make up the expression of its specific relative combined uncertainty $u_{rel-c}(X)$, the multiplying factors which are characteristic to each physical quantity involved according to its typical statistical distribution, and with the multiplying factors determined when the law of uncertainty propagation is applied.

In order to take into account all terms and factors a general expression of u_{rel-c} of $X(X_i)$ was established by Cincu[7], where the type A and type B uncertainty components are clearly distinguished and all the mentioned factors are included.

That general expression of $u_{rel-c}(X)$, is the following:

$$u_{rel-c}(\mathbf{X}) = \sqrt{\sum_{i_A} c_{i_A}^2 \cdot (f_{i_A}^p)^2 \cdot u_{rel}^2(X_{i_A}) + \sum_{i_B} c_{i_B}^2 \cdot (f_{i_B}^p)^2 \cdot u_{rel}^2(X_{i_B})} \tag{6}$$

where:
- $\mathbf{X}_{i\,(A,\,B)}$ are the quantities of type A, B on which the physical quantity $\mathbf{X}$ depends;
- $c_{i_{A,B}}$ are the coefficients of influence obtained when the partial derivatives of the function

  $X(X_i)$ to each component $\mathbf{X}_{i\,(A,\,B)}$ are calculated,
- $f_{i_{A,B}}^p$ are the amplifying factors specific to the probability distribution (Poisson, Gauss,

  rectangular, triangular or trapezoidal) characterizing each quantity $\mathbf{X}_{i\,(A,\,B)}$,
- $u_{rel}(\mathbf{X}_{i\,(A,\,B)})$ is the relative standard uncertainty associated with each quantity $\mathbf{X}_{i\,(A,\,B)}$.

The specific formula in each case (combined uncertainty of the radionuclide activity or element concentration) is established by applying the law of uncertainty propagation to the expression of the physical quantity under investigation.

We illustrate the procedure for the case of evaluating u_{rel-c} of the **Cr** concentration (C_{Cr}^x) in an unknown sample of Stainless Steel material analyzed by INAA. The basic formula for determining Cr concentration using an external standard (CRM material) is:

$$C_{Cr}^x = C_{Cr}^{ref} \times \frac{m_{ref}}{m_x} \times \frac{A_{spec}^x}{A_{spec}^{ref}} \tag{7}$$

where: C_{Cr}^{ref} is the Cr concentration in the reference standard; $m_{ref,\,x}$ is the mass of the reference and unknown samples, respectively, and A_{spec}^x and A_{spec}^{ref} are the activities of the unknown and standard/reference samples, respectively.

By applying the uncertainty propagation law to eq.(7), and taking into account the other factors indicated in eq.(6), one obtains:

$$u_{rel-c}(C_{Cr}^x) = \sqrt{(u_{A_{spec}^x}^2 + u_{A_{spec}^{ref}}^2 + u_{C_{ref}}^2) + \left(\frac{1}{\sqrt{3}}\right)^2 \left[(u_{m_{ref}}^2 + u_{m_x}^2)\right]} \tag{8}$$

Table 4 is an illustration of how to use in practice the spreadsheet corresponding to eq.(8). The sum of components in the last column gives the squared value $\{u_{rel-c}(C_{Cr}^x)\}^2$; the final value of u_{rel-c} determined by square root corresponds to the coverage factor $k = 1_{P*\,=\,68,27\,\%}$. The spreadsheet structure, based on Romanian and IAEA documents[8,9], proved to be very useful in practice as it takes into account all the influencing factors, including those typical to the statistical distribution of each quantity (generally omitted).

Table   4.   *Spreadsheet of the relative combined uncertainty of Cr concentration determined by INAA in the Stainless Steel AISI 316 sample*

i	Entry data The physical quantity X_i	The relative Uncertainty $u_{(Xi)}$ (%)	Mode of evaluation	Amplifying factor according to the type of stat. distribution f_i	The influence coefficient $c_i = \dfrac{\partial Y}{\partial x_i}$	Contributions to the: uncertainty $(u_{rel\text{-}c})^2$ $\sum_i (c_i)^2 \times f_i^2 \times (u_{rel\,(Xi)})^2 \times [10^{-4}]$
0	1	2	3	4	5	6
1	C_{ref} (%)	1,5	From Certificate	1	1	**2,250 00**
2	m_x (mg)	0,02	Experimentally determined	$1/\sqrt{3}$	1	**0,000 13**
3	m_{ref} (mg)	0,02	Experimentally determined	$1/\sqrt{3}$	1	**0,000 13**
4	A_{spec}^{x} (Bq)	2,18	Determined by calculation	1	1	**4,7524**
5	A_{spec}^{ref} (Bq)	2,22	Determined by calculation	1	1	**4.9284**

$(u_{c\text{-}rel})^2$ =11,931

x 10^{-4}

$u_{rel\text{-}c}$ = $\sqrt{(11,931 \times 10^{-4})}$ =

3,45 %

4 CONCLUSIONS

We pointed out in the paper that the current definition of the performance criterion used for evaluating laboratory performance in ILC / PT exercises needed to be expanded so as to include both metrology and routine laboratories, as -generally- their specific limits of accepted results are different. To this end, we suggested combining the 5 performance classes of $U_{test/score}$ and the E_n specific condition in a unique parameter, the 6 performance classes of which could cover the different limits proper to the metrology laboratories and routine laboratories of which experimental combined uncertainties always include components provided by metrology laboratories or other authorized certification bodies.

It was also shown that the meaning of "**Accuracy**" as defined in the metrology Vocabulary[4], is closest to the expression of the E_n numerator than any other parameter. For that reason, as well as because a mathematical expression/formula giving the numerical value of 'Accuracy' is strongly required in order to quantify in practice the performance of a technique, procedure/method, equipment we suggested to consider the difference between the reference and experimental values just the definition of the 'absolute Accuracy' (eq.4); that value divided by the value of the reference quantity gives the 'relative Accuracy'.

By a short study concerning the experimental uncertainty influence on the E_n parameter it was revealed that improvement of the experimental performance can determine surpassing of the critical E_n value, if the E_n numerator (Accuracy) is higher than a certain limit

A more detailed investigation on the experimental uncertainty influence on the E_n value would be useful; a correction of the performance criterion expression would be also necessary so that it might correctly reflect any improvement of the performance operated by reducing the experimental uncertainty of the investigated physical quantity.

With regard to the contradictory performance situations when the E_n value is good, but the associated experimental uncertainty is high, we suggest that the performance evaluation takes into account an additional criterion - the 'performance uncertainty class'- by which the reported experimental uncertainty be also evaluated as function of its correspondence to one of the several possible classes of values that have to be established (we suggest 5 %, 10 %, 20 %, and over 20 %).

As a final general conclusion we consider that the quite old Standard[2] regarding evaluation of the laboratory performance by ILC / PT exercises needs to be revised, completed and updated, so that the new content be able to cover all the problems found in practice.

References

1. ISO/IEC 17025:2000(2005) Standard *"General requirements for Competence of testing and calibration laboratories"* .

2. ISO/IEC Guide 43-1:1997 Standard *"Proficiency testing by interlaboratory comparisons - Part 1: Development and operation of proficiency testing schemes"*

3. C.J. Brookes, I.G. Betteley and S.M. Loxston, *Significance tests. Fundamentals of Mathematics and Statistics,* New York, John Wiley Ed., 1979, 369-377.

4. International Vocabulary of Basic and General Terms in Metrology, ISO, 1995

5. Em. Cincu, D. Barbos, Ioana Manea, V. Manu, approved for publication in the 'MTAA (Modern Trends in Activation Analysis) 11' Proceedings – JRNC (December 2006).

6. International Atomic Energy Agency, information extracted from the NuDat Database, via http://www-nds.iaea.org/nudat2/index.jsp, October 14, 2005

7. Em. Cincu, *Metrologie, Vol.L (noua serie),* 2003, no. 3, 71

8. Gh.P.Ispăşoiu, *„Evaluation of the Measurement Uncertainty and Reporting Results of Physical Quantities",* Application Guide of the GUM Standard, ASRO, Bucharest, 2002

9. IAEA TECDOC 1401, *'Quantifying uncertainty in nuclear analytical measurements',* AIEA, Viena, July 2005 (the previous draft of which was used).

COLLABORATIVE STUDY FOR PESTICIDES RESIDUES DETERMINATION IN WATER SAMPLES (METHOD 5060 APAT–IRSA CNR) - PROJECT 4B L. 93/01

M. Antoci[1], S. Barbizzi[2], B. Bencivenga[3], D. Centioli[2], S. Finocchiaro[4], M. Fiore[5], F. Fiume[6], V. Giudice[7], M. Lorenzin[8], M.C. Manca[9], M. Morelli[10], E. Sesia[11], M. Volante[12]

[1] ARPA SICILIA - DAP Ragusa, viale Sicilia 7 - 97100 RAGUSA;
[2] APAT – AMB.LAB - Via Castel Romano 100 - 00128 Roma
[3] ARPA LAZIO - Sezione di Roma, Via G. Saredo 52 - 00173 Roma;
[4] ARPA SICILIA - DAP Catania, Via Carlo Ardizzone 35 - 95100 Catania;
[5] ARPA SICILIA - Direzione Generale, Via Ugo La Malfa 169 – 90146 Palermo;
[6] ARPA PUGLIA - PMP Chimico, Via Caduti di tutte le guerre - 770126 Bari;
[7] ARPA SICILIA - DAP Palermo, Via Nairobi 4 - 90129 Palermo;
[8] APPA TRENTO - Settore Laboratorio e Controlli - Via Lidorno 1 - 38100 Trento;
[9] ARPA CAMPANIA - via Don Bosco 4F – 80141 Napoli;
[10] ARPA EMILIA ROMAGNA - Corso Giovecca 169 - 44100 Ferrara;
[11] ARPA PIEMONTE - P. Alfieri 33 –14100 Asti;
[12] ARPA LOMBARDIA - U.O. Laboratorio Dipartimento Lecco Via 1° Maggio 21 Oggiono

1 INTRODUCTION

In this paper we report the results of an inter-laboratory collaborative study to verify the performances of a multi residual method for the analysis of pesticides in water.

The collaborative study was developed in the framework of the Italian National Project 4B 93/01 in order to create the Italian network of reference laboratories for research of pesticide residues in all environmental matrices.

The aim of this network is the analytical harmonization among Italian laboratories.

Starting from this idea, participant laboratories had stated:

- common and shared operative rules
- inter-laboratory normalization
- inter-laboratory experimentation of reference methods for residue determinations in environmental matrices

Presently, applying the stated rules, we studied the performances of a multi residual method for analysis of pesticide residues in water.

The collaborative study was performed by the ten laboratories participating to the project.

The multi residual method under investigation was: Method 5060[1].

The substances were selected according to an integrated approach[2,3]. The method was applied on 18 residues or active substances that are prioritaries in Italy.

The study will defined precision (repeatability and reproducibility).

Because of, in this field, although there were same official methods often they did not report validation data. These data represent the base to evaluate laboratory comparability and performances.

2 METHOD

2.1 Selection Criteria for Active Substances

Pesticides are important for food production and can also be useful in many other ways. About 400 chemical compound are approved for use in Italy and they vary widely for use, chemical and physical properties, potential impact on environmental pollution.

These high number and variety of substances are detected using a lot of analytical method. For these reasons, in order to face up to a pesticide monitoring plane, is important to state the criteria that can help to have a better results and lower costs.

First approach of the project was to use a priority index to select the active substances with higher environmental risk. The selected substances were used in inter-laboratory experimentation of a multiresidual method for analysis of pesticides in fresh and ground water.

The rules were:
1. Application of Priority Index (I.P)[2,3] based on: sales data, use, degradation of active substances, distribution model in aquatic compartment;.
2. Results of previous monitoring activities of Environmental Agencies (ARPA, APPA)
3. List of dangerous substances (D.M. 367/03)
4. Historical persistent substances
5. Information on pesticide metabolites.
6. Applicability of the selected substances to chosen multi residual method (5060 – APAT- CNR IRSA[1]).

From first rule we obtain a list of selected substances at decreasing order. In the interlaboratory study were analysed the first 15 of them, i.e. the substances with the higher environmental risk, detectable by method APAT-CNR IRSA: Alachlor, Chlorpyrifos, Lindane, Linuron, metalaxyl, metolachlor, molinate, oxadiazon, oxadixyl, pendimenthalin, prometryne, propyzamide, simazine, terbumeton, terbuthylazine.

At this list we add Atrazine and two metabolites: desethylatrazine and desethylterbutylazine. Atrazine was selected because often present in previous monitoring activities performed by Italian Environmental Agencies and furthermore enclosed in dangerous substances (D.M. 367/03).

In table 1 are reported the selected pesticides and their physical-chemical properties.

Table 1 - *Chemical-physical properties of the substances under investigation*

Pesticides		M.W.	Water solubility (mg/L)	v.p. (Pa)	logK$_{ow}$	DT50 (d)
ALACHLOR	HERB	269.8	170.31	2.10E-03	3.09	30
ATRAZINE	HERB	215.7	33	3.85E-05	2.5	41
CLHORPYRIFOS	INS	350.6	1.4	2.70E-03	4.7	120
LINDANE	INS	290.8	8.52	4.40E-03	3.5	400
LINURON	HERB	249.1	63.8	5.10E-05	3	100
METALAYIL	HERB	279.3	8400	7.50E-0.4	1.75	70
METOLACHLOR	HERB	283.8	488	4.20E-03	2.9	46
MOLINATE	HERB	187.3	990	0.746	2.88	21
OXADIAZON	HERB	345.2	1.0	1.00E-04	4.91	180
OXADIXYL	FUN	278.3	3400	3.30E-06	0.65	270
PENDIMETHALIN	HERB	281.3	0.3	4.00E-03	5.18	488
PROMETRYNE	HERB	241.4	33	1.65E-04	3.1	90
PROPYZAMIDE	HERB	256.1	15	5.80E-05	3.1	60
SIMAZINE	HERB	201.7	6.2	2.94E-06	2.1	180
TERBUMETON	HERB	225.3	130	2.70E-04	3.04	300
TERBUTHYLAZINE	HERB	229.7	8.5	1.50E-04	3.21	60
	minimum	187.3	0.3	2.94E-06	0.65	21
	maximum	350.6	8400	0.746	5.18	488

2.2 Analytical Method of Collaborative Study

There are many analytical methods for determination of pesticide residues in water. They differs each other in extraction, purification or identification techniques.

In order to understand which were methods used in Italian laboratories, we asked them to give information about their analytical techniques.

The most used was the method 5060 – APAT- CNR IRSA[1]. The method is reported in figure 1.

Ethion was used as internal standard.

Because of some laboratories used Fenchlorfos as process standard in order to follow the extraction performances, in the study this option was left free. Therefore, if a laboratory belonging to the project, used a process standard, it had to use the Fenchlorfos.

Anyway, concentration of Ethion (internal standard) and Fenchlorfos (process standard) was 0.5 mg/l in both cases.

Determination of pesticides residues. Method with solid phase extraction and gas-chromatographic analysis by selective detectors.

- **CARTRIDGE ACTIVATION**

C18 Cartridge (500mg/6ml)

or

Polymeric Cartridges (divinylbenzene/ N-vinylpyrrolidone) (60mg/3ml)

Wash the cartridge with, in the order (at a flux of 8 ml/min):

➢ Ethyl Acetate 5 ml

➢ Methyl alcool 5 ml

➢ Water 10 ml

Leave an over layer of one or two millimeter.

- **EXTRACTION**

Water sample (500-1000 ml) (eventually filtered on glass fiber) is added with MeOH (5 ml/l)

The sample is collected on the SPE actived column

The cartridge is eluted with 10 ml of EtOAc and residual water is removed putting in serie together, with the first cartridge a cartridge containing anhydrous Na_2SO_4 or diatomaceous earth

If the lab use an automathic system, residual water is removed drying the cartridge under N_2 flux.

After elution, the eluate is dried with rotative evaporator and/or under N_2 flux, at T≤40°C

Eluate is dissolved with 500 µl of a solution of internal standard (ethion 0.5 mg/l) in EtOAc.

Final concentration of the sample is 1:1000 by volume.

- **ANALYTICAL DETERMINATION**

The sample is analysed by GC (ECD; NPD and/or GC-MS)

Figure 1   *Scheme of Analytical Method 5060 APAT–IRSA CNR*

2.3 Collaborative Study Protocol

One laboratory of the network prepared the analytical material for this study. The material was divided in 10 parts, stored in refrigerated box and delivered to the other laboratories within 24h.

All laboratories had to strictly observe a set of detailed instructions to handling standards and analytical material. The instructions were agreed among laboratories before starting experimental phase. This in order to minimize all the variable and obtained data useful for validation.

The method under investigation required a solid phase extraction of pesticide residues in water matrix. We considered two kind of SPE:

✓ C18 Cartridge

✓ Polymeric Cartridge (DVB/NVP)

Analytical material was a solution containing all selected substances at a concentration around 10 mg/l in EtOAc (Sample A or parent solution).

Solution had to be stored at T≤4°C up to the analysis. It was indicated also the maximum execution time of analysis.

Calibration curves of each substance were performed using a dilute solution (1mg/l) obtained from Sample A and defined Sample B in EtOAc.

From Sample B were obtained calibration solutions in the range 0.02 – 0.8 mg/l. The following concentrations were the points the of the calibration curves: 0.02mg/l, 0.05mg/l, 0.1mg/l, 0.2mg/l, 0.5mg/l, 0.8mg/l. All calibration curves were in EtOAc or Isooctane.

From Sample A were obtained also the two spiking solutions, one at 0.1 µg/l (level 1) and the other at 0.5 µg/t (level 2), acetone was the solvent used for both solutions, according to the protocol.

Each laboratory produced 10 litres of spiked water for level 1 (0.1 µg/l) and 10 litres for level 2 (0.5 µg/l). Each fortified water sample was divided in 10 different and independent samples of 500 ml. Extraction had to be done in the same way with the same analytical apparatus, with the same lot of material (solvent, cartridges, etc..) and all at the same time.

Results were given using fitting tables properly done for this experiments. Each laboratory had to report:

1. Laboratory data
2. Extraction method:
    - Use of process standard
    - Cartridge and lot
    - Extraction method manual or automatic system
3. Instrumental method
    - GC injector and working temperature
    - Temperature program
    - GC detector and working temperature (ECD/NPD or MS)
    - Column
    - Carrier gas and flux
4. GC-MS parameters (if used)
    - Target ion and two qualifier ion
    - Relative ratio
5. Analytical results of level 1
6. Analytical results of level 2

Some laboratories did the analysis with both kind of cartridges (i.e. they analysed the two level both with C18 cartridges and with polymeric cartridges). Some laboratories performed other lot of analysis with small differences from analytical method.

These last data will be compared in a second time respect to the strictly required data.

2.4 Statistical Data Analysis

Statistical data analysis was done on the basis of UNI EN ISO 5725[5,6].
Statistical elaboration phases had been:
- Check for normal data distribution (were applied two different tests: Shapiro –Wilk and Kolmogorov-Smirnof)[7];
- Graphic elaborations (scatter plot, box plot, h and k Mandel plots);
- Statistical tests (as reported in UNI EN ISO 5725[5,6]): Grubbs test and Cochran test;
- Evaluation of statistical parameters obtained after elimination laboratories not responding at reported tests.

Data from C18 cartridges and data from polymeric cartridges were treated separately according to the described procedures. This is to verify if all data belong or not to the same population. If this was true they were treated like a unique bulk. This was done by comparison (f-test, t-test)[8] on variance and mean. After verified their belonging to a unique population, the data were treated all together

3 RESULTS

Here we report some results obtained. The data were relative to five substances, so selected:

> A metabolite (Desethylterbutylazine)
> Chemical-Physical properties (Pendimenthalin: lowest solubility; Oxadixil: highest solubility, lowest K_{ow}).
> Presence in a proficiency test (organized by APAT) and performed during same period from the same laboratories (Chlorpyrifos, Metolachlor)

All pesticides were at a concentration of 0,1 µg/l (this value is closer to the MRL for ground water). Only for Oxadixil were showed the results at both concentrations (i.e. 0,1 and 0,5 µg/l).

3.1 Normal Data Distribution

The check for normal data distribution was performed by Shapiro–Wilk and Kolmogorov-Smirnof tests, looking for data obtained from both used SPE (the data from polymeric cartridge were identified as SPE1, while the data from C18 cartridge as SPE2).

All studied residues showed a normal distribution with exception of simazine, oxadixyl and desethylatrazine at concentration 0,1 µg/l and Pendimenthalin at 0,5 µg/l. In any case these residues showed a unimodal distribution, that is the necessary condition for statistical treatments according to ISO 5725 rules.

3.2 F-test and t-test

F-test (on repeatability and average of variances) and t-test were performed on residues indicated above in paragraph 3 of this paper. The test were applied separately at data belonging to the group SPE1 and data belonging to the group SPE2. Results are showed in table 2. Analysing all results, only for pendimenthalin t-test and f-test showed great differences, moreover variances appeared homogeneous. All the other data were homogeneous for averages and variances. It was possible to say that the data belonging to the same population, so they were treated as a unique data set.

Table 2 - Significativity statistical tests (p-level < 0.05)

Concentration 0.1 ug/l						
Residue	F-test (repeatability)		F-test (average)		t-test (average)	
	F sper	F tab	F sper	F tab	t sper	t tab
Chlorpyrifos	1,62	1,49	1.25	4.82	1.64	2.18
Metolachlor	1.42	1.52	1.67	4.82	0.61	2.17
Oxadixil	1.76	1.77	12.34	19.35	0.82	2.36
Pendimenthalin	3.07*	1.54	1.20	4.39	3.62*	2.26
Desethyl terbutylazine	1.01	1.54	2.70	4.88	0.94	2.20
Concentration 0.5 ug/l						
Residue	F-test (repeatability)		F-test (average)		t-test (average)	
	F sper	F tab	F sper	F tab	t sper	t tab
Oxadixil	1.51	1.52	1.46	4.82	0.31	2.20

* test not passed

3.3 Graphical Representation of Data

Before to discard outlier laboratories, we represented all results using scatter plot, box plot, h and k Mandel plots. Scatter plot and box plot were useful to check outlier data within the data set of each laboratory. This information was not used with the aim of did not conditioning the following elaborations.

Information from Mandel plots were useful to put in evidence outlier data of single residue from each laboratory. Anyway, their presence were confirmed by Cochran and Grubbs tests for elaboration according to ISO 5725.

As example we reported graphic of Metolachlor at concentration 0.1 µg/l (Figures 2, 3, 4 and 5).

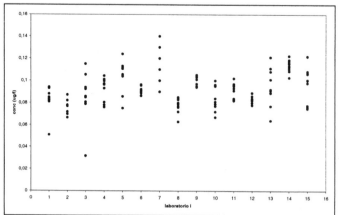

Figure 2 *Metolachlor 0.1ug/l Scatter plot*

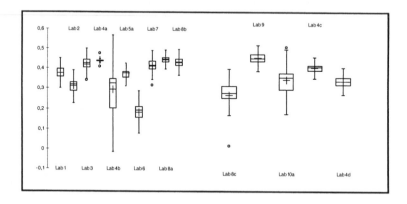

Figure 3 *Metolachlor 0.1ug/l Box plot*

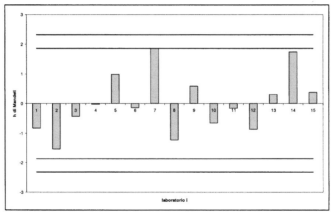

Figure 4 *Metolachlor 0.1ug/l Mandel h plot*

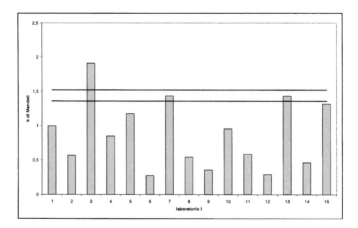

Figure 5 *Metolachlor 0.1ug/l Mandel k plot*

3.4 Elaboration Results

In table 3 were reported results obtained from elaboration of the data after elimination of outliers laboratories for specific residue.

In particular we reported:

✓ Number of laboratories which data were used for elaboration,
✓ averages,
✓ recovery percentage,
✓ repeatability and reproducibility standard deviation.

In table 4 were reported data obtained by separated elaboration of data set SPE1 and SPE2,

Table 3   *Elaboration results*

	Chlorpyriphos	Metolachlor	Oxadixil	Oxadixil	Pendimetalin	Desethyl terbutylazine
Concentration (ug/l)	0,103	0,103	0,101	0,507	0,100	0.102
Number of laboratories	15	14	11	12	14	15
Repeatability (ug/l)	0,031	0,032	0,034	0,144	0,025	0,030
Reproducibility (ug/l)	0,057	0,046	0,074	0,269	0,050	0,050
Averages (ug/l)	0,075	0,093	0,101	0,462	0,076	0,091
Recovery (%)	73	90	100	91	76	90

Table 4 – Results from SPE1 and SPE2

	Chlorpyriphos	Metolachlor	Oxadixil	Oxadixil	Pendimetalin	Desethyl terbutylazine
Concentration (ug/l)	0,103 ug/l	0,103 ug/l	0,101 ug/l	0,507 ug/l	0,100 ug/l	0.102 ug/l
SPE1						
Repeatability (ug/l)	0,027	0,038	0,033	0,186	0,015	0,027
Reproducibility (ug/l)	0,058	0,053	0,085	0,317	0,047	0,060
Averages (ug/l)	0,081	0,094	0,094	0,457	0,087	0,095
Recovery (%)	78	91	93	90	87	93
SPE2						
Repeatability (ug/l)	0,035	0,032	0,025	0,228	0,026	0,026
Reproducibility (ug/l)	0,058	0,043	0,033	0,335	0,054	0,041
Averages (ug/l)	0,067	0,090	0,086	0,441	0,065	0,088
Recovery (%)	65	88	85	87	65	86

4 CONCLUSION

Analysis of data elaboration put in evidence some important points here reported.

Evaluation of outlier data. There were full correspondence among results obtained by statistical elaboration of Grubbs and Cochran tests and the estimation by graphical technique on coherence (h and k Mandel test). For example, in figure 5 was evident that the third laboratory gave outlier data respect to the data from the other laboratories for the same residue (metolachlor 0.1ug/l). This, because of its data showed the greatest variability of replied results. This difference were confirmed by Cochran test that showed anomalies due to variance. The third laboratory did not used in statistical elaboration of Metolachlor 0.1 ug/l.

Comparison between SPE used for extraction. Results obtained from F-test and t-test, useful for evaluate date comparability, confirmed the hypothesis of equivalence between

both phases (polymeric and C18). The hypothesis was based on experience of analyst participating to the study and on method indications. Anyway, the table 4 suggested better performances in recovery percentages of the polymeric phase, at least for studied residues.
Analytical method. First results here showed confirmed how method fits for residues with very different chemical-physical properties. Repeatability values were in agreement with some specific rules (i.e. D.Lgs. 31/2001 or Directive 98/83/CE for quality of drinkable water)

References

1 APAT – IRSA CNR – Manuals and Guidelines 29/2003: Water Analytical Methods
2 M. Lorenzin, S. Coppi, A. Franchi, E. Sesia – Istisan Reports 04/35 further information on site: http:/www.appa.provincia.tn.it/slc/FrAttiviAAAF.htm
3 Internal Technical Report ANPA RTI AMB-MON 3/2000
4 The Pesticide Manual – Thirteen Edition – Editor: C D S Tomlin – 2003 BCPC (British Crop Protection Council)
5 UNI ISO 5725 – 2004
6 Technical report ISO/TR 22971 (First Edition 2005-01-15)
7 S.P. Millard, N.K. Neerchal, "Environmental Statistics with S-Plus", 2001 CRC Press, USA
8 UNICHIM "Study and Validation of Proof Methods" June 2004 – by N. Bottazzini

GROSS α/β MEASUREMENTS IN DRINKING WATERS BY LIQUID SCINTILATION TECHNIQUE: VALIDATION AND INTERLABORATORY COMPARISON DATA

I. Lopes and M. J. Madruga

Nuclear and Technological Institute, Department of Radiological Protection and Nuclear Safety, Estrada Nacional 10, Apartado 21, 2686-953 Sacavém, Portugal

1 INTRODUCTION

The importance of applied a quality assurance (QA) programme to improve the quality and reliability of the analytical results is nowadays well known (..."*the increasing need for reliable radioanalytical data creates a concomitant need for a QA system to support the acquisition of precise and accurate data..*")[1]. For that an internal and external quality control should be defined and applied.

A quality assurance programme has been implemented at the *Department of Radiological Protection and Nuclear Safety* (DPRSN) of *Nuclear and Technological Institute* (ITN) in order to guarantee the quality of the results and for the further accreditation of the Environmental Radioactivity laboratories based on ISO 17025 standards. Some management and technical requirements have been carried out. New methodologies have also been developed and their validation performed to verify if those methods are acceptable for our intended purpose.

The validation of internal methods includes studies on accuracy, precision, specificity, detection and quantification limits, linearity, range and/ or robustness. This validation requirements depends on the type of methodology and is an interactive process (revalidation may be necessary). Performing a through method validation could be a tedious process and often the time constraints do not allow for sufficient method validations. However, the consequences of invalid methods and the amount of time required to solve the problems exceeds what have been spent initially, if the validation studies had been performed properly[2].

The accuracy of the method, express the closeness of agreement between the measured value and the reference value. It can be assessed by analyzing a sample of known concentration and by comparing the measured value to the true value, using reference materials (CRM). The accuracy of the method can also be evaluated through the participation in interlaboratory exercises or it can be assessed by comparing test results from the new method with the results from an existing alternative method that is known to be accurate.

Liquid Scintillation Counting (LSC) technique is one of the most practical methods for determination of alpha and beta activities, in water samples[3-7]. The samples are homogeneously mixed with the scintillation cocktail, the counting geometry is essentially

4π, which provides high efficiencies and eliminates matrix effects. A method based on this methodology was developed and applied. The preliminary results have already been reported[8]. To ensure the confidence level of the data and the method accuracy, in 2004, the laboratory participated in an international analytical intercomparison trial[9]. The results obtained in this exercise and the calibration and validation tests performed recently to improve the data quality of the LSC method are presented and discussed in this paper.

2 METHOD AND RESULTS

2.1 Optimization of the Pulse-Decay Discriminator (PDD)

For gross alpha and gross beta measurements in water samples a low-level Tri-Carb 3170 TR/SL (Packard) equipment was used. The counter has a Bismuth Germanium Oxide (BGO) guard detector that completely surrounds the sample vial and delivers a lower background. This counter has the ability to discriminate between alpha and beta particles by pulse decay analysis (PDA). The percentage of events incorrectly classified can be determined and recorded as the percentage of misclassification. The best value of pulse decay discrimination (PDD) is the point at misclassification is minimum[10]. To select the optimum counting condition for alpha/ beta particle detection, an α and β emitter standards were measured separately at different PDD settings. After counting the alpha emitter source, the percentage of alpha events counted as beta events were plotted on the discriminator setting. Similarly, the events recorded as alpha emissions in the counting of the pure beta emitter source were plotted also in percentage of total counts against the discriminator setting in the same graphic.

Firstly, ^{241}Am and ^{36}Cl standards sealed vial sources (SURRC) were used and the optimum PDD value obtained was 136 (the spillover curve was reported in previous work)[8]. Afterwards, standard sources of ^{241}Am and ^{90}Sr were prepared using standardised solutions from Amersham at the same conditions of the samples to analyse (vial and cocktail type, acid concentration and ratio sample/ cocktail volume). The alpha (^{241}Am; 24 Bq L^{-1}) and beta (^{90}Sr/ ^{90}Y; 23 Bq L^{-1}) standards, with a quenching level of tSIE $\approx$300 (transformed Spectral Index of the External standard) prepared with Ultima Gold LLT scintillation cocktail (Packard) were used for the determination of the optimum PDD value. In these conditions, the minimum interference occurred for a pulse-delay discrimination parameter set at 127. However, it is known the quenching level could affect the value of PDD[11-13].

To study the influence of the quench on the optimum PDD setting for the alpha and beta separation, standards of ^{241}Am and ^{90}Sr with the same activity, quenched with different amounts of CCl$_4$ (quenching agent) were prepared in same conditions as the samples and measured at different PDD settings. The percentage of spillover versus optimum PDD value at different quenching levels is presented in Figure 1. The optimum PDD value was changed from 127 (0μl CCl$_4$) to 96 (80μl CCl$_4$). The α/ β misclassification under related optimum PDD settings changed from 5.8% (α)/ 5.2% (β) at a tSIE $\approx$ 300 to 17.1% (α)/ 17.7% (β) at a tSIE $\approx$ 100. The decreasing of the optimum PDD value with the increasing of the quenching is shown in Figure 2. The tSIE $\approx$ 100 corresponds to a higher quenching level while the tSIE $\approx$ 300 corresponds to a lower quenching level. The tSIE parameter, which gives the quenching level measurements, ranges from 0 to 1000, where 1000 indicates an unquenched sample and 0 indicates a most quenched sample[14].

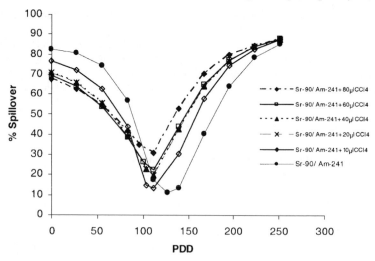

Figure 1 – *Total interferences versus PDD value, using* $^{241}Am/\,^{90}Sr$ *standard sources prepared with Ultima Gold LLT cocktail and* CCl_4

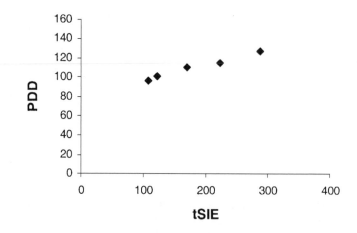

Figure 2 – Optimum *PDD value versus tSIE quenching parameter, using* $^{241}Am/\,^{90}Sr$ *standard sources prepared with Ultima Gold LLT cocktail and* CCl_4

2.2 Counting Efficiency Determination

For the determination of the counting efficiency, alpha (^{241}Am) and beta (^{90}Sr/ ^{90}Y) standard sources were prepared with activities ranging from 1 to 25 Bq L^{-1}, using 10 ml of Ultima Gold LLT scintillation cocktail and 10 ml of 0.1M HCl distilled water, in order to achieve the same chemical and physical conditions as the environmental samples to analyse. The background samples were prepared with distilled water and the same amount of cocktail and acid loading as the standards. The mean value of the counting efficiency for ^{241}Am and ^{90}Sr/ ^{90}Y sources was 95.8% and 90.5%, respectively, using the PDD value set at 127.

The influence of the quenching level on the counting efficiency was also studied. The alpha (^{241}Am; 24 Bq L^{-1}) and beta (^{90}Sr/ ^{90}Y; 23 Bq L^{-1}) standards, prepared with different CCl$_4$ amounts were measured using the same value of PDD setting at 127. The volume of CCl$_4$ used (from 10 µl to a maximum of 90 µl) produced tSIE range values between 300 and 100. The quench curve for ^{241}Am (Figure3) shows that the alpha particle efficiency remains higher at lower quench ($\approx$ 95%) and drop off towards lower values ($\approx$ 50%) as the quench increases. The effect of quenching on the beta particle efficiency is less pronounced, the efficiency values ranging from 87.6% to 95.5% (Figure 4).

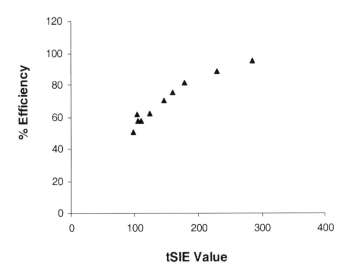

Figure 3 – *Alpha particle efficiency versus quench (tSIE) using PDD set at 127*

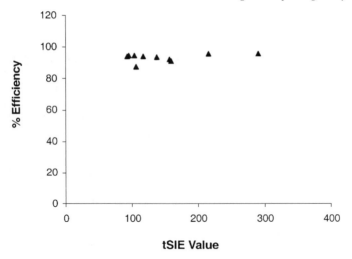

Figure 4 - *Beta particle efficiency versus quench (tSIE) using PDD set at 127*

2.3 Validation Test

Validation test based on the recovery of spiked samples was performed using the same procedure as in the sample analyses. Tap water samples (1000 ml) spiked with alpha (^{241}Am) and beta (^{90}Sr/ ^{90}Y) standard sources, were prepared in triplicate with activities ranging from 1 to 25 Bq L^{-1}. The samples were acidified with 1 ml HCl (pH>2) to avoid losses due salt precipitation in the glass beakers. Afterwards, the samples were pre-concentrated ($\approx$50-100 ml) by slow evaporation on a hot plate (pH drops to pH <1) and one aliquot of 10 ml of this aqueous concentrate was withdrawn and added to 10 ml of cocktail. Tap water samples without spike were used as background samples. The same counting time of 240 minutes was used for the spiked and background samples at PDD of 127. Table 1 summarizes the results of the samples used in this validation test.

The nominal activities and the results obtained for the measurements of spiked samples are in good agreement. The percentage of recovery was calculated. For gross alpha, the recovery ranges from 84.5% to 113.4%, with a mean value of 101.7%, higher than the result obtained for gross beta, which recovery ranges from 88.6% to 94.0 % with a mean value of 90.6%. The mean recovery percentage was in range 80-120% which indicated the accuracy of the methodology[2].

Mean recovery values of 94.2% and 103.4% for gross alpha and gross beta respectively, were obtained in previous validation test[8] performed with Ultima Gold AB cocktail at a PDD of 136.

Table 1 - *Alpha and beta activities (Bq L^{-1}) for water samples spiked with ^{241}Am and ^{90}Sr, using Ultima LLT scintillation cocktail.*

Alpha Activity (^{241}Am)	
Known[a]	Measured[b]
48.3	46.7 ± 2.4
24.1	23.9 ± 0.8
12.1	12.4 ± 0.5
5.8	4.9 ± 0.2
0.097	0.11 ± 0.04
Beta Activity (^{90}Sr)	
Known[a]	Measured[b]
22.5	20.2 ± 1.7
11.3	10.0 ± 2.2
5.5	5.0 ± 0.3
1.8	1.7 ± 0.1

(a) Uncertainty of the nominal value of activity is 0.47 % and 0.88% for ^{241}Am and ^{90}Sr, respectively.
(b) Average of the measurement of three replicate samples with the same nominal activity.

2.4 Intercomparison Data

The LSC procedure was applied to determine the alpha and beta activities of a drinking water matrix in an international intercomparison exercise performed between environmental radioactivity laboratories and organized by Consejo de Seguridad Nuclear (CSN), Spain[9].

The water sample (1000 ml) was pre-concentrated (80 ml) by slow evaporation on a hot plate. Five aliquots of 10 ml of this aqueous concentrate was withdrawn and added to five vials (22 ml glass scintillation vial, Packard) with 10 ml of Ultima Gold AB (Packard). The background samples were prepared with distilled water using the same volume of cocktail as the samples. Counting time of 240 minutes was used for the background and water samples, with an open window of 0-1000 keV for alpha and of 0-2000 keV for beta energy range. The detection efficiency was evaluated through the measurements of ^{241}Am and ^{90}Sr/ ^{90}Y standard sources prepared with Ultima Gold AB scintillation cocktail. The efficiency for alpha and beta emitters was 97.2 % for ^{241}Am and 92.8% for ^{90}Sr/ ^{90}Y, respectively, with a PDD value set at 136 using the ^{36}Cl and ^{241}Am standard sources as described in section 2.1. Table 2 shows the results obtained in this intercomparison exercise.

Table 2 *CSN/ CIEMAT intercomparison results*

	Gross Alpha Activity (Bq L^{-1})	Gross Beta Activity (Bq L^{-1})
DPRSN/ ITN results by LSC[a]	0.135 ± 0.017	1.698 ± 0.055
CSN/ CIEMAT reference values	0.091 ± 0.020	1.333 ± 0.227

(a) Mean value

The intercomparison trial have indicated that the LSC method provides acceptable result (Z = 2.21)* for alpha activity and more accurate result for beta activity (Z= 1.62)*. The criterion proposed to the Z classification[9] was |Z| ≤ 2 accurate, 2<|Z| <3 acceptable and

$|Z| \geq 3$ not acceptable. It was verified that uncertainty associated to the LSC results are subestimate (En >1)[**]. The reference value is outside uncertainty range for alpha and for beta activities.

$$^{(*)} Z = \frac{|x - X|}{\sigma_p} \qquad\qquad ^{(**)} En = \frac{|x - X|}{\sqrt{U^2_x + U^2_x}}$$

x = results obtained by the laboratories; U_x = uncertainty of laboratory results
X = reference activity value; U_X = uncertainty of reference value
σ_p= standard deviation.

3 CONCLUSION

The results of the calibration indicated that the increasing of the quenching level originated a decreasing of the optimum PDD value and also a decreasing of the counting efficiency, mainly for gross alpha emitters. So, in the LSC methodology a suitable quenching correction should be performed to obtain the optimum PDD value and a good α/ β separation.

The validation test results show a good agreement between the spiked samples recovery and the nominal activities.

Acceptable results were obtained on the intercomparison exercise using the cocktail Ultima Gold AB. However, more accurate data will be expected in further intercomparison exercises using the optimization described in this paper with Ultima Gold LLT cocktail.

More studies will be necessary in order to examine thoroughly the accuracy of this methodology and to test others requirements (specificity, precision, etc.) to obtain a properly validation of the method.

References
1 M. Betti, L. Heras, *J. of Environ. Radioactivity*, 2004, **72**, 233.
2 J. M Green, *Anal. Chem. News & Features*, 1996, 305 A.
3 J. A. Sanchez-Cabeza, L. Pujol, J. Merino, L. León, J. Molero, A. Vidal-Quadras, W.R. Schell, P.I. Mitchell, *Radiocarbon*, 1993, 43.
4 J. A. Sanchez-Cabeza, L. Pujol, *Health Phys.*, 1995, **68**, 674.
5 W. C. Burnett, J. Christoff, B. Stewart, T. Winters, P. Wilbur, *Radioact. and Radiochem.*, 1999, **11**, 26.
6 R. I. Kleinschmidt, *Appl. Rad. Isot.*, 2004, **61**, 333.
7 C. T. Wong, V. M. Soliman, S. K. Perera, *J.Rad. Nucl. Chem.*,2005, **264**, 357.
8 I. Lopes, M. J. Madruga, F. P. Carvalho, *Proceedings of an Int. Conf. Naturally Occurring Radioactive Materials (NORM) IV*, Polónia 16-21 Maio 2004. IAEA-TECDOC-1472, 305.
9 CSN/ 2004 "Intercomparación analítica entre laboratories de radioactividade ambiental 2004: Evaluación de la intercomparación analítica de radionucleidos en muestras ambientales.
10 Introduction to Sample Preparation for Alpha/ Beta LSC, Packard Company, 1993.
11 Quench, Application Note BTB-001, Packard Instrument Company
12 J. M. Pates, G. Cook, A. Mackenzie, C. Passo Jr., *Analyst*, 1998, **123**, 2201
13 D. Yang, in: *Liquid Scintillation Spectrometry* 1994, ed. G. T. Cook, D.D. Harkness, A. B. Mackenzie, B. F. Miller and E. M. Scott, *Radiocarbon*, 1996, 339.
14 Thomson, J. Liquid Scintillation Counting, Application Note LSC-007, Packard BioScience Company.

IMPROVEMENT OF A RADIOCHEMICAL LABORATORY THROUGH FOURTEEN YEARS OF PARTICIPATION IN AN INTERCOMPARISON PROGRAM

M.H.T Taddei

Laboratório de Poços de Caldas, Comissão Nacional de Energia Nuclear
P.O.Box 913, Zip Code 37701-970 - Poços de Caldas - MG, Brazil

1 INTRODUCTION

CNEN's radiochemistry laboratory in Poços de Caldas has been participating in the 14 years National Intercomparison Program (NIP) for the analysis of radionuclides in environmental samples. The program is conducted by the Radioprotection and Dosimetry Institute of CNEN.

Three times a year, this program distributes simulated environmental samples which contain one or more radionuclides and the results of the analyses are sent for statistical and analytic performance evaluation. Each round, a report that allows performance to be evaluated and procedural or instrumental problems to be identified and corrected is sent to the participating laboratory. Participation in the program has improved analytical techniques in the determination of radionuclides and has introduced advancements in laboratory practices which have reduced many sources of uncertainty. Statistical parameters are applied to the analytical data in order to show the performance of the laboratory. The variable coefficient of three measurements that have been performed is related to standard deviation and the accuracy is precisely related to the reference value.

During the period of the study, various changes in analysis methods and the acquisition of new reagents were introduced, not just to achieve better performance in the exercises but also to develop methodologies for other radionuclides which are necessary for regulatory control of the nuclear fuel cycle and are offered during the intercomparison program rounds. The improvements have been pointed out in the graphs that are included here.

The data base was made up of 43 runs from 1992 to 2005, with 730 radionuclide analyses, ^{226}Ra, ^{228}Ra, ^{210}Pb, natural uranium and its isotopes, thorium, ^{90}Sr, ^{241}Am, total alpha and beta, approximately 10 gamma emitters and $^{238/239}$Pu, in water, vegetation, soil and aerosol filter.

The evolution of our results which were demonstrated by the results of the intercomparison allowed us to implement radiochemical analysis training programs for other CNEN laboratories and for research laboratories of both universities and private enterprises.

2 METHODS AND RESULTS

The samples are sent through the mail three times per year in the moths of April, August, and December. The deadline for the delivery of the results is three months after receiving the samples.

In 1992, we began to participate in the program solely with samples of water for which determinations of alpha and beta, natural uranium, ^{226}Ra, ^{228}Ra, and ^{210}Pb were performed. At that time the lab was only equipped with a thallium activated sodium iodine alpha counter and a Geiger Müller beta counter. These detectors were used to determine ^{226}Ra, ^{228}Ra, and ^{210}Pb after radiochemical separation by coprecipitation with sulfate, barium/radium, and lead chromate. The laboratory already had a well-established procedure for natural uranium which used solvent extraction to purify the uranium followed by UV-VIS spectrophotometry. As the NIP evolved, more radionuclides and matrices were made available in the rounds, and for the purpose of CNEN's inspection of nuclear and radioactive facilities, new techniques were developed and more specific equipment was appropriated.

The Figure 1 shows the improvements introduced in the laboratory from 1992 to the present.

	1992	1993	1994	1995	1996	1997	1998	1999	2000	2001	2002	2003	2004	2005
alpha-beta	Total Counting				Ultra Low Level Counting									
^{226}Ra	Coprecipitation + alpha Total Counting				Coprecipitation + Ultra Low Level Counting							Micropreciptation + alpha Spectrometry (^{228}Ac)		
^{228}Ra	Coprecipitation + beta Total Counting				Coprecipitation + Ultra Low Level Counting							Micropreciptation + gamma Spectrometry (^{228}Ac)		
^{210}Pb	Coprecipitation + beta Total Counting				Coprecipitation + Ultra Low Level Counting							Sr spec resin purification + Liquid Scintillation Counter ^{210}Pb		
^{90}Sr	Solvent Purification + beta Total Counting ^{90}Y				Solvent Purification + Beta Ultra Low Level Counting							Sr spec resin purification + Liquid Scintillation Counter ^{90}Sr		
U isotopes					Anion exchange resin purification + alpha Spectrometry									
Th isotopes					Anion exchange resin purification + alpha Spectrometry									
^{241}Am					TRU spec resin purification + alpha Spectrometry									
Pu isotopes					Anion exchange resin purification + alpha Spectrometry									
Gama emitters					Gamma Spectrometry HPGe 15%		Gamma Spectrometry + HPGe 45%							

Figure 1. *Evolution of analytical techniques during the period of the study.*

Statistical criteria from reference 1 were used in order to evaluate performance in terms of precision and accuracy. In this way, analysis results were classified using normalized standard deviation D, defined as the difference between the average value obtained in the three readings, $\overline{X}$, and reference value U, normalized to the standard deviation of reference value, s_U, divided by the square root of the number of determinations, n:

$$D = \frac{(\overline{X} - U)}{s_U / \sqrt{n}} \tag{1}$$

If D results in ± 2, laboratory performance is good, if ±3< D < ±2, laboratory performance is acceptable and for results in which D≤±3, laboratory performance is unacceptable.

The accuracy of each analysis can be expressed by the variable coefficient:

$$VC = \frac{s}{\overline{X}}\%\qquad(2)$$

In which $\overline{X}$ is the average value of the three determinations and s is the corresponding standard deviation.

The Figures 2 (2.a, 2.b, 2.c, 2.d) show the evolution of the results for the radionuclides that were analyzed in the intercomparison program in terms of the normalized standard deviation for the different types of matrices and techniques during the period of the study.

Soil Gamma

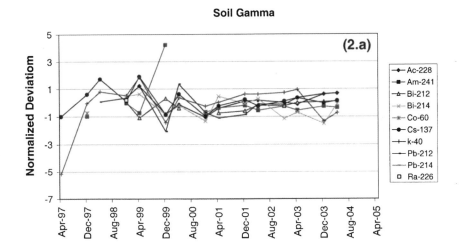

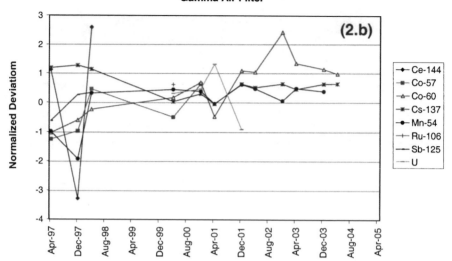

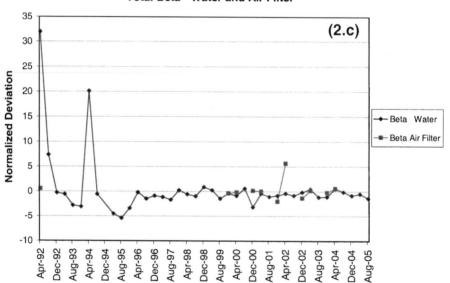

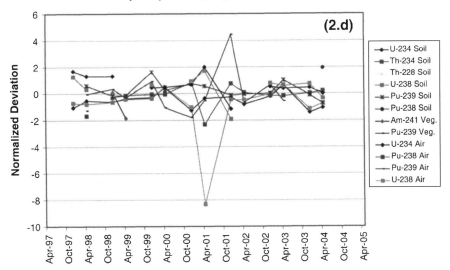

Figure 2 (2.a, 2.b, 2.c, 2.d). *Evolution of the results for the radionuclides in differents types of environmental samples and techniques.*

The figure 3 shows the evolution of the results for frequency of occurrence at each interval of percentage in relation to the adjusted standard deviation of the period.

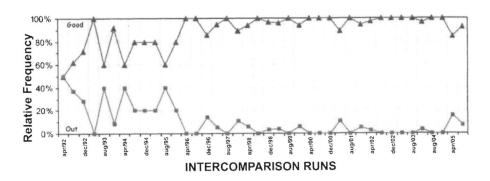

Figure 3.. *Variation of percentage for correct readings in each of the rounds from 1992 to 2005.*

3 CONCLUSION

The graphs in Figure 2 show:
For the samples that were analyzed through direct determination, that is, without the parts of the procedure that call for radiochemical separation, there was a notable improvement in the determination of total alpha and beta due to the introduction of low-background equipment, and for gamma radionuclide emitters due to the acquisition of more sensitive equipment.

In the case of radionuclides that undergo radiochemical purification and separation, improvements were observed in the results due to the introduction of new purification methods, though it appears that a need to standardize operational procedures within the laboratory certification criteria in order to reduce sources of error in the analytical procedures.

Even though some of the determinations weren't what was expected, figure 3 shows that participation in the program significantly improved the laboratories analytical performance with a few small drops which were caused by lack of operational control, which needs to be improved.

Participation in the intercomparison program made it possible to verify the weaknesses in the methods that were being used, introduce improvements in the procedures and improve the number of radionuclides that are being analyzed through the free distribution of reference standards by the coordinating body of the NIP. With these standards, it became possible to offer training to researchers and technicians for inspection agencies and companies involved in mining and processing of minerals that contain uranium and thorium.

References

1. L.Tauhata, M.E.C.Vianna, A.E.Oliveira, C.C.S.Conceição, Metrological capability of the Brazilian laboratories of analyses of radionuclides in environmental samples.Applied Radiation and Isotopes, 2002, **56,** 409-414.
2. A.N.Jarvis,L.Siu, Environmental Radioactivity Laboratory Intercomparison Studies Program, 1981, EPA-600/4-81-004, EMSL-USEPA, Las Vegas.

Acknowledgements

The author wish to thank CNPq and PBMQ for their financial assistance and the staff of CNEN/COLAB (in special Ms Maria Rego Monteiro Gomes for their valuable contribution in the realization of this work).

Subject Index